The Foundations of Experimental Physics

Jacopo Parravicini

The Foundations of Experimental Physics

Unraveling the Premises of Physical and Scientific Knowledge

Springer

Jacopo Parravicini
Department of Physics and Astronomy
University of Florence
Sesto Fiorentino (Firenze), Italy

ISBN 978-3-031-55451-3 ISBN 978-3-031-55452-0 (eBook)
https://doi.org/10.1007/978-3-031-55452-0

The translation of the original texts was done by the author. Where available, some English translations were consulted, but the exposed text is an own translation unless explicitly stated otherwise. Professor V. Sanvito's consultancy was also used, in particular for German and Russian texts.

This Springer imprint is published by the registered company Springer Nature Switzerland AG
The registered company address is: Gewerbestrasse 11, 6330 Cham, Switzerland

Paper in this product is recyclable.

Πολλὰ τὰ δεινὰ κοὐδὲν ἀνθρώπου
δεινότερον πέλει.

Sophocles, Antigone

Visibilia regenda, invisibiles regentes:
interroga ista.
Respondent tibi omnia: Ecce vide, pulchra
sumus.

Saint Augustine, Sermones

To my wife Viola and my children Tecla, Martino, Clara

To my mother Mirella and my father Giambattista

Foreword

What is the scientific method today, roughly 400 years after Galileo and 100 years after the quantum physics revolution? What are the logical, philosophical, mathematical, and physical patterns and ingredients underlying the experimental physics of our time? Jacopo Parravicini answers these important questions with a beautiful journey along the Foundations of Experimental Physics, a book that crosses all the founding aspects of today's scientific method in an original and stimulating way. The book was born as the text of a very successful university course with the same title, started 3 years ago by Jacopo Parravicini for Ph.D. students in Physics and Astronomy at the University of Florence. However, the text does not assume a deep knowledge about experimental physics, rather it guides readers in an in-depth but also exciting and often amusing way to discover the many involved conceptual aspects, which go far beyond the typical preparation of physics students.

This book addresses the important relationship between physics and mathematics in the research and verification of physical laws, and it shows that the scientific method has gradually evolved with the introduction of new scientific methodologies, new physical laws, and in general with the evolution of culture and knowledge at a global level. The text discusses important epistemological aspects, making them understandable with practical examples of scientific approaches drawn from the past and the present. Aspects of scientific research that we all certainly know, such as the difference between theoretical and experimental research, the units of measurement, the uncertainties of measurement, are for the first time explained and connected to each other in a way that makes us grasp their relationships and importance. The book also addresses much less known aspects, such as that of scientific frauds, describing them in a fascinating way and discussing the role of the scientific community, both in the past and in the present, with all its related open issues.

The journey along the Foundations of Experimental Physics by Jacopo Parravicini—an Experimental Matter Physicist at the Department of Physics and Astronomy of the University of Florence—is certainly aimed at a much wider audience than the Ph.D. students in physics for whom it was born. I am sure that many researchers and teachers, and more generally all the public with a passion for scientific research, will find this book instructive and stimulating as I did.

Florence, Italy
September 2023

Giovanni Modugno

Preface

A discussion of the nature of any intellectual effort is difficult per se—at any rate, more difficult than the mere exercise of that particular intellectual effort.

Johan von Neumann

Those who have a passion for Physics and scientific research usually focus their energies towards the specific content of the discipline, i.e. the study of natural phenomena. Whether this research is carried out as a job or that it is a simple curiosity, it is very rare to spend time to think about the methods and meaning of this intellectual enterprise. Looking at university courses in scientific branches, we notice how it is very difficult for a student to face, during his training course, a specific study regarding the foundations of sciences, what we see when experiments are carried out, what is the meaning of the claim that natural sciences are based on experiments. In short, in the curricula of the scientific faculties there is no a Method Course, where one learns how to "do science". What the scientific method is and how it is applied can only be learned through practice, there is no systematic treatment about it, not even broadly speaking. The text here presented aims at addressing this lack, proposing a reflection path on the general principles of Physics, as the discipline at the basis of all natural sciences.

There are many textbooks having, in their respective languages, titles such as "Fundamentals of Physics" or "Principles of Physics" or similar ones. Almost all of these deal with the main, or elementary, specific contents of the physical sciences, such as, for example, classical mechanics, thermodynamics, electromagnetism. Similarly, the addition of the "experimental" adjective ("Principles of Experimental Physics", etc.) makes the content a reasoned exposition of the main techniques and of the most important experimental methods. Allow me a metaphor, which I will recall throughout the text. Let physical science be a great building. The "foundations" or "principles" referred to by the titles of the aforementioned books are the structural elements of the building (mechanics, electromagnetism, etc.), those general concepts on which it is possible to build a peculiar and detailed explanation of a large number of natural phenomena. However, those same pillars and architraves, which support the weight of the building and give it shape, require solid foundations, the deeper and wider the bigger the building. These foundations are made not to be seen and not to be directly exploited by the users of the building. However, it is clear that they are the prerequisite for the whole construction. Well, the subject of this text is the foundations of the Physics building, not the pillars and lintels. The subject will be that which is not immediately before the eyes.

It is therefore clear that, although the present text has a very similar title to the aforementioned ones, it will deal with quite different subjects. The proposed path will address the study of what lies at the basis of science, in general, and of Physics in particular. Therefore, we will treat those elements which, in the practice of this discipline, are generally "taken for granted". We will address elements such as the relationship between the natural phenomena that Physics studies and the mathematical language it uses. I will highlight the main hubs of the conceptual development of science, also outlining the conceptual and practical instruments gradually conceived for the purpose of investigating nature. The life of every human being takes place between a concrete dimension, made up of body and matter, and an abstract one, made up of mind and thought. Similarly, Physics insists between a material world of experiments, highly concrete, and an ideal one made by mathematical entities and models, often highly abstract. In the middle there is "scientific practice", i.e. that set of proven, more or less empirical, more or less codified customs and methods, stabilized by successive approximations, which today determine the life of what is called the "scientific community", which determines the advancement of knowledge with its own work.

These and other highly heterogeneous elements contribute to the construction of knowledge in Physics. It is with all these elements that both those who practice research and those who benefit from its results deal, often unconsciously. If the small group of scientists, especially physicists, belongs to the first category, everyone belongs to the second one. Everyone, in the modern era, has known or used the fruit of scientific knowledge. Never as in our times we have been confronted with that powerful method of investigating reality which is called "science", a word, a principle, a concept, alas, often more abused than used. For this reason, in general, I don't like to talk about "science", but about "scientific research". During the text, however, I will be forced to use the word "science" several times, but I will always take care that the specific meaning attributed to it in that specific context is clear. As my personal contribution to clarity and truth based on facts, in this text I intend to describe the functioning and the main mechanisms of that sort of "machine", or rather, of *organism*, whose result is scientific knowledge in the discipline of Physics.

The here proposed path comes from a class for the Ph.D. course in Physics and Astronomy that I held at the University of Firenze. Therefore, it gives as notes the main notions of Physics and Mathematics that are learned in the first years of the technical-scientific faculties and, with them, the fundamental concepts of so-called modern physics. However, I believe that this text is within the reach of any university student or professor who, despite not having studied Physics in detail, intends to know the question of its foundations. Concepts that I take for granted non-physicists will easily find in suitable introductory textbooks.

As a general method criterion, in my text I will "make the great scientists speak", reporting as much as possible the quotations of those who were the builders of the discipline, Physics in particular, but not only. Similarly, to illustrate the concepts of the scientific method, I will propose the cases of some important paradigmatic discoveries, analysing their circumstances and procedures. I chose this approach not because of a historicist inclination, but because I don't consider myself entitled to explain "what Physics is" or, worse, "what science is". I will therefore ask the great ones to take me on their shoulders and I will lend them my voice so that whoever reads me can hear theirs.

Of course, I make no claim to exhaustiveness. On the one hand, anyone will certainly find works on the subject by great scientists that have escaped me and have not found a place here. In the text I have necessarily made a choice, but, in case, I ask my reader to report me further pertinent quotations. The intrinsic non-exhaustiveness of this book also derives from the fact that each of the individual considered topics would alone deserve a much more extensive treatment. For each of the topics that will be here treated, it is possible to find various texts that go into it in much larger detail than here. There are several books that deal in detail with subjects such as the planning of an experiment, or the features of measurement units, or the structure and principles of Mathematics, or scientific revolutions, or the features of the scientific community. This text, on the other hand, aims at highlighting a common thread between all these elements, showing how they all come together to support the adventure of scientific research in the field of Physics. Therefore, each element will necessarily be exposed in broad terms, favouring its role in the general framework of science rather than its particular articulations. With this I prevent possible criticisms for excessive conciseness and simplification or, worse, superficiality: my aim is not to provide an analysis, but a synthesis. I intend to propose an itinerary where the correlation of each element with respect to the others is highlighted, in the belief that in the "science phenomenon" we can find a great plurality and, in it, a unity.

The text consists of three parts. A *first part* (I), *Concepts and Instruments*, deals with the intrinsically phenomenological nature of Physics and its most evident peculiarities (Chap. 1), highlighting the relationship between mathematical concepts and experimental data (Chap. 2), then discussing the two pillars on which physical quantities are defined, the mathematical one in requesting models, and the experimental one in requesting operating procedures and units of measurement (Chap. 3); the section concludes by briefly reviewing the development of scientific thought in relation to the language of Mathematics (Chap. 4). The *second part* (II), *Structure of Scientific Knowledge*, deals with the status of the knowledge obtained from Physics: a purely epistemological chapter (Chap. 5) is followed by the discussion of the hierarchy and the bases of knowledge, with the great issues of reductionism (Chap. 6) and the foundations of Mathematics (Chap. 7); finally, it will deal with the solidity of physical models and theories, related to measurement processes and the meaning of the notions of truth in scientific practice (Chap. 8). The *third part* (III), *Practice of Knowledge*, intends to expose the effective modalities of scientific inquiry, as a product of men who, with their behaviour towards research, teach how to carry out (Chap. 9) or not (Chap. 10) scientific investigation, and facing the problem of the features of the so-called scientific community. At the end of the discussion, I want it to emerge that the practice of experimental sciences is a *profoundly human activity* (Chap. 11). By this I mean that it is not given once and for all to be practiced with immutable rules; on the contrary, it has been and is the subject of continuous study and rethinking.

We can find two opposite attitudes (with all intermediate gradations) towards science. On the one hand, there is a *radical positivist* attitude, according to which the only real knowledge would be scientific knowledge, the one that goes under the horrendous name of *hard sciences*, and all the other disciplines would not be real, objective knowledge. On the opposite side there is the approach that we could call *radical humanist*, according to which the natural sciences can perhaps be considered knowledge, since they have an objective utility, but they cannot be defined as *culture*, their learning would not be considered necessary for spiritual maturation and cultural background of an individual with a deep education. It seems to me that both attitudes are highly reductive, if not completely incorrect. This text aims at overcoming the first attitude by showing that experimental science is an activity in which everything of man is involved, that is, in which history, philosophy, genius, creativity, morality are involved, categories generally considered external to the field of mathematical, physical, natural ("hard") sciences. On the other hand, the intention is to overcome the second attitude by highlighting the *cultural dignity* of humanity's scientific experience.

The great mathematician Ennio De Giorgi spoke of the *sapiential value* of the "mathematical, physical and natural sciences", which he compared to that of the "human sciences". Science is not only technique, it is not a sort of well-thought-out instruction manual, it is instead a building of conquests of human thought whose essential features should become part of all those who educate themselves, who should learn the foundations of experimental sciences (and of mathematics which is its language) just as they learn the rudiments of history, geography, grammar, literature.

I hope that the reader, at the end of this itinerary, has built a sufficiently clear picture of what lies at the foundation of physics and its experimental method. I wish for succeeding in proposing an agile path, which may be able to provide the essential coordinates of this great activity, whose instruments humanity has developed over centuries and whose results are among the greatest achievements of mankind.

Florence/Milan, Italy Jacopo Parravicini
July 2023

Acknowledgements

The reader has certainly already understood that there have been many cues that contributed to this book. At the origin of these suggestions, teachings, and reflections, there are many masters that here I intend to briefly recognize.

My first and most important thanks go to my two teachers in physics and science: *Giambattista Parravicini* and *Eugenio Del Re*. They taught me curiosity, method, ability to judge and reflect, articulated thought, a method for evaluating scientific texts and communities, passion, and the human and the cultural value of the discipline.

Among the physicists working at "Sapienza" University of Rome, and at the University of L'Aquila, I thank Profs. *Paolo Di Porto*, *Bruno Crosignani*, as well as Prof. *Claudio Conti* for their stimulating scientific discussions and for the experimental works carried out together.

In the Mathematics and Physics Departments of the State University of Milan, I thank first of all Profs. *Luigi Galgani* and *Andrea Carati*, for arousing my interest in the foundations of physics, for their attention to the original sources, and for teaching me intellectual nonconformism in scientific research; I also thank Prof. *Marcel Di Vece* for discussions and suggestions regarding the modern scientific community.

Moreover, I thank and remember *Alessandra Tomaselli* for her great teaching on carrying out laboratory activities and for her professional and human support she gave me in important moments of my career: she was an experimental physicist and engineer, teacher, researcher, and scientist with a humble personality but of great value, but unfortunately she prematurely died.

For their attention to a multidisciplinary approach, I would like to mention in particular Prof. *Elio Sindoni* from the University of Milano-Bicocca and the Rev. Prof. *Alberto Strumia* from the "F. Severi" High Mathematics Institute: they are the ones who taught me to range between the various branches of culture without worrying about the artificial subdivision between humanistic and scientific disciplines.

Still within the context of the University of Milano-Bicocca, I would like to remember Prof. *Francesca Della Volta* from the Department of Mathematics, who helped my reflections to take a first coherent path, as well as Profs. *Leonardo Colzani*; *Matteo Luciani*, *Matteo Pavan*, and *Daniele Zanotti* for our fruitful intellectual exchanges about the problems of the foundations of physics and mathematics; Profs.

Maurizio Acciarri and *Simona Binetti* from the Department of Materials Science, with whom I shared a deep and formative experimental work; and Prof. *Marco Fanciulli* for our stimulating discussions.

Among colleagues from the Physics and Astronomy Department of Florence, I would like to thank the director Prof. *Duccio Fanelli*, the Ph.D. coordinator Prof. *Giovanni Modugno* and his predecessor Prof. *Raffaello D'Alessandro*, the President of the Degree Course Council Prof. Vitaliano Ciulli, who showed great appreciation for my work and put me in the best conditions to carry out. Moreover, for being valuable and stimulating, both intellectually and scientifically, colleagues, I want to remember Profs. *Francesco Biccari*, *Giuseppe Luca Celardo*, *Alessandro Cuccoli*, *Leonardo Fallani*, *Massimo Gurioli*, *Gabriele Pasquali*, and *Anna Vinattieri*.

Other important intellectual contributions came to me from teachers and scholars external to the specific physical-mathematical studies: I thank the Rev. Profs. *Gabriele Mangiarotti*, H.M.R.E. *Luigi Negri*, and Prof. *Stefano Tasselli* for the stimuli on philosophical, historical, logical topics, and Prof. *Giraldi* for introducing me to Aristotelian and symbolic logic. To them are added friends and colleagues from very different disciplines, who stimulated me to an in-depth investigation of scientific knowledge: Prof. *David Conversi* for psychology, Profs. *Nicola Schiavone* and *Mariano Bizzarri* for biomedical sciences, and Profs. *Dario Altobelli* and *Barbara Osimani* for sociology and philosophy of scientific research.

In conclusion, I thank Prof. *Marta Benedetti* for the English translation revision and Prof. *Viola Sanvito* for her help in analysing and translating the original texts.

Jacopo Parravicini

L. D.

Contents

Part I
Concepts and Instruments

This part provides some answers to the question of what the scientific method is in general, and in particular what is physics. The discussion starts with the following issues:

What is scientific knowledge?
What is physics?
What is the relationship between physics and mathematics?
What is a measure?
What is an experiment?
What are the conceptual instruments that make it possible to carry out an objective measurement?
What are the assumptions that make it possible to carry out an experiment?
How have we arrived at the current discipline of physical science?

Chapter 1
The Bases of Scientific Knowledge

Δῆλον ὅτι καὶ τῆς περὶ φύσεως
ἐπιστήμης πειρατέον διορίσασθαι
πρῶτον τὰ περὶ τὰς αρχάς.
Aristotle

1.1 Phenomena

There is no instruction manual that describes how that intellectual challenge called "science" is done. Science intended as a subset of the larger enterprise of human knowledge. However, the great men who made the greatest contributions to the construction of this building have often provided insights into the structure of scientific research. In this text I rely on these men to explain what science is and what physics is.

It is a fact that the essential traits on which scientific research is based have gradually been revealed, coming from the work and reflection of so many men, of various inclinations, heterogeneous formations and different epochs. Therefore, it is all the more remarkable that, looking at the edifice of science as a whole, one can notice a surprising unity whereby each element, although of different origin and importance, relates to the others in a balanced way. Thus, we are facing a building similar to the ancient cathedrals, whose construction has often lasted for centuries, but whose result is equally harmonious and unitary, even in the variety of individual elements. This unity in plurality shows an aspect: at the base of scientific research there is a common factor, identical for all scholars and scientists. All the cultists of the discipline must confront themselves with it identically as it transcends the diverse and heterogeneous contexts. This makes it possible to reach that unity that has produced the great achievements of knowledge of recent centuries. A principle always identical to itself. What is the name of this item?

J. Parravicini, *The Foundations of Experimental Physics*,
https://doi.org/10.1007/978-3-031-55452-0_1

Since, with good approximation, the problem of the foundations of physics is largely overlappable with that of the foundations of science, we can begin by asking ourselves what is modernly called science. Albert Einstein writes

> (1.1) Science is the attempt to make the chaotic diversity of our sense-experience correspond to a logically uniform system of thought. In this system single experiences must be correlated with the theoretic structure in such a way that the resulting coordination is unique and convincing.[1]

A feature of many of Einstein's texts is the large conceptual density exhibited in a clear and seemingly simple form. The items which he finds here in the scientific enterprise are

1. sensible experience;
2. the apparent chaos of such experience;
3. thought;
4. logical coherence;
5. the mental construction of a system;
6. the internal uniformity of such a system;
7. the necessity of a correlation between mental construction and experience;
8. the necessity of a sufficiently robust ("convincing") resulting balance;
9. the nature of attempt of the scientific investigation, for which it is inherently incomplete.

It is clear that the elements listed above are many and each of them deserves a discussion on its own. For the sake of our argument, we begin talking about the first element: sensible experience. The starting point is therefore the sensitive experience of each one. Einstein continues

> (1.2) The sense-experiences are the given subject-matter.[2]

What is given, which is the subject of science, is what reaches our senses, i.e. what is shown to us. This is "the data", this is the subject of scientific investigation. What is shown to us is told *phenomenon*. Phenomena are the bases of all science.

The word phenomenon comes from the Greek verb φαίνειν, which means "to show", "make evident", "make clear". "Phenomenon", in particular, is the substantiated neuter gender of the middle-passive participle:

τὸ φαινόμενον

It has got a simultaneously passive and reflexive meaning. Therefore, it means at the same time "that which is showing itself" and "that which is shown"; it can also be translated as "that which is evident", "that which is made evident", or "that which becomes evident". Summarizing, we can state that:

Definition 1.1 (*General object*) The object and starting point of all scientific research is what reaches our senses from the world around us.

[1] Albert Einstein in Einstein (1940).

[2] A. Einstein in Einstein (1940).

Approaching the specific topic of this discussion, we refer to the natural world. Thus, we reformulate the above statement as

Definition 1.2 (*Specific object*) The object and starting point of all scientific research are NATURAL PHENOMENA.

The sciences that have got the phenomena of nature as their object are therefore called *natural sciences*. Consequently, we may set a first fixed point of our reasoning: scientific research (particularly physics) *does not* arise from reasoning, but from OBSERVATION. We find this concept in the thought of Alexis Carrel[3]:

> (1.3) *Observer est moins facile que raisonner. Comme on le sait peu d'observation et beaucoup de raisonnement conduisent à l'erreur; beaucoup d'observations est peu de raisonnements, à la vérité. Mais il y a un plus grand nombre d'esprits capables de faire des syllogismes que de saisir exactement le concret. [...] Pour ne pas se tromper dans la poursuite du réel, il importe de se baser non sur les vues de l'esprit mais sur les résultats de l'observation et de l'expérience.*[4]
>
> Observing is less easy than reasoning. It is known that little observation and much reasoning lead to error; much observation and little reasoning to truth. But there are many more souls capable of doing syllogisms rather than grasping exactly the concrete [...] In order not to be mistaken in pursuing the real, one must rely not on the visions of the soul but on the results of observation and experience.

The above statements are obviously not an invitation to deny the use of reason. On the contrary, they are a reminder of a correct use of reason, as an instrument that has got a precise purpose, namely, the knowledge of reality, and not an end in itself. Somehow this can be interpreted as a modern version of the principle "adequatio rei et intellectus"[5] enunciated by St. Thomas Aquinas centuries earlier. Carrel's affirmation also invites us to always prefer the observation, which in scientific research is condensed into what are called *data* (which are discussed later), to any mental construction. In the research process, the practice of the natural sciences must constantly be guided by observation, and must constantly return to observation.

Given the phenomenological basis of the natural sciences, we may now take a closer look at how the scientific approach to phenomena is specifically structured. According to Richard P. Feynman , three actions can be identified that form the basis of the scientific approach:

1. OBSERVATION,
2. REASONING,
3. EXPERIMENT.

To develop scientific knowledge these actions in themselves are not sufficient: they must be carried out in a peculiar way. Specifically, in order to develop a scientific discourse, it is necessary:

---

[3] 1912 Nobel Prize in Medicine.

[4] A. Carrel in Carrel (1950).

[5] "Correspondence between reality and intellect".

1. to observe the *right things* in the *right way*;
2. to ask the *right* questions;
3. to consider the experiment as a true *judge*.

At this point our discussion focuses on the features of these two groups: the first contains a list of actions, the second the peculiarities that these actions must present.

The starting point of scientific knowledge is, precisely, a given phenomenon, which is *observed* and recorded. After such a process, our mind employs *reason* in order to construct a logical-rational scheme of what has been observed. Such scheme determines logical implications with the features that they can, at least in part, be proven through an *experiment* in the real world. It is worth noting that in the first action (observation) nature (which produces the phenomenon) is active and man (that records the phenomenon), in some way, is passive: what is observed is a given.

Man has only to be open and listen to reality. In the other two steps the man plays a markedly active part. Reasoning must be properly carried out in such a way that it can be tested by experience. The *experiment* is as decisive as it is conceptually delicate. On the one hand, it is a pure construction of the scientist, where he can exercise his imagination as well as his expertise and technical ability in properly interrogating nature. On the other hand, it is nature itself that responds to the experiment, with its own answers, which can confirm or refute, totally or partially, the reasoning previously conducted by the researcher.

With regard to the features that the aforementioned actions must have, first of all it is necessary to understand what is meant by "right". If our aim is to build knowledge of natural phenomena, the meaning of the adjective "right" must be practical and utilitarian: it is what makes the knowledge of natural phenomena develop to be considered right. Outside this criterion, it is not possible to establish a mechanical procedure to understand what are actually the *right things* to observe and what is the *right way* to do it. This is in the realm of experience and intuition. Some examples can be provided. If I wanted to study the motion of a macroscopic object by neglecting its mass and concentrating on its colour, evidently I would not observe the *right thing* about it. If I wanted to describe a tree by trying to give instantaneously the number of atoms of which it is composed, it would *not* be equally *right*. Similarly, if you pretended to measure the size of an atom, or the distance between the earth and the sun, with an ordinary ruler, this would obviously be *not the right way*. The *reasoning* step requires asking *the right questions*. Again, *right* according to the criterion of advancing knowledge. Among the right questions we can include that about the predictions and the measurable implications of the model that is being developed.[6] Finally, the experiment is said to be the *judge*. It is constructed by the researcher and the researcher submits himself to it. Thus, the experiment is the judge chosen by the defendant himself. Consequently it is noteworthy how the honest researcher builds with his own hands that tribunal which is potentially able to prove him wrong

[6] On this topic, it is worth making a distinction: it is not necessary that every single part of the developed model is measurable, but it is necessary that there are sufficient measurable elements, verifiable through experiments, so that the degree of correctness of the model can be resolved. This issue is discussed in more detail in Sects. 8.2.1 and 8.2.2.

and, nevertheless, submits himself to it.[7]. The distinguishing factor of what is called *science* has been seen in this:

(1.4) The principle of science, the definition, almost, is the following: *The test of all knowledge is experiment.* Experiment is the *sole judge* of scientific "truth."[8,9]

It is worth noting that the need to accept the real data, even taken to the extreme—even when they do not confirm the research's reasoning and thus constitute a judge to the work—represents a strong ethical teaching. The scientific method is a test bench on which whoever wants to progress in knowledge must humbly accept being judged, sacrificing, if necessary, the mental schemes he has built to accept real data. As mentioned above (quote 1.3), in many cases it is not easy to abandon one's mental schemes in favour of reality: in the daily practice of scientific research various obstacles are introduced, which we explore in Chap. 10.

(1.5) *Certes, [ce qu'on étudie l'expérimentation] c'est une nature simplifiée, préparée, parfois mutilée en fonction de l'hypothèse préalable, que l'expérimentation interroge; il n'empêche qu'elle garde en général les moyens de démentir la plupart des hypothèse. Einstein faisait remarquer que la nature, aux questions qu'on lui pose, réspond le plus souvent* non, *et, parfois,* peut-être.[10]

Certainly, [what experimentation considers] is a simplified nature, it is prepared, sometimes mutilated according to prior hypothesis, which experimentation investigates. However, nature generally maintains the ability to refute most hypotheses. Einstein pointed out that nature most often answers *no* and sometimes *perhaps* to the questions we ask.

To explain in more detail how scientific investigation is carried out and the conceptual strategies that are normally employed, Feynman uses the metaphor of a chess game[11], which is a vivid example. Let us suppose that the set of natural phenomena is a chess game that is taking place before our eyes, and also suppose that we know nothing about chess except that it is a game. In these conditions, our goal is to deduce the rules, but we cannot do anything else but attend the game. The investigation of nature is very similar to this condition: an attempt to reconstruct the rules of the great chess game in which we are present exclusively by observing it, without anyone from outside explaining the rules to us. You can make the first hypotheses on how the game works by watching the match. Typically the verification of such hypotheses takes place using different strategies.

1. SIMPLIFYING the situation, the phenomenon, the problem. For example, we can neglect the colour of the pieces and concentrate on the different shapes.

---

[7] These concepts on the conduct of scientific research are developed more extensively in Chap. 8 and in Part III.

[8] R. Feynman Sect. 1–1 in Feynman et al. (2005a).

[9] A precise enunciation of the experiment is provided in Chap. 4, especially in the definition 4.2, distinguishing between a more generic concept of *Observational investigation* of natural phenomena, defined in 4.1 (editor's note).

[10] I. Prigogine et al., Chap. 1 of Prigogine and Stengers (1986).

[11] Section 2–1 in Feynman et al. (2005a).

2. In LIMITING the situation one evaluates a single phenomenon in simpler circumstances, trying to sketch partial rules. For example, we can evaluate only the movement of the pieces and not the act of capturing them.
3. If we APPROXIMATE, only a part of the involved elements can be considered. In the example of chess, we can limit ourselves to what happens around a single piece.

Simplifying, limiting, and approximating are the typical strategies that are employed in scientific research. They are the result of the experience, and *trying and trying again*[12] has proved they are fruitful strategies: they are *right* in the aforementioned sense.

All the above considerations concern the phenomenological foundations of scientific research and of the scientific method and are valid for all the natural sciences. So what distinguishes physics from the other natural sciences?

1.2 Peculiarity of Physics

At this point the question arises as to what is physics. To start with, the etymology of the word "physics" is provided:

φύσις

which comes from ancient Greek and literally means "nature". The correlated adjective, which literally means "natural" is φυσικός, but the word "physics" comes from the noun plural neutral, i.e.

τὰ φυσικά

which literally means "natural things" or rather "what is related to nature": the concept is rendered with a plural neuter. Physics is therefore by etymology everything that is found in nature and the term already has in itself the concept of plurality. Recalling Einstein's quotation 1.1, we underline the fact that what pertains to physics is only and exclusively what is perceived by our senses. Therefore, physics has nothing to do with reality outside our senses[13] such as, for example, abstract mathematical or logical systems. In short, being based on the experience of our senses, by definition physics does not deal with everything that is μετὰ τὰ φυσικά, that is, "beyond what is related to nature".

[12] "*Provando e riprovando*", motto of the Accademia del Cimento (1657–1667), the first scientific association in the modern sense, founded by the disciples of Galileo Galilei. This Italian syntagm is taken from Dante Alighieri's *Divine Comedy*, the III Canto of Paradise; in that context the words mean "demonstrating and refuting". Therefore, the motto assumed by the Academy has a double meaning: a more modern and literal one, stating the need for experimentation—"*trying and trying again*"—and an original and ideal one—"*demonstrating and demonstrating again*"—stating the need for demonstration, including counterfactual reasoning.

[13] Chapter 1 of Lindsay and Margenau (1957), Part I Chap. 1 of Duhem (2016).

(1.6) Physics is the most fundamental and all-inclusive of the sciences, and has had a profound effect on all scientific development. In fact, physics is the present-day equivalent of what used to be called *natural philosophy*, from which most of our modern sciences arose.[14]

What we call physics, understood as a discipline at the foundation of all studies concerning the investigation of natural phenomena, has sprung from what for centuries has been called

philosophia naturalis,

that is, "natural philosophy" or, better, "philosophy of nature". *Philosophia*, of course, should also be understood in an etymological sense as "love for" -and therefore "practice of"- knowledge. We here see that at the beginning there was no dichotomy between that knowledge which today we would call humanistic, the tout-court philosophy, and scientific knowledge, physics. The former is the philosophy today called theoretical (including what are classically called ontology and "metaphysics"), the second one is the philosophy of nature. Both of them were and are parts of a single discipline, *philosophia*, the background of the one who loves knowledge and wants to devote himself to it[15].

The aforesaid considerations have anticipated some concepts that we now need to systematize. To say that physics deals with phenomena, and natural phenomena specifically, does not allow us to fully identify physics as a discipline. In fact, it is necessary to specify further details within a scheme that is more systematic.

To precisely identify the specific features of this branch of knowledge, we may refer to the most classic and tested system of classification of disciplines, which provides us with a precise and relatively simple scheme.[16] The purpose of any discipline is to increase knowledge through a specific investigation of reality. On the basis of which criteria, therefore, is it possible to clearly specify the discipline here considered? Referring to the aforementioned scheme, we assume the *subiectum* ("subject") of the investigation as the starting point; moreover, we identify four specific features that describe the conceptual content of the discipline considered. We could make an analogy with an origin and a four-dimensional Cartesian axis system. Through these coordinates, which are defined starting from subject and object, the discipline is uniquely determined[17]:

0. *subiectum*, i.e. the *subject* of the investigation;
1. *obiectum materiale*, i.e. the *material object*, the actual matter that the discipline deals with;

---

[14] R.P. Feynman Sect. 3–1 in Feynman et al. (2005a).

[15] It is worth noting that in the same Aristotelian classification the investigation of nature, τά φυσικά, anticipates the theoretical investigation: nowadays the latter is considered a discipline belonging to philosophy alone as it is called *metaphysics*, from the expression μετὰ τὰ φυσικά, literally "after what is related to nature".

[16] We here refer to the classical epistemological classification (Strumia 2017; Basti 2002).

[17] (Strumia 2017).

2. *obiectum formale*, i.e. the *formal object*, the form that the aforementioned material assumes for investigation;
3. *obiectum formale quo*, i.e. the *formal object from which* [we investigate], the specific *point of view* of the discipline;
4. *obiectum formale quod*, i.e. the *formal object which* [is investigated], the aspects of the subject that are inspected by the investigation as determined by the aforementioned point of view.

The *first item*, the SUBJECT, is the origin of our reference system, it is what science studies. We have thus already identified the subject in "what is related to nature" (τὰ φυνικά), making physics part of the complex of natural sciences.

The other four items, the coordinates, are all theses, doctrines, specific concepts of the discipline. As far as physics is concerned, we have already introduced the first three coordinates. The MATERIAL OBJECT are the aforementioned phenomena. The FORMAL OBJECT is correlated with the methodology employed, i.e. the investigation of an experimental nature, and coincides with what can actually be grasped of the subject: these are the observations and the corresponding experiments. Specifically, the results of observations and experiments.

This first pair of coordinates, material and formal objects, are common to all natural sciences. Conversely, the last two coordinates allow us to identify physics with precision, grasping its specific features.

Definition 1.3 (*Einstein's definition of Physics*)

> (1.7) What we call physics comprises that group of natural sciences which base their concepts on measurements, and whose concepts and propositions lend themselves to mathematical formulation.[18]

Here, in addition to synthetically recalling the formal object ("measurements"), Einstein clearly defines the FORMAL OBJECT QUO and the FORMAL OBJECT QUOD.

The formal object *quo* is the point of view of the investigation. It is the key feature that distinguishes physics from other disciplines. Specifically, the formal object *quo* is mathematics.

The formal object *quo*, we said, determines the *quod*, i.e. what is actually studied and investigated of the subject. In the specific case of physics this coincides with those "concepts and propositions [that] lend themselves to mathematical formulation". Among these are the *physical quantities* (which are discussed in more detail in Chap. 3), i.e. all those measurable elements of reality that can be placed in a correspondence one-to-one with mathematical entities; however, there are also all the concepts related to the terms of relation and extension (studied for example by topology).

Thus, Table 1.1 summarizes the specific coordinates we have just discussed. In brief, we can reduce these coordinates to the two essential pillars of the discipline:

[18] A. Einstein in Einstein (1940).

Table 1.1 Coordinates of physics discipline

General feature	Specific feature
Subject	Nature
Material object	Natural phenomena
Formal object	Observations and experiments
Formal object *quo*	Mathematics
Formal object *quod*	Mathematizable quantities

natural phenomena and reality. Paraphrasing Einstein's Definition 1.3 (quotation 1.3), we can therefore state that

Definition 1.4 (*Field of investigation of physics*) Field of investigation of physical science coincides with the mathematizable characteristics of natural phenomena.

Proceeding along this line of thought, the next chapters of I part are written from a perspective aimed at specifically highlighting the features and the complex relationship between mathematics and physical science.

References

G. Basti. *Filosofia della Natura e della Scienza*, vol. 1, 1a edn. (Pontificia Università Lateranense, Roma, Città del Vaticano, 2002)

A. Carrel, *Réflexions sur la conduite de la vie* (Librairie Plon, Paris, France, 1950)

P. Duhem. *La théorie physique. Son objet, sa structure*, 2nd edn. (ENS Éditions, Lyon, France, 2016). http://books.openedition.org/enseditions/6077

A. Einstein, Considerations concerning the fundaments of theoretical physics. Science **91**, 487–492 (1940)

R.P. Feynman, R. Leighton, M. Sands, *The Feynman Lectures on Physics*, vol. 1, 2nd edn. (Addison-Wesley, Boston, USA, 2005)

R.B. Lindsay, H. Margenau, *Foundations of Physics*, 2nd edn. (Dover, New York, USA, 1957)

I. Prigogine, I. Stengers, *La nouvelle alliance-Metamorphose de la science*, 2nd edn. (Gallimard, Paris, France, 1986)

A. Strumia, *Dalla filosofia della scienza alla filosofia nella scienza*, 1a edn. (ESC-Edizioni Santa Croce, Roma, Italia, 2017)

Chapter 2
The Role of Mathematics

Principia mathematicæ sunt applicabilia naturalibus rebus.
St. Thomas Aquinas

2.1 Use of Mathematics in Physics

We have so far discussed the phenomenological empirical foundation of physics, which is shared by this discipline with all the natural sciences. Then, we have specified that the peculiar characteristic of physics lies in the possibility of a mathematical formulation of its concepts (according to Einstein's definition 1.3). Regarding what we mean by "mathematics", we here need to provide a specification of what we mean in the context we are dealing with. There are many definitions of what mathematics is[1]. Many of them are related to the problem of the foundations of the discipline. As far as our discussion is concerned, closely related to experimental physical science, we limit ourselves to choose a useful definition that is specifically practical to our discussion. Therefore, we provide the following:

Definition 2.1 (*Mathematics for physics*) What we mean by *mathematics* is the *language* which describes the reality according to the categories of *quantity*, *relation*, and *extension*.[2,3]

It is worth emphasizing how, on the basis of the aforementioned definition 1.3, Einstein defines mathematics as a language ("formulation") capable of describing that part of reality which is suitable for such a representation. Thus, we introduce the

[1] Cf. e.g., (Sakkurai 2005; Dorato 2005; Basti 2002; von Neumann 1995a; Wilder 2012; Strumia 2017), etc.

[2] Cf. e.g., (Strumia 2017) and (Basti 2002).

[3] Actually, the category of "extension" could be reduced to the first two, but we would enter into a philosophical discourse that goes beyond our present discussion.

J. Parravicini, *The Foundations of Experimental Physics*,
https://doi.org/10.1007/978-3-031-55452-0_2

term *mathematizability* as the characteristic of "being suitable for a mathematical formulation". This concept requires a precise definition in relation to the specific issue of the *language* to which it is connected. We say that

Definition 2.2 (*Mathematizability*) We call *mathematizable* an aspect of natural reality that can be described and/or treated using mathematical language.

The problem of *language*, specifically of mathematics understood as language, to which the concept of *mathematizability* as defined in 2.2 is closely related, plays an essential role as a founding element of the natural sciences. Already in ancient and medieval ages there were many intuitions regarding the suitability of mathematics to describe natural phenomena: these contributions, among which the epigraph at the beginning of this chapter is an important example, are mentioned in Chap. 4. Nonetheless, it is universally recognized that the fundamental intuition of the systematic mathematizability of natural phenomena is provided by Galileo Galilei, in a famous passage from *Il Saggiatore*. We take this passage as the turning point where not only the suitability, but even the need to use a mathematical language to understand nature is openly stated.

> (2.1) *La filosofia [naturale] è scritta in questo grandissimo libro che continuamente ci sta aperto innanzi gli occhi (io dico l'universo), ma non si può intendere se prima non s'impara a intender la lingua, e a conoscer i caratteri ne' quali è scritto. Egli è scritto in lingua matematica, e i caratteri sono triangoli, cerchi, ed altre figure geometriche, senza i quali mezi [sic] è impossibile a intenderne umanamente parola; senza questi è un aggirarsi vanamente per un oscuro laberinto.*[4]
>
> [Natural] philosophy is written in this huge book which is continually open before our eyes (I mean the universe), but it cannot be understood unless one first learns to understand its language and to know the letters in which it is written. It is written in mathematical language, and the letters are triangles, circles, and other geometric figures, without which it is humanly impossible to understand a word of it; without these it is a vain wandering through a dark labyrinth.

The above statement can be considered the birth declaration of modern physics,[5] because it precisely identifies its form, specifying the main ones among the characterizing elements that we have previously stated (cf. Table 1.1). One could say that the idea expressed in this statement is the *founding concept* of physical science. It is interesting to note that this embryonic affirmation already raises the question of mathematics understood as a language. For what Galilei wants to do, in fact, mathematics is not a discipline but a language, a tool aimed at understanding and describing a specific reality. To analyse this item, we summarize the problem of the relationship between mathematics and nature (that is, the problem underlying physical science) by asking the following question:

---

[4] G. Galilei, Chap. 6 of Galilei (1623).

[5] I define this affirmation as a "declaration", i.e. a "birth certificate", as it is in all respects, but I deliberately do not affirm that physics is actually born through this assertion. Galilei's statement specifies the fundamental characteristics of physics, but it does not determine the event of its birth, which is a much debated issue (cf. inter alia Sect. 4.1).

Why is mathematics able to describe such a large number of natural phenomena in such detail?

The possible answers belong to two categories: those of the *a priori*, called *Platonic*, and those of the *a posteriori*, called *antinaturalist*.[6]

Beyond the huge historical significance of the 2.1 statement, whose importance we do not want to underline further, it is necessary to analyse its meaning. With these words not only does Galileo recognize that mathematics is suitable for describing natural phenomena, but he even goes one step further in assuming that the very structure of the universe is mathematical. We call this assumption *Galilei's postulate*, noting the fact that, since it is a *postulate* that Galilei considers a presupposition for the development of physical science, it does not belong to the latter one: in the etymological sense, it is a *metaphysical* affirmation (μετὰ τὰ φυσικά, cf. Sect. 1.2). Actually, in modern terms this statement belongs more to philosophy than to science. One side of this postulate, assumed as a "working hypothesis", certainly led to great scientific results. Among many others, Isaac Newton held it to be entirely valid and on the basis of this criterion he accomplished a large part of his scientific and philosophical work. Therefore, Galilei's answer to the question we have asked belongs to the category of *a priori*. We can paraphrase it as follows:

Mathematics provides a good description of natural phenomena because the very structure of nature is mathematical; indeed, mathematics itself is the principle of all nature.

Therefore, this interpretation considers mathematics an *a priori* because it believes that mathematics underlies all natural phenomena which somehow come after it. This interpretation is undoubtedly charming, and several great physicists were or are convinced of its veridicity. However, it should be noted that, although being important and fruitful, *Galilei's postulate* is not empirically and phenomenologically founded, mainly for two reasons. First of all, the experimental sciences, as we saw, cannot be based on anything other than the investigation of phenomena. No natural phenomenon, no experiment is able to *display* relationships, structures, mathematical laws; these are always abstraction and description created by the human mind. Secondly, Galilei's postulate carries a risk for the development of knowledge. In fact, if we assume that the universe is "written" in mathematical language, and that the only suitable instrument to its investigation is mathematics, then we introduce the forcing that any development in knowledge requires making mathematical any form of knowledge. In other words, developing real knowledge about some aspect would coincide with mathematizing that aspect. Galileo does not directly claim this, but after him several people take this perspective on a scientific, philosophical, and epistemological level. The problem with this is that nothing guarantees us that everything that is knowable can be expressed in mathematical language. After centuries of progress in scientific knowledge, Einstein reverses Galilei's perspective. According to Galilei, we practice physical science because the universe is written in mathematical language. Einstein, in some way, reverses this vision: for him the practice

[6] Cf. e.g., (Sakkurai 2005; Dorato 2005).

boundaries of physical science exactly delimit those parts of the universe that can be described in mathematical language, as stated in the quote 1.7. Starting precisely from the relationship with mathematics, it is also possible to specify the application field of physical science.

(2.2) [The] realm [of physics] is accordingly defined as that part of the sum total of our knowledge which is capable of being expressed in mathematical terms.[7]

Compared to Galilei, Einstein can exploit three additional centuries of knowledge, of recorded phenomena, of developed physical and mathematical concepts. Then he may provide a more balanced definition of the use of mathematics in the natural sciences. In this passage, Einstein traces the specific boundaries and thus the field of investigation of physical reality, consequently stating that, a priori, not all knowledge (not even that referring to natural phenomena) is expressed, nor should it be, in "mathematical terms".

At this point the picture looks like this. On the one hand there are natural phenomena and on the other there is mathematics. These two areas coexist quite separately. It has been empirically noted that mathematics is able to describe a large part of natural phenomena in a detailed, elegant, powerful, and predictive way (cf. Sect. 2.3). The cause of this possibility can be investigated by specific disciplines such as philosophy, but it is not a matter of investigation of physical science, which is limited to exploit it. In this sense, Nicolas Bourbaki[8] states:

(2.3) *Il est en tout cas une constatation qui pourrait, sur ce point, incitér à l'avenir les philosophes à plus de prudence: avant les développements révolutionnaires de la physique moderne, on a dépensé beaucoup de peine our vouloir à toute force faire sortir les mathématiques de vérités expérimentales, notamment d'intuitions spatiales immédiates; mais d'une part la physique des quanta a montré que cette intuition "macroscopique" du réel couvrait des phénomènes "microscopiques" d'une toute autre nature, relevant de branches des mathématiques qui n'avait certes pas été imaginées en vue d'application aux sciences expérimentales; et d'un autre côté, la méthode axiomatique a montré que les "vérités" dont on voulait faire le pivot des mathématiques n'etait que des aspects très spéciaux de conceptions générales qui n'y limitait nullement leur porté. Si bien qu'en fin de compte, cette intime fusion dont on nous faisait admirer l'harmonieuse nécessité, n'apparaît plus que comme un contact fortuit de deux disciplines dont le liens sont beaucoup plus cachés qu'on ne pouvait le supposer* a priori.[9]

There is in any case an observation which could, on this item, incite philosophers to be more cautious in the future: before the revolutionary developments of modern physics, a lot of effort was spent trying to force mathematics to be obtained from experimental truths, notably of immediate space insights. But, on the one hand, quantum physics showed that this

[7] A. Einstein in Einstein (1940), continuation of quote 1.7.

[8] Pseudonym given to a group of important mathematicians, mainly of the French school, who, starting from the 1930s, developed the project of exposing the most important results of modern advanced mathematics in a series of encyclopedic volumes. Members of this group were, among others, Henri Cartan (1904–2008), Jean Dieudonné (1906–1992), Szolem Mandelbrojt (1899–1983), André Weil (1906–1998). Through its work this group had an enormous impact on the development of mathematical science in the twentieth century.

[9] N. Bourbaki, in Bourbaki (1948).

> "macroscopic" insight of the reality would cover "microscopic" phenomena of a completely different character, relating to branches of mathematics which had certainly not been imagined in view of application to the experimental sciences. On the other hand, the axiomatic method showed that the "truths", which we wanted to make the pivot of mathematics, were only very special aspects of general concepts which in no way limited them. Consequently, in the end, this intimate fusion, about which we were made to admire the harmonious necessity, only appears as a fortuitous contact between two disciplines whose links are much more hidden than we could *a priori* suppose.

From the point of view of physics, mathematics would be a construction with a specific aim. Given physical phenomena, the necessary mathematics would be used or constructed to suitably describe them in the most appropriate language. Therefore, the "mathematical language" would not reside in nature at all, but exclusively in the human mind as its construction. Considering the aforementioned question on why mathematics is able to describe so numerous natural phenomena is possible to carry a summary. According to the position just explained, which is purely phenomenological, the answer to our question paints mathematics as an *a posteriori*.

Mathematics provides a good description of natural phenomena because it has been built and refined by the human mind for this exact aim; therefore, mathematics is not the foundation of natural phenomena, but natural phenomena are the foundation of the construction and structure of mathematics.[10]

The relation between physics and mathematics would thus be "constructive".[11] Asking why mathematics is suitable to describe nature would be like asking why a car is able to transport people: beyond the technicalities, it was built exactly for that purpose, so the question is trivial.

This state of affairs appears satisfactory to a certain extent, but it is an evident simplification. In fact, both the *a priori* and the *a posteriori* perspectives of mathematics have got their strengths and weaknesses.

It is true that experiments *themselves* are unable to show any mathematical structure or concept; these ones are abstract and what the first ones evaluate are only the concrete natural phenomena. Therefore, it would seem impossible to experimentally demonstrate that there is an actual mathematical structure underlying natural phenomena. In short, *Galilei's postulate* would remain only a suggestive hypothesis. Furthermore, much of mathematics, ancient and modern, is born for supporting physics. This has occurred since the *protophysics* of ancient peoples, where the goal of measuring objects and extensions led to the development of geometry, namely, the "measurement of the earth".[12] This is confirmed in classical modern physics, for which the tools of analytic geometry, mathematical analysis, vector spaces, etc. were

[10] An enunciation of this kind of position is provided and supported, among others, by the Italian mathematician G. Israel (1945–2015), cf. (Israel 2015). In a less radical way it is also the position of J. von Neumann, cf. (von Neumann 1995b) as stated in the quote 8.11, which is somehow the opposite of Galilei's 2.1 statement.

[11] This term should not be confused with "mathematical constructivism" (Wilder 2012).

[12] Literal translation of the original Greek term γεωμετρία, from γῆ, "earth", and μετρία, "act of measuring", "measurement".

developed as tools for a rigorous formulation of classical mechanics and electromagnetism.[13]

However, it is equally true that the development of mathematics has not always been driven by that of physics. Indeed, the great development that the mathematical sciences had since the nineteenth century led to an almost opposite evolution.

> (2.4) The effect of mathematics on physics was, however, unexpected. It turned out that theories created for purely mathematical reasons were applicable to physics. That surprising result is still a puzzle, although we may surmise some partial answers. For one, the mathematical theories invented by pure mathematicians without any consideration for their possible applications, give schemes of reasoning and models of inference that are suited to the human mind; the scientist then may find the phenomena that fit the paradigm. This does not, however, solve the puzzle of why the fit is so good. Eugene Wigner (1902-[1995]), the distinguished mathematical physicist who did so much to introduce group theory in modern physics, has written an essay, "The Unreasonable Effectiveness of Mathematics in the Natural Sciences," [...] Whatever the reason, the application of theories created for purely mathematical reasons, and without any idea of their applicability to physics, is a striking phenomenon [...] In modern times, group theory, tensor analysis, Lie groups, noncommutative algebras, matrix theory, and fiber bundles are only some of the chapters of pure mathematics first developed as such and later applied to physics. The physicist finds himself in need of progressively more abstract and sophisticated mathematics, as well as more refined observational instruments.[14]

In the next section we discuss in more detail some of Wigner's ideas here mentioned. The argument developed by Segrè highlights the fact that many mathematical concepts have been developed *completely independent* on natural phenomena, without any idea of a possible use in explaining anything about nature. Just later these concepts, already available, have been used to powerfully describe several natural phenomena.[15] Therefore, these considerations raise doubts about the mere "constructive" hypothesis of mathematics and about the position of the *a posteriori* correlated to it: affirming that the mathematical language works in describing nature because it is a construction planned for this purpose evidently does not hold up in the several cases where this construction was prepared for completely different purposes. In these cases, with much greater awareness and reasonableness compared to Galilei's time, the hypothesis that *concretely there is something mathematical* in nature itself appears again: at a certain point the human mind formulates this idea in correspondence with concepts abstractly developed.

There is not a definitive answer to this issue. Certainly, the great usefulness and power of the mathematical language in physics suggested the application possibility to branches very far from the natural sciences, in particular to economics. In these areas the use of mathematics seems to provide a psychological guarantee of rigour and certainty. However, the results are still very far from being comparable to those obtained in the field of physical sciences.[16] The specification by Einstein in 2.1 is

---

[13] A brief overview is provided in Sect. 4.2.

[14] E. Segré, Chap. 3 in Segrè (1984).

[15] Quick examples are provided in Sect. 4.3.

[16] References to deepen these aspects, which go beyond the present discussion, can be found in Israel (2015) and in Ingrao and Israel (1987).

therefore crucial, where he traces the perimeter of physics by stating that what is suitable of being expressed in mathematical terms is only a part of our knowledge. What this part is, and what the corresponding part of reality that can be profitably mathematized, is a matter that must be evaluated with care and intellectual honesty, for which there is no a definitive criterion. However, it is still important to identify where the use of mathematics is appropriate and to state to what extent and under what assumptions it reasonably works. Therefore, human and even ethical elements begin to emerge, which we discuss later.[17]

Ultimately, the issue of the relationship between mathematics and reality, as posed in the above question regarding the origin of this intimate connection, remains in between the two extremes of *a priori* and *a posteriori*, between *Galilei's postulate* and the *constructive* perspective. As regards our treatise, specifically focused on experimental physics, a general interpretation can be given starting from the definition of mathematics that we have provided in Sect. 1.2, as a language describing reality according to the categories of quantity, relation, and extension. Therefore, it is reasonable to infer that the presence of these elements in natural reality generates the suitability of the mathematical language in describing physical phenomena. However, the intimate details of this remain a complex and still open question.

2.2 Mathematics as a Language

2.2.1 General Features

When we analyse the founding elements of mathematical thought and discourse, which constitute the basis of the entire building, three pairs of abstract objects can be identified.[18]

1. LOGIC and INTUITION,
2. ANALYSIS and CONSTRUCTION,
3. GENERALITY and INDIVIDUALITY.

This list is an illuminating and effective synthesis, but it is also evident that these "fundamental elements" are not exclusive to the mathematical discipline, instead they are shared, to varying degrees, by all HUMAN LANGUAGES. As a matter of fact, on closer inspection even the everyday language, when used to rationally describe a huge number of reality elements, has equally logical and intuitive, analytical and constructive, general and individual features. This shows how mathematics can be considered a full-fledged language. It is a language that has got its own peculiarities, and it is chosen because it alone has got the ability to describe *some large regions of nature* with *sophistication*, *elegance*, and deep *effectiveness*. In this regard, in the quote 2.4 Emilio Segrè recalls the opinion of Eugene Wigner on the "unreasonable efficacy of mathematics":

[17] Chapter 8 and Part III.

[18] Introduction of Courant and Robbins (1941).

(2.5) The miracle of the appropriateness of the language of mathematics for the formulation of the laws of physics is a wonderful gift which we neither understand nor deserve. We should be grateful for it and hope that it will remain valid in future research and that it will extend, for better or for worse, to our pleasure, even though perhaps also to our bafflement, to wide branches of learning.[19]

In his famous text Wigner shows that he embraces two important concepts of scientific-mathematical thought. The *first* one is what we are discussing: the adequacy of mathematics in describing nature laws as well as the depth of the agreement, found in many sectors, between its concepts and the entities that we find in nature, as we infer them from their phenomenology. It is also worth recalling the heart of Wigner's research, mainly aimed at studying the *fundamental symmetry principles*[20] applied to nature. These symmetry principles are a much less intuitive and evident aspect of mathematics than the usual "quantitative" considerations, i.e. correlated with numbers, which cover a large part of the usual mathematical science. The discovery of a symmetry in nature, in other words, gives the impression of facing an order that is in some exotic way, very little evident, and therefore very deep. This characteristic probably plays a role in the *second* expressed concept, that of the mystery or the gift, that conveys a conception of the physical-mathematical sciences that borders on mysticism.[21]

2.2.2 Logical Formalism, Semantic, and Structure

Beyond these philosophical, or even "mystical" implications, the listed-above founding elements allow us to glimpse a special richness, which the mathematical language seems to share with all other human languages, all capable of conveying analysis and synthesis, stringent reasoning and formidable intuition, creativity and constructiveness. Many of these features belong to several other human languages, both the ordinary language and, to some extent, the specific symbolic or even artistic languages.

At this point our discussion analyses in more detail the peculiarities that the mathematical language shares with other human languages.

1. *Analytical-predictive feature*. Thanks to this peculiarity, if we demonstrate that the mathematical language is able to suitably describe some physical phenomena, it is possible to carry out an abstract, purely mathematical reasoning, in order to predict a phenomenon that has not yet been measured, but which is known to be connected, through the mathematical model, to what has been measured.[22]
   To clarify this feature it is useful to consider a simple example. Let us assume that we measured that a given body moves with uniform rectilinear motion at

[19] E. Wigner in Wigner (1960).

[20] He was awarded with a Nobel Prize in 1963 precisely for this study.

[21] Israel (2015).

[22] This is discussed in Sect. 8.2.1 in the discussion on *physical theories*.

speed $v(t) = v_0$ along the axis x during time t. We know that the distance from the origin is given by

$$x(t) = x_0 + v_0 t$$

where x_0 is the distance of the considered body from the origin in the instant $t = 0$.

Let us assume that we measure the body position every 10 seconds and we see that it agrees, within our error, with the aforementioned relationship, to then finish our measurement after a number of recordings. If the same conditions persist in physical reality, it is reasonable to predict, through mere mathematical reasoning, the position of the object that is moving. This procedure, which today seems trivial because it is by now completely customary, requires that the mathematical language has got the characteristic of adhering to the reality it describes in such detail (analyticity) that it can formulate predictions (predictability).

In order to correctly set up the analytic-predictive feature of the mathematical language in the landscape of the various languages, it is important to note that it shares this characteristic with the classic *logical-deductive* reasoning, which is expressed in a language that uses the *current language* (English, Latin, Italian, Greek, etc.). The common feature of mathematical structures and logical-deductive structures are so deep that, starting from the nineteenth century, it was possible to develop *propositional calculus* as a formalization of the classic inferential and syllogistic structures in a symbolic language of a mathematical nature,[23] giving life to the branch of *mathematical logic*.[24] However, to be clearer and more incisive, here we choose to neglect, at least for now, the symbolic-propositional writing and to employ the usual natural language.[25] Therefore, to fix ideas, let us consider the most classic syllogism,[26] that of *mortal Socrates*.[27] Premises are

All men are mortal.
Socrates is a man.

Suppose we know these premises from having experienced. There are sufficient reliable experimental data that can reasonably be described by claims that Socrates is a human being and that human beings are mortal. This is analogous to saying that we had enough experimental data to affirm that the motion of a given object

[23] In particular, starting from the work of George Boole (1815–1864) and Augustus De Morgan (1806–1871), who was later mainly developed by Gottlob Frege (1848–1925), Giuseppe Peano (1858–1932), Alfred N. Whitehead (1861–1947) and Bertrand Russell (1872–1970), cf. (Boyer 1968; Bottazzini et al. 2016) and, more specifically, A. Church in Chap. 2 of Church (1956).

[24] Rautenberg (2009), Enderton (2001), Cole Kleene (1967), Church (1956).

[25] We limit here to briefly analyse a parallelism with the world of logical-deductive classical reasoning, neglecting the formalism of propositional calculus and metamathematical languages that go beyond our discussion. Very brief hints of propositional calculus and metamathematics are given in Sect. 7.1.

[26] DARII syllogism of *first figure* (cf. Sect. B.2.2). About this syllogism a possible simple writing is $A \rightarrow B, C \rightarrow A \vdash C \rightarrow B$.

[27] Copi et al. (2019); Lindsay and Margenau (1957).

is well described by the law of uniform rectilinear motion. Given the aforementioned premises, we can deduce a prediction without needing further experimental data, accomplishing a purely abstract logical reasoning, which leads us to conclude that:

Socrates is mortal.

This conclusion leads us to do a prediction about reality, a prediction that is true even without having passed through experiential data, i.e. without actually "measuring" the death of Mr. Socrates, who could very well still be alive and in excellent health! We note that the premises are given by experiential data; thus, their truth is not determined by the correctness of the formal reasoning but exclusively by the correspondence between the statement and the fact inferred from an experiment. The truth of the conclusion, *if premises are true*, i.e. if the statements correctly describe the phenomenon, will be exclusively determined by the reasoning correctness. Conversely, if I said "Socrates is a man" and Socrates were not a man, but rather an angel and therefore immortal, the second premise would be syntactically correct, but it would be semantically false because it would not correctly describe reality. Then, the conclusion of the syllogism, although formally correct, is false due to a fallacy of one of the premises.
Summarizing, the *analytical-predictive feature* of the language, either *mathematical* or *classical logical-deductive*, shares the possibility of building up predictions through:

(a) analysis of suitable experience (phenomenological) data,
(b) their formulation through the syntax of the employed language,
(c) construction of the result by a (abstract) reasoning which follows the specific language rules.

In this sense, mathematics is a language, because it shares the typical analytical-predictive capacity of a rational language. There is a descriptive and semantic feature, which may or may not be correct in relation to a phenomenon of reality; then, there is a syntactic feature, which is completely abstract and which requires the development of a reasoning with certain stringent rules.

2. *Structural feature*. The previously introduced statement that mathematics is a language capable of explaining reality, according to the categories of quantity, relation, and extension (definition 2.1), implies that the *structure* plays an essential role.

> (2.6) *Les mathématiciens n'étudient pas des objects, mais des relations entre les objects; il leur est donc indifférent de remplacer ces objects par d'autres, pourvu que les relations ne changent pas.*[28]
>
> Mathematicians do not study objects, but relations between objects; therefore, it is indifferent to them to replace these objects by others, provided that the relations do not change.

[28] H. Poincaré, Chap. 2 in Poincaré (1902).

As a matter of fact, the concept of *structure* stands as a backbone between the three aforementioned concepts. Indeed, it is precisely due to its structure that mathematics is able to provide the language adopted by physical science. Nicolas Bourbaki[29] underlines the concept of structure as the cornerstone connecting mathematics to the natural sciences.[30]

(2.7) *Le trait commun des diverses notions désignées sous ce nom générique [de* structure mathématique*], est qu'elles s'appliquent à des ensembles d'éléments dont la nature* n'est pas spécifiée[31]

The common feature of the various notions designated under this generic name [of *mathematical structure*] is that they apply to sets of elements whose nature *is not specified.*

This structure is itself a correct, well-thought-out instrument, whose truth and whose specific meaning are given by the interpretation that is given through the correlation with physical reality. However, what do we mean by "elements whose structure is not specified"? The relations mentioned by Poincaré develop in the mathematical discourse forming *structures*. These relationships can relate very different elements to each other, elements which are not *a priori* specified.
The purely structural aspects of language were also brought to light by recent studies about the syntax of current languages, which are worth briefly outlining. The current languages, such as English, and Italian, simultaneously have got semantic, morphological, syntactic, phonetic, etc. features. If, for example, we utter the sentence

Many clouds are moving in the sky.

beyond the specifically linguistic-syntactic areas, other areas are activated in the brain, such as, for example, the sensory areas dedicated to vision. The most characterizing element of human language, however, is the syntax.[32] For this reason, there is a specific research line aimed at performing studies exclusively on the syntactic aspect, i.e. the most markedly structural one of the language. These studies demonstrated the possibility of preparing sentences which are *syntactically correct but without semantics*, in such a way as to (almost) exclusively activate the brain areas dedicated to syntax.[33] Therefore, one can use sentences like[34]

---

[29] Cf. above, note 8.

[30] The Bourbaki group vision, for which mathematics is above all a science of structures, largely dominant until half a century ago, is now joined by other visions, favouring concepts such as "model" and "interaction" (Bottazzini et al. 2016). We do not enter into such diatribes, which go beyond our purposes, and we limit to consider the useful elements for the discussion we are developing.

[31] N. Bourbaki, *L'architecture des mathématiques* in Bourbaki (1948).

[32] Moro (2008).

[33] Moro et al. (2001).

[34] The example is taken from (Moro 2008) (Sect. 2.2), a text which gives an in depth description of these studies, here summarized briefly.

Many grapots atted.
Every blick has milened the flust.
The coives were featided.[35]

For our discussion about structure, it is interesting to note that the aforementioned sentences are, syntactically speaking, correct, even though they mean absolutely nothing. They can assume a specific meaning only when we specify the meaning of each noun (as they cannot be found in a dictionary). Therefore, such linguistic studies illuminate the parallelism between mathematics and languages.
Returning to mathematics, it is therefore clear how focusing on the structural aspect allows us to use mathematics with flexibility and freedom. Indeed, it follows that it is possible to use different parts of mathematics according to the different parts of reality that we want to describe, just as in the current language we use the singular and the plural, the masculine or the feminine, the present tense or the past depending on what is meant to describe.

Within the great debate on non-Euclidean geometries, Poincaré exactly provides an interpretation of geometry that is based on the concept of structure and language, an interpretation which for our discussion we extend to the entire mathematical language.[36]

(2.8) Les axiomes géométriques ne sont donc ni des jugements synthétiques à priori ni des faits expérimentaux.

Ce sont des conventions*; notre choix, parmi toutes les conventions possibles, est* guidé *par des faits expérimentaux ; mais il reste* libre *et n'est limité que par la nécessité d'éviter toute contradiction. C'est ainsi que les postulats peuvent rester* rigoureusement *vrais quand même les lois expérimentales qui ont déterminé leur adoption ne sont qu'approximatives [...]*

Dès lors, que doit-on penser de cette question: La géométrie euclidienne est-elle vraie?

Elle n'a aucun sens.

Autant demander si le système métrique est vrai et les anciennes mesures fausses; si les coordonnées cartésiennes sont vraies et les coordonnées polaires fausses. Une géométrie ne peut pas être plus vraie qu'une autre; elle peut seulement être plus commode.[37]

Geometrical axioms are therefore neither a priori synthetic judgements nor experimental facts.

These are *conventions*; our choice, among all the possible conventions, is *guided* by experimental facts; but it remains *free* and is limited only by the need to avoid any contradiction. This is how the postulates can remain *rigorously* true when even the experimental laws which determined their adoption are only approximate [...]

Therefore, what should we think about this question: is Euclidean geometry true?

It makes no sense.

[35] In Italian they are roughly "Molte grapotte amionarono", "Ogni ditra ha milenato il fiommo", "Le corle furono featide", respectively (from Moro (2008)).

[36] Actually, Poincaré had not the opinion that his geometry conception could be extended to arithmetic and numerical mathematics, but we here omit this distinction for the purpose of our discussion.

[37] Henri Poincaré, part, II, Chap. III in Poincaré (1902).

> As well as asking whether the metric system is true and the old units false; whether Cartesian coordinates are true and polar coordinates are false. One geometry cannot be truer than another; it can only be more convenient.

In summary, we can state that physics finds mathematics an useful language, which provides, through "necessary demonstrations", *convenient* and varied structures; however, the specific meaning of these structures is provided by reality, whose natural phenomena are investigated through "sensible experiences".[38] From the point of view of physical science, this language, in itself detached from reality, has not got a particular meaning and has not got a particular truth content, except for the correspondence to the rules of the structure, exactly like the aforementioned sentence "Many grapots atted" does not report any truth except for its correspondence to the syntactic rules of the English language.[39]

2.2.3 Axioms, Interpretations, Models

At this point, let us reconstruct the actual procedure by which a given mathematical structure is correlated to a given physical reality, a process producing what is called *interpretation*. The *interpretation* concept was formalized within the mathematical discourse and thus it is worth finding through it a formally correct definition. In this exposition we follow a modern thought line, neglecting the historical aspects and those related to the debate about the origin of mathematics.[40] Modern mathematics exploits the so-called *axiomatic method*, already theorized and practiced with great results by the Greeks, in particular Euclid.[41] On the basis of this method, as a starting point of the *discussion* a certain number of statements are *assumed to be true* without proof: these are called *axioms* or *postulates*.[42] The axioms are required to be *consistent*, i.e. not contradictory, so as to form an *axiomatic system* (a.s.), which we conventionally indicate with Σ. Moreover, in their formulation these axioms recall

---

[38] "Regarding to natural conclusions, we must esteem the necessary demonstrations and sensible experiences", G. Galilei, letter to Christine of Lorraine, 1615.

[39] It is here proposed the specific perspective that may be of interest to a physicist, or to a natural scientist, without entering into a dispute regarding the proper meaning of mathematics from the perspective of a pure mathematician.

[40] We deal with it more extensively in Chap. 7 and in particular in Sect. 7.1.

[41] *Euclid's elements* were the most influential mathematics textbook and, we can say, scientific book of all time. Despite the vast majority of scientific results exposed in ancient texts, such as those in Aristotle's *Physics*, are by now completely outdated or even refuted, it is noteworthy that the content of Euclid's treatise still remains valid (Boyer 1968).

[42] According to a classical line of thought, the distinction between *axioms* and *postulates* consisted in the fact that the former ones were assumed to be true because they were considered *self-evident* (for example, the *First Postulate of Euclid*, which reads: "From any point to any point it is possible to draw a line".), while the latter ones were assumed to be true even without being perceived as immediately evident (for example, the famous *Euclid's Fifth Postulate*) (Basti 2002). In modern mathematical practice there is a tendency not to consider this distinction significant, and we speak of axioms or postulates in an almost identical way (Wilder 2012).

some *indefinite terms*, whose interpretation can be later specified.[43] In summary, at the basis of a given mathematical construction we have got necessarily two elements:

1. an AXIOMATIC SYSTEM (a.s.)
2. some INDEFINITE TERMS

Briefly, the concepts of abstractness and structure correlated with this axiomatic method satisfy the condition of separation from the specific semantics, in a completely analogous way to what it has been shown in the previous section regarding current languages.

> (2.9) La forma logica che si vuole dare ai postulati è precisamente quella di *relazioni aventi un significato indipendente dal particolare contenuto dei concetti*, cioè di relazioni affatto generali che possono sussistere fra "enti astratti".[44]
>
> The logical form that we want to give to the postulates is precisely that of *relations having a meaning independently on the specific content of the concepts*, i.e., they should be quite general relations that can exist between "abstract entities".

To get a concrete idea of this method, an example is provided. The most immediate example is provided by Euclidean geometry[45]: in a flat simplified form, we could list some axioms.

Axiom. $\exists$ 2 points, p, q.
Axiom. $\forall L$ is a collection of points.
Axiom. q and r are points and $q \neq r, \Rightarrow \exists!\, L$ s.t. $\{q, r\} \in L$.
Axiom. L is a line $\Rightarrow \exists\, p$ s.t. $p \notin L$.
Axiom. L is a line and $p \notin L \Rightarrow \exists! M \parallel L$ s.t. $p \in M$.

Assuming the above ones, we have got the first element of our mathematical construction, namely, the *system of axioms* Σ. These axioms, however, assume the existence of two terms that are *undefined*, i.e.

Indefinite terms: *point*, *line*.

The heart of the interpretative process lies here: according to the meaning we attribute to indefinite terms, the specific meaning of the considered axiomatic system changes. For the example here proposed Euclid provides the concepts of *point* and *line* stating them as *definitions*[46]:

> (2.10)
> α'. Σημεῖον ἐστιν οὗ μέρος οὐθέν.
>
> β'. Γραμμὴ δὲ μῆκος ἀπλατές.

---

[43] Wilder (2012).

[44] F. Enriques in Chap. III of Enriques (1922).

[45] Example taken from Chap. 1 of Wilder (2012).

[46] In the original text ὅροι, from Euclide (2007).

1. A point is that of which there is no part.
2. A line is a length without width.

We note that the above "definitions" maintain something indefinite in the terms "line" and "dot", to which we can give a meaning using different criteria. In standard geometry these terms correspond to the intuitive concepts of *point* as an *object without extension* and *line* as a *minimum distance between two points*. However, many other interpretations are possible. For example, it is known that, given the *functional spaces*,[47] "points" are mathematical functions with certain properties. Nevertheless, the interpretations can also be not strictly mathematical: by suitably redefining the various terms, the "points" can be objects of any set (coins, books, etc.).[48] For instance, if we assumed that the term "point" is equivalent to "a human being " and the term "line" is equivalent to "two human beings", then the five axioms listed above would turn out, even intuitively, to be "true" in relation to our concrete experience (even the definition of "point" given by Euclid would be intuitively satisfied), and therefore all the demonstrable theorems starting from these postulates would maintain their validity.

This can be fixed by providing formalized mathematical definitions.[49]

Definition 2.3 (*Interpretation of an a.s.*) Given an axiomatic system Σ, then we call *interpretation* I of Σ a meaning assigned to the *indefinite terms* of Σ such that the axioms become all statements true at the same time for all values of their variable terms.

Definition 2.4 (*Model of an a.s.*) Given an axiomatic system Σ and its interpretation I, we call *model* resulting from I, which we denote by $\mathfrak{M}(I)$, the set of concepts which originate from this interpretation.

The above definitions were constructed as auxiliaries of the theory of logical foundations of mathematics (and of sets) and they represent a concise version[50] of what is much more widely developed in the *model theory* of mathematical logic.[51] In this simplified form, these definitions can also be used to classify the relationship between the mathematical structure world and that of physical systems. With the help of the concepts of *interpretation* I and *model* $\mathfrak{M}(I)$, according to the aforementioned definitions, it is possible to reconstruct the process through which a given mathematical structure is chosen, based on certain postulates, and is applied to explain some aspects of reality.[52]

[47] Cf. e.g., (Kolmogorov and Fomin 1999; Robinson 2003), etc.

[48] Chapter 1 of Wilder (2012).

[49] Chapter 1 of Wilder (2012), Chap. 7 of Brewer (2022).

[50] According to the formulation of Chap. 1 of Wilder (2012).

[51] For a more precise idea of this theory, see, e.g., (Hodges 1997).

[52] We do not deal with such concepts in the context of *formal languages*, which require to deepen the distinction between *syntax* and *semantics* (cf. e.g., (Rautenberg 2009; Enderton 2001) and also (Brewer 2022; Hodges 1997)), but here we intend to highlight only the *empirical* aspects related to I and $\mathfrak{M}(I)$ useful for application in the physical description of nature.

Finally, we can now give an example in physics. The mathematician, through his work on structures, provides one or more purely mathematical objects. We assume, indicatively, the homogeneous *wave equation*, such as

$$\nabla^2 \mathcal{F} - \frac{1}{v^2}\frac{\partial^2 \mathcal{F}}{\partial t^2} = 0$$

and its general integral.[53] These mathematical objects are based on the usual axioms of the real number filed $\mathbb{R}$, or of the complex ones $\mathbb{C}$, and on the consequent usual theorems of mathematical analysis. Among the variables therein present, the physicist is able to give different *physical* interpretations, formally indicated as I_1, I_2, I_3, in order to, respectively, describe the sounds (assuming $\mathcal{F} :=$ *pressure*) or light (assuming $\mathcal{F} :=$ *electric (magnetic) field*) or the motion of water on the surface of a lake (assuming $\mathcal{F} :=$ *height of water surface*). From them, we obtain general models, respectively, $\mathfrak{M}_1 = \mathfrak{M}(I_1)$, etc., which conceptually and mathematically describe the typical noticed wave phenomena. Each *practical* model can name differently the same mathematical object. The same mathematical entity is thus called *height of sound* according to $\mathfrak{M}_1$, *color of light* according to $\mathfrak{M}_2$, and *wave motion frequency* according to $\mathfrak{M}_3$.

2.2.4 Choice of Useful Elements

Accordingly, we have got the necessary mathematical and linguistic structures. We have a syntax and (mathematical) words to build "sentences" to reproduce, predictively and structurally, natural phenomena, within the approximation allowed by our observations.[54] Thus, we must move towards constructing. This construction of the mathematical language is necessarily driven by facts. Says Poincaré:

> (2.11) *Comparant les données brutes de nos sens et ce concept extrêmement complexe et subtil que les mathématiciens appellent grandeur, nous sommes bien forcés de reconnaître une divergence; ce cadre où nous voulons tout faire rentrer, c'est donc nous qui l'avons fait; mais nous ne l'avons pas fait au hasard, nous l'avons fait pour ainsi dire sur mesure et c'est pour cela que nous pouvons y faire rentrer les faits sans dénaturer ce qu'ils ont d'essentiel.*[55]
>
> If we compare our raw sensory data with this extremely complex and subtle concept that mathematicians call quantity, we will be forced to recognize a discrepancy; this framework where we want everything to fit in, it has been build by us; but we did not it randomly, we did it, so to speak, tailoring and that is why we can fit the facts into it without distorting what is essential.

---

[53] Where, as usual, $\mathcal{F}$ is a generic perturbation, t the time variable and $v = \lambda\nu$ is the parameter obtained from the product between the wavelength and its frequency.

[54] In this regard, we deepen the issue in Sects. 8.3 and 8.2.2.

[55] H. Poincaré, Introduction of Poincaré (1902).

As far as physics is concerned, the axioms of mathematics we discussed in the previous section are not arbitrary, but their selection and formulation is guided by experience. Hence, the choice of language elements strongly depends on *what we want to investigate* and is determined by it. A perspective of this kind is an *open perspective*, since it does not start from the assumption of bending the phenomenological data to mental structures. On the contrary, a *closed perspective* (claiming to bend the phenomenological data to the representations of the mind accomplished through a given language) would lead to discarding facts that do not bend to a mathematical description. Conversely, the *open perspective* finds a discrepancy between language and correlated model, between language and phenomenon, not surprising. If this discrepancy widens, it will be the scientist's task either to limit his investigation field, or to broaden the language (mathematics, in this case), or even to change the language. Therefore, from analysing the choice of the axiomatic system and of the model built on it, we arrive at posing the more general problem of the language choice. In the case of physics, the *closed perspective* leads to downplaying those phenomena that cannot be represented through the chosen language, which is a given form of mathematics. This was the weak point of the eighteenth-century mechanistic approach, exaggerated by Pierre-Simon Laplace, and of a large part of nineteenth-century positivism: making the knowledge of phenomena coinciding with their mathematizability, the claim that all phenomena can be mathematizable. This led to the belief, which we know today as erroneous, that mathematics were able to explain all phenomena.[56]

Today we may recognize that in the modern understanding of scientific disciplines, the characteristic of mathematizability is not a necessary condition to define a discipline as "science", and even less as "knowledge". That is "one does science" even without choosing a mathematical language to describe the phenomena (think about disciplines such as medicine or geology) or by exploiting a language that is only partially mathematical (for example, geophysics or chemistry). Poincaré reminds that the choice of a given axiomatic system or a given model is driven by the utility and convenience regarding the study of phenomena (quote 2.8); the choice of language is based on the same utility and comfort criteria, as well as elegance. In other words, mathematics is a useful, powerful, and elegant language, but it is not always the mandatory one, or even the best one. A famous statement by Einstein reported by Max Born summarizes this concept well:

> (2.12) *Wenn man alles auf physikaische Gesetzmässigkeiten zurückführen würde, wäre das eine Abbildung mit inadäquaten Mitteln, so als ob man eine Beethoven-Symphonie als Luftdruckkurve darstelte.*[57]

> It would be possible to describe everything scientifically, but it would make no sense. It would be a description without meaning as if you described a Beethoven's symphony as a variation of wave pressure.

In summary, the development of a language to study natural phenomena consists of a series of choices, constantly driven by the comparison with the experience data. The

---

[56] Cf. quote 6.6.

[57] A. Einstein reported in Born (1966) by M. Born.

first choice is precisely that of the kind of language, which in the case of physics is mathematics. Subsequently, the most suitable structures for the development of the discipline are chosen, according to specific criteria of language, rationality, and experience: axiomatic systems, indefinite terms, interpretations, and models. By exploiting this conceptual construction, from the comparison with the experience data, we enunciate what are called *physical laws*.

2.3 Physical Laws

The interpretation of natural phenomena according to mathematical criteria, which is obtained through a more or less articulated construction, is usually summarized in what we name *physical laws*.

> (2.13) All the laws of nature are conditional statements which permit a prediction of some future events on the basis of the knowledge of the present [...] Naturally, we do use mathematics in everyday physics to evaluate the results of the laws of nature [...] In order that this be possible, the laws of nature must already be formulated in mathematical language.[58]

These constructions, the physical laws, have got an abstract nature, since they are expressed mathematically. They have got two fundamental characteristics: the ability to describe phenomena and that of predicting their trend. This dual nature leads us to formulate the following[59]

Definition 2.5 (*Physical laws*) Physical laws are a *highly predictive* description of natural phenomena performed through a mathematical language.

It is clear that the validation of a given physical law, i.e. the affirmation that it is *true*, is a complex process, aimed at assessing *to what extent* it accurately describes and predicts a given class of natural phenomena.[60]

> (2.14) To see what is general in what is particular and what is permanent in what is transitory is the aim of scientific thought [...] This possibility of disentangling the most complex evanescent circumstances into various examples of permanent laws is the controlling idea of modern thought.[61]

However, it should be noted that, although physics is a discipline based on the observation of natural phenomena, investigated through suitable experiments,[62] the purely mathematical nature of the *physical laws* means that they cannot be found in experimental investigation. In other words, physical laws are a mere mental construction:

---

[58] E. Wigner in Wigner (1960).

[59] Cf. e.g., Part I Chap. 1 and Part II Chap. 5 of Duhem (2016).

[60] This topic is explored later in Chap. 8.

[61] Alfred. N. Whitehead in Chap. 1 of Whitehead (2017).

[62] Cf. Table 1.1.

> (2.15) [It] is not at all natural that "laws of nature" exist, much less that man is able to discover them.[63]

In symbols we may synthetically state that

$$\exists \textit{ physical laws}? \text{ No, } \exists! \text{ phenomena!}^{64}$$

Here we must recognize that Galilei (see the quote 2.1) and probably also Newton (see Sect. 4.2) would probably object both to this last synthetic statement, and to the definition 2.5. Following Galileo's essentially Platonic conception, the mathematical relations are the *true reality* of nature. On the other hand, we also must agree that, on a purely experiential level, only phenomena can be investigated through experiments; so, we can provide experimental evidence only of them, and not of the physical laws, which are mathematical constructs.

On these aspects there are often misunderstandings even among scholars. Therefore, it is worth dwelling on the most frequent problems that often invalidate a correct understanding of the physical law meaning.

First of all, in the introductory and elementary level classes, we often find a statement that through experiments we *see* the *physical laws*. As mentioned above, this is false: what is seen through the experiments are only the phenomena. Actually, through experiments it is not possible to visualize any relationship, structure, or mathematical entity. In short, it is not possible to see any *physical law*. Experiments make it possible to see natural phenomena, whose appropriate features are measured (the *physical quantities* which we discuss in the next chapter); these features are summarized and systematized within specific models through mathematical relationships.

Secondly, the *physical laws* are provided according to some synthetic statement: they are considered *valid*, i.e. able to describe "well"[65] a class of phenomena, under certain *hypotheses*, within a given *interpretation* and a given *model* and within a given *precision degree*. These assumptions are largely determined by the *simplification*, *limitation*, and *approximation* strategies discussed in Sect. 1.1. Outside of these assumptions, a physical law is a virtually meaningless statement. This is well specified by Federigo Enriques:

> (2.16) [R]esta associata al concetto di una *legge fisica*, l'idea di qualcosa di *necessario*, che *debba* essere all'infuori del verificarsi o meno dei fatti contingenti, di cui essa esprime il rapporto: così, p. e., quando si pensano le leggi geometriche e meccaniche, che pur esprimono alcuna proprietà del movimento dei corpi, come indipendenti dall'esistenza di questi. [...] [Al] sentimento della necessità [...] si collega l'*illusione* che certe rappresentazioni costruite *a posteriori*, sulla base di osservazioni o esperienze preliminari, abbiano un valore *a priori*, e si estendano quindi illimitatamente al di là del campo cui queste si riferiscono. Importa perciò di rilevare che una siffatta necessità di ordine psicologico, resta sempre un

---

[63] E. Wigner in Wigner (1960).

[64] The symbol $\exists$ is here used to indicate the existence in the real world, i.e. the phenomenological one, investigable through experiments, differently from mathematics.

[65] In this point I deliberately keep this concept in a vague form, which is treated in more detail later on (see Chaps. 5 and 8).

sentimento, il cui valore è puramente subiettivo, cioè che non può essere invocato come prova di sussitenza obiettiva.[66]

The idea of something *necessary* is still associated with the *physical law* concept and it *must* exist in spite of the occurrence or the contingent facts whose relations it describes: so, e.g., when geometric and mechanical laws are thought of, although they express some property of the body movement, as independent on their existence. [...] [The] feeling of necessity [...] is linked to the *illusion* that some representations constructed *a posteriori*, on the basis of observations or preliminary experiences, have got an *a priori* value, and then they extend indefinitely beyond the field to which they refer. It is thus important to note that such a psychological necessity always remains a sentiment, whose value is purely subjective, that is, it cannot be invoked as proof of objective subsistence.

In other words, besides a given physical law, stated within an organic theory, there are several elements and hypotheses that must be specified, without which the enunciation of the law is meaningless.

References

G. Basti, *Filosofia della Natura e della Scienza, Pontificia Università Lateranense*, vol. 1, 1a edn. (Roma, Città del Vaticano, 2002)

M. Born, *Physik im Wangel meiner Zeit*, 4th edn. (Vieweg, Braunschweig, Deutschland, 1966)

U. Bottazzini, E.A. Marchisotto, P. Miller, *Hilbert's Flute: The History of Modern Mathematics* (Spriunger-Verlag, Berlin, Deutschland, 2016)

N. Bourbaki, L'architecture des mathématiques: la mathématique ou les mathématiques?, in *Les grandes courants de la pensée mathématique*. ed. by F. Le (France, Lionnais (Cahiers du Sud, Paris, 1948)

C.B. Boyer, *A History of Mathematics*, 2nd edn. (Wiley, New York, USA, 1968)

W.D. Brewer, *Kurt Gödel. The Genius of Metamathematics*. (Springer, Cham, Schweiz, 2022)

A. Church, *Introduction to Mathematical Logic* (Princeton University Press, Princeton, New Jersey, USA, 1956)

S. Cole Kleene, *Mathematical Logic*. (Dover, Mineola, New York, USA, 1967)

I.M. Copi, C. Cohen, V. Rodych, *Introduction to Logic*, 15th edn. (Taylor & Francis, New York, USA, 2019)

R. Courant, H. Robbins, *What is Mathematics? An Elementary Approach to Ideas and Methods* (Oxford University Press, Oxford, UK, 1941)

M. Dorato, The laws of nature and the effectiveness of mathematics, in *The Role of Mathematics in Physical Sciences*. ed. by G. Boniolo et al. (Springer, Berlin, Deutschland, 2005)

P. Duhem, *La théorie physique. Son objet, sa structure*, 2 edn. (ENS Éditions, Lyon, France, 2016). http://books.openedition.org/enseditions/6077

A. Einstein, Considerations concerning the fundaments of theoretical physics. Science **91**, 487–492 (1940)

H.B. Enderton, *A Mathematical Introduction to Logic*, 2nd edn. (Harcourt Academic Press, San Diego, USA, 2001)

F. Enriques, *Problemi della Scienza* (Zanichelli, Bologna, Italia, 1906)

F. Enriques, *Per la storia della logica-I principii e l'ordine della scienza nel concetto dei pensatori matematici* (Zanichelli, Bologna, Italia, 1922)

Euclide, Euclid's elements of geometry (2007). ISBN 978-0-6151-7984-1

[66] F. Enriques, Chap. II Sect. 13 in Enriques (1906).

G. Galilei, *Il Saggiatore*. (Giacomo Mascardi, Roma, Italia, 1623)
W. Hodges, *A Shorter Model Theory* (Cambridge University Press, Cambridge, UK, 1997)
B. Ingrao, G. Israel, *La mano invisibile. L'equilibrio economico nella storia della scienza*. (Laterza, Roma-Bari, Italia, 1987)
G. Israel, *La matematica e la realtà* (Carocci, Roma, Italia, 2015)
A.N. Kolmogorov, S.V. Fomin, *Elements of the Theory of Functions and Functional Analysis* (Dover, New York, USA, 1999)
R.B. Lindsay, H. Margenau, *Foundations of Physics*, 2nd edn. (Dover, New York, USA, 1957)
A. Moro, *The Boundaries of Babel-The Brain and the Enigma of Impossible Languages* (The MIT Press, Cambridge, Massachusetts, USA, 2008)
A. Moro, M. Tettamanti, D. Perani, C. Donati, S.F. Cappa, F. Fazio, Syntax and the brain: Disentangling grammar by selective anomalies. NeuroImage **13**, 110–118 (2001)
H. Poincaré, *La Science et l'Hypothèse* (Flammarion, Paris, France, 1902)
W. Rautenberg, *A Concise Introduction to Mathematical Logic*, 3rd edn. (Springer, New York, USA, 2009)
D. Robinson, *An Introduction to Abstract Algebra*. (de Gruyter, Berlin & New York, Deutschland & USA, 2003)
S. Sakkurai, Revisiting the 'unreasonable effectiveness' of mathematics. Curr. Sci. **88**, 415 (2005)
E. Segrè, *From Falling Bodies to Radio Waves: Classical Physicists and their Discoveries* (W.H. Freeman & C., New York, USA, 1984)
A. Strumia, *Dalla filosofia della scienza alla filosofia nella scienza*, 1a edn. (ESC-Edizioni Santa Croce, Roma, Italia, 2017)
J. von Neumann, Method in the physical sciences, in *The Neumann Compendium, volume 1 of World Scientific Series in 20th Century Mathematics*. ed. by F. Bródy, T. Vamós (World Scientific, Singapore, 1995), p. 627
J. von Neumann, The mathematician, in *The Neumann Compendium, volume 1 of World Scientific Series in 20th Century Mathematics*. ed. by F. Bródy, T. Vamós (World Scientific, Singapore, 1995), p. 618
A.N. Whitehead, *An Introduction to Mathematics*. (Dover Publication, Mineola, New York, USA, 2017). unabridged republication of the work originally published in 1911
E. Wigner, The unreasonable effectiveness of mathematics in the natural sciences. Commun. Pure Appl. Math. **13**, 1–14 (1960)
R.L. Wilder, *Introduction to the Foundations of Mathematics*, 2nd edn. (Dover Publications, Mineola, New York, USA, 2012)

Chapter 3
Conceptual and Empirical Structure of Physical Quantities

Non più per immaginazione, ma per sensata esperienza e necessaria dimostrazione.
Galileo Galilei

3.1 Two Cornerstones: Reality and Mathematics

At the beginning of this analysis we have recognized phenomena as the starting point of all natural sciences. Then we have identified the specificity of physical science in the use of mathematics, whose role we have analysed under various aspects. The concept of *physical law* provides the passage between the purely mental world of mathematical concepts, whose truth is demonstrated through pure reasoning, and the world of phenomena, whose truth is demonstrated by observations and experiments. The "physical law" object, despite being a concept and not a phenomenon, does not belong to mathematics, but it arises from the specific development of the physical science discipline. The main rational steps leading to the construction and enunciation of the *physical laws* can be schematized and summarized as follows:

- We understand, or intuit, that mathematics is a suitable language to describe some features of nature, according to Galileo's great insight (quote 2.1).
- If you want to undertake this path, you will need to create a bridge between reality, i.e. natural phenomena, and mathematics.
- Therefore, it is necessary to find, among phenomena, entities that can be described through mathematics, i.e. entities that can be put into correspondence with mathematical entities.

Such entities should be present in the *physical laws* and consequently they should have a mathematical nature, but, above all, they should be somehow verifiable in natural phenomena, i.e. experimentally measurable. They are the PHYSICAL QUANTITIES. The precise identification of the distinctive feature of a physical quantity was a key element that allowed the birth of modern science.

J. Parravicini, *The Foundations of Experimental Physics*,
https://doi.org/10.1007/978-3-031-55452-0_3

(3.1) *Ma se vorremmo fermarci nell'apprensione di alcune affezioni, non mi par che sia da deperar di poter conseguirle anco ne i corpi lontanissimi da noi, non meno che nei prossimi [...] Voglio per tanto inferire, che se bene indarno si tenterebbe l'investigazione della sustanza [...], non resta però che alcune loro affezioni, come il luogo, il moto, la figura, la grandezza, l'opacità, la mutabilità, la produzione ed il dissolvimento, non possino da noi esser apprese.*[1]

But if we would like to stop in the understanding of some affections, it does not seem to me that we should despair of being able to obtain them even in bodies very distant from us, no less than in the closest ones [...] Then I want to deduce that, if we vainly attempted the investigation of their substance [...], however, it would remain that some of their affections, such as place, motion, shape, size, opacity, mutability, arising and dissolution, can be known by us.

What Galileo calls "affections" are, precisely, the *physical quantities*: since they are quantitatively measurable, therefore, they are, in other words, mathematizable.[2]

It becomes here appropriate to provide a concise definition of *physical quantity*. Actually, the problem of a formal specific definition is rather complex, giving rise to many discussions as it is correlated with logical, mathematical, epistemological, phenomenological concepts.[3] For this reason, while adhering to the purposes of this discussion, we limit ourselves to provide an intuitive and, precisely, *synthetic* definition. In order to achieve a clear form, this definition should highlight the *dual nature*, mathematical and phenomenological, of these quantities. Referring to Table 1.1, we recognize that it is a matter of constructing the *quod* formal object proper to the physics discipline.

Definition 3.1 (*Physical quantity*) A *physical quantity* is a feature of a natural phenomenon that is measurable and can be correlated to a mathematical quantity.[4]

This definition is provided from the phenomenological point of view of physical science. If we move on to the mathematical point of view, the question arises about which are the mathematical entities that can be associated with measurable physical quantities. This problem has turned out to be particularly important during the historical development of sciences. As a matter of fact, at the dawn of modern physics, at Galilei's and Newton's time, it was necessary to develop a suitable mathematics to explain the new concepts of mechanics, since the mathematical entities conceived, or investigated, had been largely unsuitable or insufficient until then.[5] By way of example, it is worth recalling some of the main classes of mathematical objects capable of explaining aspects of natural reality:

---

[1] G. Galileo, Third Letter in Galilei (1613).

[2] In spite of being useful in physics, there are some mathematical concepts not specifically or immediately quantitative in nature (for example, some algebraic structures, symmetries, etc.). Following Galilei, our discussion develops referring to the quantitative aspects, bearing in mind that the discussion is easily generalized.

[3] de Boer (1995), Basti (2002), Strumia (2017).

[4] Cf. e.g. (Bureau international des poids et mesures 2008).

[5] Whether mathematical entities are either *constructed* or *discovered* objects is still a debated issue, which mainly belongs to the philosophy of mathematics, cf. Chap. 2 as well as (Boyer 1968; Courant and Robbins 1941; Bottazzini et al. 2016; Israel 2015).

Functions explain *changes*, especially on time and space.
Vectors describes directionality.
Tensors the combination of several directionalities.
Lebesgue integrals are the bases of a rigorous definition of *probability*.[6]
Sets allows us, through the *specification axiom*, to group objects of reality as a function of shared properties.[7]
Symmetry operations enable to describe regularities of nature, and to predict new ones.

Although this list is by no means exhaustive, these cases highlight that there has been a mutual exchange between mathematical entities and the notion of physical quantity which has led to the mutual development of both disciplines: some hints of this exchange are given in Chap. 4.

Concerning the Definition 3.1, Maxwell stresses the *dual nature* in identifying a general criterion which a physical quantity must satisfy. It is important to note that Maxwell does not provide a definition in the proper sense of the term, but a *practical* description of the concept which takes into account its actual use. It can be considered as an *operational definition* (a concept that is introduced in the next section).

> (3.2) Every expression of a Quantity consists of two factors or components. One of these is the name of a certain known quantity of the same kind as the quantity to be expressed, which is taken as a standard of reference. The other component is the number of times the standard is to be taken in order to make up the required quantity. The standard quantity is technically called the Unit, and the number is called the Numerical Value of the quantity.[8]

In mathematical terms we may write[9]

$$Q = \{Q\}[Q], \tag{3.1}$$

where quantity Q is expressed as $\{Q\}$ such the numerical value of the quantity multiplied by a suitable unit of measurement $[Q]$.

Maxwell's statement makes explicit the intrinsically composite nature of a *physical quantity* Q. The object $\{Q\}$ is a numerical value: in this we find its being a mathematical object, i.e. its submission to the articulated theoretical construction where results such as theorems, corollaries, etc. can be recognized *true*. According to Maxwell, this $\{Q\}$ is *multiplied* (mathematical operation) by an object of a completely different nature, $[Q]$. This is the *unit of measurement*, which is *necessarily* defined in relation to reality, i.e. in a non-mathematical way. Maxwell develops the argument by noting that the description provided by (3.1) has the advantage that the Q physical quantity is invariant with respect to the used unit of measurement.[10]

---

[6] Cf. (Kolmogoroff 1933).

[7] We discuss set theory in more detail in Chap. 7, especially in Sect. 7.2. Concerning the axioms of set theory, see, e.g. (Lolli 2011; Wilder 2012; Levy 1979).

[8] James C. Maxwell in Maxwell (1873).

[9] (de Boer 1995; Bich 2019).

[10] (Maxwell and Jenkin 1865).

In short, the definition (3.1) expresses a "product" between a mathematical object and a non-mathematical object. A formalist mathematician could object that this statement is meaningless: how could the "product" be defined with a non-mathematical entity? From a formal point of view, specific algebraic structures have been developed to provide a suitable logical-mathematical support for such kind of operations, theories that have their roots in the formulation of logic according to Boole's algebra.[11] Without going into a specific analysis of these topics, we can note that the specificity of physical science precisely lies in these conceptual difficulties, which defines notions proper to the discipline. Therefore, the notion of "physical quantity" is developed as an amphibious being, different from the concepts of mathematical science, although intimately intertwined with them. In particular, H. Helmholtz arrives to identify $\{Q\}$ (Maxwell's "numeric value of a physical quantity") with the expression of "concrete number" (*benannte Zahl*),[12] which is effectively an oxymoron. Consequently, the construction of physical science requires the use of physical quantities; these need to be defined equally through mathematical and empirical concepts. The starting concepts appear independent of each other; however, like two pillars, their coexistence, in symbols represented by (3.1), is a necessary condition for the construction of the discipline, as understood and enunciated by Galilei.[13]

3.2 Operational Definitions

A physical quantity itself must be a characteristic of the natural phenomenon under investigation which can be mathematically described. Using a different language, we could say that a physical quantity is an entity of reality that is isomorphic, *at least approximately*, to a mathematical entity. To build a passage between physical and mathematical reality it is necessary to go through the operating procedure of *measurement*. Here we give a provisional formulation, i.e. correct but approximate[14]:

Definition 3.2 (*Measurement (preliminary definition)*) We define *measurement* as the process of discrimination, quantification, correlation, recording of a physical quantity, which can be found within one or more natural phenomena under investigation, through some reasoned experimental procedure.

The introduction of the MEASUREMENT concept transfers our discussion into the experimental world. This requires identifying the practical features of physical quantities, specifying the criteria by which they can be uniquely determined in practice. We focus here on quantities according to the notion (3.2) introduced by Maxwell,

[11] (Bottazzini et al. 2016; Copi et al. 2019; Boyer 1968; Lolli 2011; Rautenberg 2009).

[12] Quoted in de Boer (1995).

[13] Cf. quote (3.1).

[14] A second formulation, equivalent but more precise, is provided in Sect. 8.1, Definition 8.1.

i.e. that they are expressible according to the relation (3.1). Therefore, a physical quantity necessarily satisfies some criteria.[15]

1. It provides comparable data. In other words, for a given quantity it is necessary to be able to verify whether there is a relationship of equality between different measurements.
2. It supplies data that can be added together. In other words, for a given quantity it is necessary to be able to evaluate whether there are differences between different measurements.
3. It is constantly compared to a unitary sample that is the same for all.

We note that the key feature of *objectivity* follows from the 3 assumption, implicit in the formulation (3.1): the fact of having to compare oneself with a unitary sample that is the same for all gives independence to the recorded data no matter who is investigating reality, i.e. the experimenter.

Thus, we can collect the specific elements and characteristics that provide us with a precise method to practically identify a specific physical quantity. The measurement process according to Definition 3.2 plays a key role in this practical determination. Especially, it is assumed that the intention is to compare, add, relate, and evaluate. In a word, it means measuring. This leads us to state that a physical quantity is necessarily defined in relation to the measurement process. Indeed, the measurement procedure itself defines a given physical quantity.

Definition 3.3 (*Operational definition*) A physical quantity is experimentally determined by means of an *operational definition*: this consists of explaining how to identify and measure a given phenomenological object which can be associated with a mathematical quantity or entity.

It is evident that a OPERATIONAL DEFINITION is completely different from a *mathematical definition*. The latter one assumes the existence of certain objects, either directly from postulates or through demonstrations, so it provides them with a name or category based on certain characteristics of interest. In the phenomenal reality there are neither postulates nor demonstrations, but only evidences; experiences; and, under certain conditions, experiments. Thus, the operational definition is based on a reasoned experiment, i.e. a measurement process.

> (3.3) *Quand on dit que la force est la cause d'un mouvement, on fait de la métaphysique, et cette définition, si on devait s'en contenter, serait absolument stérile. Pour qu'une définition puisse servir à quelque chose, il faut qu'elle nous apprenne à* mesurer *la force; cela suffit d'ailleur, il n'est nullement nécessaire qu'elle nous apprenne ce que c'est la force en soi, ni si elle est la cause ou l'effet du mouvement.*[16]

> When we say that force is the cause of a movement, we are engaging in metaphysics, and this definition, if we were to be content with it, would be absolutely sterile. For a definition to be useful, it must teach us how to *measure* force; that is enough, while it is by no means

[15] Cf. e.g. (Feynman et al. 2005a, b, c; Enriques 1906; Arons 1997; Poincaré 1902), etc.

[16] H. Poincaré in Chap. 6 of Poincaré (1902).

necessary for it to teach us what force is in itself, nor what is the cause or the effect of the movement.

The advantage of this approach is that it exempts us from addressing questions of an ontological nature, today called theoretical-philosophical. To clarify this point, it is worth providing a brief example.

Suppose we want to define the "distance" physical quantity, which is directly correlated with the mathematically defined "distance" function.[17] Then the experimenter asks himself the apparently trivial question:

"What is *distance*?"

The answer is by no means immediate if one tries to generically and abstractly answer. First, we need to specify that the "distance" is never defined in absolute terms, but it is always defined as the "distance between two objects". Even with this specification, however, the problem remains open. One could sketch out an answer like "The distance between two objects is the space between them", or something similar. Such a predicative statement, however, would immediately shift the burden of definition to the nominal part, i.e. the *space*. Then, the question about the nature of *distance* would become

"What is *space*?"

Answering to this question belongs more to the domain of theoretical philosophy than of physics: the latter wants to intentionally focus on the phenomenal investigation of nature, leaving out the ontological aspect. On the other hand, an operational definition makes possible to overcome these difficulties. In this process, an ontological determination of the physical quantity is left out in order to focus on the actual measurement process. In our example, the question "What is distance?" will be answered using an *operational definition*, stating:

"Given two physical objects and assuming a third extended physical object, which we call *unit*, we evaluate how many times the unit object lies between the two given objects. The numerical result of this count is the distance between the aforementioned two objects".

Of course such a definition can be enriched by adding various details and specifications; for example, it could be made explicit that among all the possible counts it is necessary to choose the one with the smallest value. However, the definition method is sufficiently precise and general, favouring a practical identification method which does not involve discussions of a theoretical kind.

The decision to leave out philosophical determinations and concentrate on experimentally constructable definitions was one of Galilei's great intuitions. This choice allows us to focus on the investigation of natural phenomena, the "affections" of Galilei's quotation (3.1). The purpose of avoiding disquisitions that today we would call "philosophical" is explicitly stated by Galileo himself.

[17] Cf. (Rudin 1987).

(3.4) *Il tentar l'essenza l'ho per impresa non meno impossibile e per fatica non men vana nelle sustanze elementari che nelle remotissime e celesti.*[18]

I believe that looking for the essence is an enterprise no less impossible to attempt and an effort no less vain in elementary substances than in the most remote and celestial ones.

Since we do not want to "look for the essence" but to concentrate on the study of *things of nature* (τὰ φυνιϰά), we identify what we are talking about, the physical quantities, through operational definitions.

Following our general method, we evaluate the implications of this operational approach by considering a concrete example of how it has influenced the development of physics. Indeed, it is worth noting that this criterion plays a decisive role in important turning points of modern physics.

(3.5) Einstein liked the practical physics laboratory, where he could see phenomena with his own eyes rather than through mathematical symbols.[19]

From specific attention to *phenomena*, Einstein (although he was a theoretical physicist) moved on to exploiting specifically the operational definitions of concepts when he set out to develop the special theory of relativity.

(3.6) Einstein attacked the time and space transformation problem with a deep, almost childlike freshness of approach. With his powerful logic he carefully analyzed the concepts of time and space using an operational method that, for each concept introduced, required a rigorous and concrete specification of how to measure the magnitudes involved. Of course he was concerned not with the instruments and their mechanics, but with the logic of the experiment.[20]

The criterion of the *operational method* is therefore not exclusively useful within a discourse on the first principles of physics. The research experience shows that the use of an *operational method* remains fruitful as a general criterion for tackling subtle physics problems and for keeping experimental physics strictly correlated with theoretical physics.

On a conceptual level, it is worth noting that a given physical quantity must necessarily and simultaneously be determined according to two quite heterogeneous ways. On the one hand, it is to be found in the reality of phenomena and identified through a clear operational definition. On the other hand, despite the definition of quantity has got an operational nature, this quantity must necessarily relate to mathematical entities, which have their purely mental and ideal definitions. This ambivalence evidently lies at the heart of the *amphibious nature* of physical science, between the world of ideas and that of phenomena, and places it in particular emphasis.

---

[18] G. Galileo, Third Letter in Galilei (1613).

[19] Emilio Segrè in Chap. 5 of Segrè (1980).

[20] ibid.

3.3 Unit of Measure

3.3.1 Unit of Measurement Features

Going back to Maxwell's 3.2 statement, and in particular to the formulation in mathematical symbols provided by (3.1), given a physical quantity Q, it is expressed as "the product of two factors" $Q = \{Q\}[Q]$. Since this relationship has got a foundational importance and a key role, we have taken the two factors of the product as two pillars. In the so far discussion we focused on one of them, the element $\{Q\}$. We are now focusing on the second pillar, $[Q]$, the second element of the relationship, i.e. the *unit of measurement*. A modern version of Maxwell's statement provides a useful concise definition of the three elements Q, $\{Q\}$, $[Q]$:

> (3.7) The value of a quantity is generally expressed as the product of a number and a unit. The unit is simply a particular example of the quantity concerned which is used as a reference, and the number is the ratio of the value of the quantity to the unit.[21]

So, we obtain the following:

Definition 3.4 (*Unit of measurement*) We call *unit of measurement* the reference unit of a given physical quantity.

A UNIT OF MEASUREMENT[22] is the basis of any measure which wants to be scientific. It is a necessary condition for scholars to understand each other, in order to build a knowledge that is universal. We note that in mathematics some objects can be assumed as a unit of measurement, but this choice is arbitrary. In geometry, for example, one can assume a segment has got arbitrarily value 1; from this, through the Cartesian construction, the entire analytic geometry is built up. However, it does not matter which segment has been assumed to be unitary. Quite different is the choice for units of measurement in the natural sciences, which must satisfy specific criteria of reasonableness and practicality. A given unitary sample must have got the following features. It should be

1. adopted by convention and universally recognized;
2. easily comparable with the quantity to be measured;
3. easily reproducible and (ideally) without errors;
4. identical to itself, in space and time.

Several related units of measurement form what is called a unit *system*. From the properties listed above, it can be deduced that a unit of measurement (or a system) is assumed taking into account a combination of factors of various kinds. The first factor is, so to speak, *social* (or *political*). The second and third ones are of an eminently technical-practical kind. The fourth one is of the *physical* kind, i.e. dependent on natural features (laws). Therefore, the whole matter has got multiple factors that are the result of a subtle balancing and are constantly subject to revisions and adjustments.

---

[21] Section 2.1 in Bureau international des poids et mesures (2019).

[22] Sometimes referred to as *unitary sample*, which however in some contexts can be misleading.

A signature of this multiplicity are the many units of measurement and the different systems that were adopted throughout history. In general, ancient systems did not correlate inhomogeneous quantities to each other, as it is the case for the present *international system* (which we deal with in the following section); conversely, in those systems were enunciated multiples and submultiples, through the reciprocal relations, of the same fundamental unit.[23] The unit of measurement which is universally considered the simplest one, in some way the most *fundamental*, has always been the length. The reason for this is intuitive: the spatial extension of objects is the prerequisite for measuring, and their extension is the most evident characteristic of their existence. Most of the historical systems were not decimal, but they were systems of base 6 or, more often, of base 12, or in their multiples.[24] These choices were due to the practical need to have got units that were simultaneously divisible by 2 and 3. In the twenty-first century, the system conventionally accepted as universal is the *international system* (IS), whose characteristics make it *unicum* (these are discussed in the following section). Outside of it, the only measurement system that is not strictly local is the *British imperial unit*, which is used in many English-speaking countries, alongside the IS, with some variations.[25]

Fundamental elements of any modern measurement system for scientific uses are the definitions of *basic units*. They can be given according to three types:

1. a standard artefact;
2. a natural standard;
3. a natural constant.

Historically, the category of artefacts was the first to be established, due to its simplicity of practical implementation. To define a unit of measurement through an artefact, it is sufficient to identify a physical feature of a given object, such as its length, and arbitrarily declare it to have got the unitary value. This can be done where there is a sufficiently strong and respected authority to impose the convention on who should use the measurement unit. Examples of standard artefacts can be found in several times, starting from Ancient Egypt.[26] The advantage of this approach is the easiness of the reference construction. The obvious disadvantage is the need to repeatedly prepare reliable replicas of the fundamental unit that be *virtually identical* to each other. This operation is particularly difficult when the required levels of precision become increasingly stringent. For this reason, the use of artefacts had been gradually fell into disuse, in order to assume natural standards or, even better, natural constants as references, so that they could be replicated everywhere without errors.[27] In this sense, the most advanced measurement unit system, which has fully embodied

---

[23] (Gyllenbok 2018).

[24] A system of such kind, which has got still an impact in our world, is the *British Imperial System*: for example, 1 fathom = 6 feet, 1 foot = 12 inches, 1 inch = 12 lines, etc.

[25] For example, *United States customary units* are widely used in the United States for everyday use.

[26] (Gyllenbok 2018; Boyer 1968).

[27] (Bich 2019).

the aforementioned criteria, is precisely the modern *international system*. Accordingly, we are going to expose its cornerstones to highlight, rather than the accidental and conventional aspects, the objective characteristics that make it the most refined *measuring instrument*[28] ever conceived.

3.3.2 Construction and Structure of the International System

To understand the structure of the International System, it is useful to keep in mind its evolution over time. The current *international system* arises from the definition of the metre, which was conceived in France starting from 1791 and legally introduced in 1795, at the climax of French Revolution. To give a name to the fundamental measurement unit, it was decided to use a word from the ancient Greek, μέτρον ("measure"), with the express purpose of pointing out the desire to provide a universally accepted unit. For the same goal, the first definition of the metre aimed at referring to a natural standard, specifically to the size of Earth: as a matter of fact, 1 m should be *one forty-millionth of the average circumference of the Earth* or, in another way, one ten-millionth of the (average) distance between the pole and the equator, measured on the earth's surface.[29] Since from a practical point of view it was necessary to get a sample that could be easily reproducible by everyone, and the measurement of the Earth's circumference was not, right from the beginning it was essential to provide a specific artefact (in French *étalon*) to which the fundamental measurement could be referred.[30]

In the nineteenth century, the use of the *metre* spread and related measurement units was gradually conceived and introduced. In 1875 there was the *metre convention*, an agreement among 17 states which established the *Bureau International des Poids et Mésures* (BIPM) as a supervisory authority,[31] until arriving in 1889 at the *I Conférence général des poids et des msures* , where, among other things, the concept of *system* was officially introduced, being the birth of the *MKS system*. The latter assumed the metre, kilogram, and second as fundamental units, with appropriate relationships between them.[32] We remark that the establishing of MKS recognized that the measurement units were inter-correlated to form a *system*.[33] Subsequently, the need to insert a fundamental unit into the system to measure electromagnetic

---

[28] I deliberately do not use the adjective "metrological".

[29] (Gupta 2010).

[30] (Bureau international des poids et mesures 2008; Gupta 2010).

[31] It is worth noting that in the official emblem of the institute there is the inscription "ΜΕΤΡΩ ΧΡΩ" ("DO MEASURE [and] DO USE"), an expression also used in the ancient Greek language, a choice evidently motivated by the same reason as the choice of the word μέτρον eighty years before.

[32] (Gyllenbok 2018; Gupta 2010).

[33] We discuss this topic more extensively in the following section.

quantities became mandatory; in 1935 the *Giorgi system*[34] was formally organized just before the beginning of WWII.[35] This led to the official birth of the *International System* (IS) in 1960, in which the relationships between the various measurement units were specified as we know today. Furthermore, on that occasion, the definitions of the *fundamental units* and that of the *metre* were modified so as to limit as much as possible the use of *standard artefacts*. Then, the SI of 1960 had *almost* all the fundamental units referred either to *natural standards* or to *natural constants*.[36] Due to a series of technical reasons, the *kilogram* remained defined as the mass of the specific artefact conserved at the BIPM headquarters. However, this conformation showed more and more limitations over the years. This is why the current definition of the International System was reached in 2018. This new formulation maintains the relations between units crystallized in the assumptions of 1960, but defines again the fundamental units so as to refer exclusively to *natural standards* and *natural constants*.[37] The definitions are also assumed in such a way that the standards are easily reproducible in any laboratory, virtually without errors.[38]

After this very brief excursus on the evolution of the present scientific measurement system, it is worth analysing in detail its current state in its definitive formulation.

The current form of the International System was adopted by the XXVI General Conference on Weights and Measures held in November 2018.[39] The essential purpose of the most recent formulation was to redefine the fundamental samples without referring to artefacts. Accordingly, the basic units are given either through a natural standard or through a natural constant. Thus, it was decided to provide a list of fundamental physical constants which are assumed as invariants. They are, by definition, *exact*, in the sense that their numerical value expressed in measurement units of the International System is *without uncertainty by definition*.

The International System (IS or SI) assumes 7 *fundamental physical quantities*, corresponding to 7 fundamental units of measurement (u. m.), following Table 3.1.[40]

Definition 3.5 (*International System (IS or SI)*) Given the physical quantities in Table 3.1, IS is defined as the system where, *by definition*, the statements in Table 3.2 hold.

In the definitions by Table 3.2 the units hertz, joule, coulomb, lumen, and watt, respectively, corresponding to Hz, J, C, lm, and W symbols, are referred to second, metre, kilogram, ampere, Kelvin, mole, and candela units, with their respective symbols, m, kg, A, K, mol, and cd, by the relationships:

[34] The name is after Giovanni Giorgi (1871–1950), an Italian physicist and engineer who devised this system.

[35] (Kennelly 1935; Gupta 2010).

[36] (Bich 2019; Gupta 2010).

[37] ibid.

[38] An extremely short picture of the historical path leading to the current system is here provided. For further information on the topic, refer to the texts in the bibliography.

[39] The official date is though May 2019.

[40] Sections 2.3.1 and 2.3.3 in Bureau international des poids et mesures (2019).

Table 3.1 Fundamental physical quantities with corresponding measurement units. *Typical symbols* are those recommended by BIPM, while the indicated *dimensional symbols* are those usually employed in the so-called *dimensional analysis*

Physical quantity	Recommended symbol	Dimensional symbol	Unit of measurement	Official symbol
Time	t	T	second	s
Length	l, x, r, etc.	L	metre	m
Mass	m	M	kilogram	kg
Electric current	I, i	I	ampere	A
Thermodynamic temperature	T	Θ	kelvin	K
Amount of substance	n	N	mole	mol
Luminous intensity	I_v	J	candela	cd

Table 3.2 Reference standard of international system

Definition	Symbol	Exact numerical value	Corresponding unit
The frequency of the hyperfine transition of the not-perturbed ground state of the Cesium-133 atom is	$\Delta\nu_{\mathrm{Cs}}$	9192631770	Hz
The *in-vacuum* speed of light is	c	299792458	m/s
The Planck constant	h	$6,62607015 \cdot 10^{-34}$	Js
The elementary charge is	e	$1,602176634 \cdot 10^{-19}$	C
The Boltzmann constant is	k_B	$1,380649 \cdot 10^{-23}$	J/K
The Avogadro constant is	N_A	$6,02214076 \cdot 10^{23}$	mol^{-1}
The luminous efficiency of a monochromatic radiation with frequency $540 \cdot 10^{12}$ Hz è	K_{cd}	683	lm/W

$\mathrm{Hz} = \mathrm{s}^{-1}$,
$\mathrm{J} = \mathrm{m}^2 \cdot \mathrm{kg} \cdot \mathrm{s}^{-2}$,
$\mathrm{C} = \mathrm{A} \cdot \mathrm{s}$,
$\mathrm{lm} = \mathrm{cd} \cdot \mathrm{m}^2 \cdot \mathrm{m}^{-2} = \mathrm{cd} \cdot sr$ (where sr is a *steradian*),
$\mathrm{W} = \mathrm{m}^2 \cdot \mathrm{kg} \cdot \mathrm{s}^{-3}$.

To build up the system we therefore need to simultaneously have got the definitions provided by Tables 3.1 and 3.2. The first of these identifies suitable physical quantities correlated to each other, allowing to construct a *system*. The relationships between the quantities are suggested by nature, as we currently know it; therefore, the construction of the 3.1 essentially belongs to fundamental physics. Conversely, the assumptions of

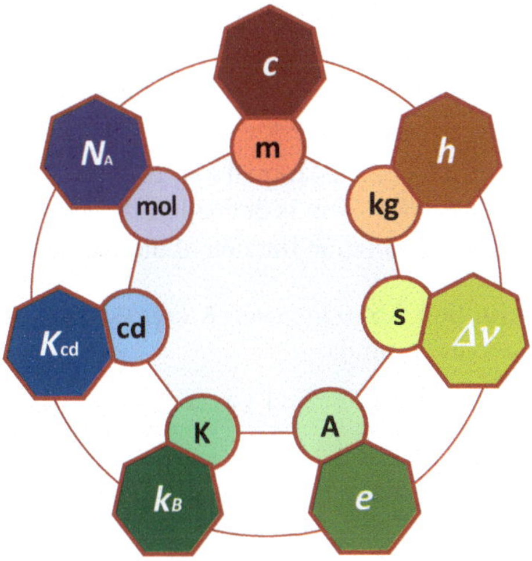

Fig. 3.1 Scheme of the seven basic units (external circle) which define the seven fundamental measurement units (internal circle) of the International System according to the formulation adopted in 2018

standards in the 3.2 are determined by application possibilities and, for this reason, they have got a high degree of arbitrariness.

(3.8) To be of any practical use, these units not only have to be defined, but they also have to be realized physically for dissemination.[41]

In this most recent formulation, which is the most precise, refined, and powerful one, the heart of the system lies in obtaining the fundamental units of measurement starting from the aforementioned physical constants. Since it was decided to maintain the distinction between *basic* and *derived* measurement units, the basic units are defined through the fundamental constants listed above, according to the scheme shown in Fig. 3.1.

The most updated definition of the IS constructs the units of the seven quantities given by Table 3.1 through the standards of Table 3.2 as it follows.[42]

Definition 3.6 (*second*) *Second* (s) is defined according to the $\Delta\nu_{\text{Cs}}$ constant (which is a *natural standard*), that, by definition, has got a fixed numerical value. So we obtain

$$1 \text{ s} = \frac{9\,192\,631\,770}{\Delta\nu_{\text{Cs}}}.$$

In other words, 1 s is defined as the time in which the radiation corresponding to the hyperfine transition of the Cs-133 atom effectuates 9 192 631 770 oscillations.

---

[41] Section 1.2 in Bureau international des poids et mesures (2019).

[42] (Bureau international des poids et mesures 2008, 2019; Bich 2019).

Definition 3.7 (*metre*) *Metre* (m), which is a unit of length, is defined according to the formula

$$1\ \text{m} = \left(\frac{c}{299\,792\,458}\right)\ \text{s},$$

that is, it is obtained form the second, through the *fixed* value of c (a *natural constant*). In other words, 1 m is defined as the distance travelled by light in vacuum during a $1/299\,792\,458$ time fraction of the previously defined second.[43]

Definition 3.8 (*kilogram*) *Kilogram* (kg), which a unit of weight, is defined according to the formula

$$1\ \text{kg} = \left(\frac{h}{6.626\,070\,15\cdot 10^{-34}}\right)\ \text{s/m}^2,$$

that is, it is specifically obtained from the fixed value of the constant h (again, a *natural constant*) expressed in the previously defined units. In other words, 1 kg is defined as $h\Delta\nu_{\text{Cs}}$, the quantum of energy which is associated to the hyperfine transition of Cs-133, divided by c^2, and then multiplied by the fraction $\frac{299\,792\,458^2}{6.626\,070\,15\cdot 10^{-34}\cdot 9\,192\,631\,770}$.[44]

The *second*, the *metre*, and the *kilogram* are the measurement units providing the reference quantities for mechanics and gravitation.

Definition 3.9 (*ampere*) *Ampere* (A), which is a unit of current, is given by

$$1\ \text{A} = \left(\frac{e}{1.602\,176\,634\cdot 10^{-19}}\right)\ \text{s}^{-1};$$

this means that such quantity is provided on the basis of the fixed value of the *natural constant* e and of the second, which is provided by the previous definitions. That is, 1 A is $\frac{1}{9\,192\,631\,770\cdot 1.602\,176\,634\cdot 10^{-19}}$ di $\Delta\nu_{\text{Cs}}$ multiplied by the constant e.[45] The ampere is the measurement unit on which the reference of the electromagnetic quantities is based.

---

[43] We also obtain

$$1\ \text{m} = \frac{9\,192\,631\,770}{299\,792\,458}\frac{c}{\Delta\nu_{\text{Cs}}} \approx 30.663\,319\frac{c}{\Delta\nu_{\text{Cs}}}.$$

[44] We also obtain

$$1\ \text{kg} = \frac{299\,792\,458^2}{6.626\,070\,15\cdot 10^{-34}\cdot 9\,192\,631\,770}\frac{h\Delta\nu_{\text{Cs}}}{c^2} \approx 1.475\,521\,4\cdot 10^{40}\frac{h\Delta\nu_{\text{Cs}}}{c^2}.$$

[45] Moreover, doing calculations, we get

$$1\ \text{A} = \frac{1}{9\,192\,631\,770\cdot 1.602\,176\,634\cdot 10^{-19}}\Delta\nu_{\text{Cs}}e \approx 6.789687\cdot 10^{8}\,\Delta\nu_{\text{Cs}}e.$$

Definition 3.10 (*kelvin*) *Kelvin* (K), which is the unit of thermodynamic temperature, is given by

$$1\text{ K} = \left(\frac{1.380649}{k}\right) \cdot 10^{-23}\text{ kg m}^2\text{ s}^{-2},$$

that is, it is based on the fixed value of k (a *natural constant*) and on the previous definitions of s, m, kg, correlated to $\Delta\nu_{Cs}$, c, h.[46]

Definition 3.11 (*mole*) *Mole* (mol), which is indicated as the measurement unit of the *quantity of substance*, is given by the formula

$$1\text{ mol} = \left(\frac{6.02214076 \cdot 10^{23}}{N_A}\right),$$

i.e. it is provided in terms of the *Avogadro number*, whose numerical value is fixed.

The *kelvin* and the *mole* are the measurement units that provide the main references of the thermodynamic quantities, therefore they have got particular interest in chemistry. Finally, there is a last quantity which is defined in relation to those previously given, the *candela*, which is the unit of luminous intensity in a given direction.

Definition 3.12 (*candela*) *Candela* (cd) is given by

$$1\text{ cd} = \left(\frac{K_{cd}}{683}\right)\text{ kg m}^2\text{ s}^{-3}\text{ sr}^{-1}.$$

In this formula the previously defined units are present, the fundamental fixed constant K_{cd} and the mathematically defined unit *steradian* (sr).

The above-listed definitions highlight that each fundamental unit is built upon its specific reference, either when it is standard (the second) or a constant (all the others). This relationship between each unit and its respective reference is graphically shown in Fig. 3.1.

> (3.9) The set of seven defining constants has been chosen to provide a fundamental, stable and universal reference that simultaneously allows for practical realizations with the smallest uncertainties. The technical conventions and specifications also take historical developments into account.[47]

The reasons for which the aforementioned complex was built lie in this declaration by the BIPM. Among the three kinds of *basic units* listed in Sect. 3.3.1, the chosen

---

[46] We also have

$$1\text{ K} = \left(\frac{1.380\,649 \cdot 10^{-23}}{6.626\,070\,15 \cdot 10^{-34} \cdot 9\,192\,631\,770}\right)\frac{\Delta\nu_{Cs}h}{k} \approx 2.266\,665\,3\frac{\Delta\nu_{Cs}h}{k}.$$

[47] Section 2.2.1 in Bureau international des poids et mesures (2019).

standards achieve the aim of not including any artefact among them. There is only one natural standard (in order to define the second); all other references are represented by natural constants.

The first advantage is the *stability* of the references: for what we currently know about nature, in every place and in every time the aforementioned standards, such as the in-vacuum speed of light, Planck's constant, the fundamental charge, etc., are maintained unchanged. Of course, the question about whether this is actually always satisfied has been posed. In particular, the constant e corresponds to a coupling constant of the electromagnetic field through the *fine structure constant* α.[48] The carried out experiments show that e and α remain identical over time and that, even if the variations foreseen by some theories were confirmed, they would be so small as to be completely irrelevant for the current experimental limits and in the near future.[49] In conclusion, the target of stability over time of all reference units is largely satisfied.

The other characteristic aimed at by the current formulation of the IS, as stated in the quote (3.9), is to obtain the most precise and effective *reproducibility*. This goal is achieved by exploiting standards whose practical realization can be accomplished with relative simplicity by any laboratory.

It has been known for more than one century that spectroscopic techniques make it possible to reach very high levels of precision. In this vein, the great precision that is achievable through atomic spectroscopy made it possible to build *atomic clocks* to provide standards for time measurement.[50] These devices and their related techniques became reliable and reproducible from the mid-50s of the twentieth century. The use of the Cesium-133 atom hyperfine transition was motivated by several practical reasons.[51] Although the definition of second adopted at the birth of the IS (1960) still assumed the astronomical references analogous to those of the old decimal metric system,[52] from 1967 the greater precision of the atomic references was officially recognized, adopting the definition of the second based on the $\Delta\nu_{\mathrm{Cs}}$ standard.[53,54] Consequently, the possibility of reproducing the standard of the time comes from the features of the Cs-133 isotope, which are the same everywhere, and from the well-known and consolidated techniques of atomic spectroscopy.

The definition of the metre is based on that of the second and on the invariance of the in-vacuum light speed: it is superfluous to recall the certainty of the c invariance in any reference system. Accordingly, the reproducibility of the metre arises from the consolidated techniques for measuring the speed of light, combined with the standard of the second.[55]

---

48 $\alpha = e^2/(2c\epsilon_0 h)$, where ϵ_0 is the vacuum dielectric constant.

49 Section 2.2.1 in Bureau international des poids et mesures (2019).

50 (Inguscio and Fallani 2013; McCarthy and Seidelmann 2018).

51 Chapter 11 in McCarthy and Seidelmann (2018).

52 Eleventh General Conference of Weights and Measures.

53 Thirteenth General Conference on Weights and Measures.

54 Chapter 13 in (McCarthy and Seidelmann 2018).

55 Appendix 2 in Bureau international des poids et mesures (2019).

The identification of suitable, easily measurable, modern standards for the definition of the units of electric charge and mass has allowed, at present, the formulation of the most powerful and precise IS. A fundamental role for the identification and replication of these standards was played by the discovery and use of the *Josephson Effect* and the *Quantum Hall Effect*.

The Josephson effect[56] is found when there are two superconducting electrodes separated by a layer of insulator. By choosing suitable materials, geometries, and temperatures, if a fixed voltage ΔV is applied to the electrodes, charges pass through the insulator due to the tunnel effect. In particular, an oscillating current $I = I(t)$ is established:

$$I = I_0 \sin[\Delta\phi(t)],$$

whose phase, unless a not-time-dependent term ϕ_0, is found to be proportional to the applied voltage, according to the relationship

$$\Delta\phi(t) = \phi_0 + \frac{2e}{\hbar}\Delta V t.$$

So, the oscillation frequency is given by

$$\frac{d}{dt}\Delta\phi(t) = \frac{2e}{h}2\pi\Delta V = K_J 2\pi\Delta V,$$

where $K_J = 2e/h$ is called *Josephson constant* and is expressed in Hz/V. The remarkable peculiarity is that, once the Josephson effect has been established, the value of the oscillation frequency of $I(t)$ does not depend on anything other than the applied voltage. This is indicated by the fact that within K_J there are only the fundamental charge and Planck's constant. Thus, an apparatus for Josephson effect measurement provides, via K_J, a way to directly access the e/h ratio. More importantly from a practical point of view, the constant K_J is able to provide a practical representation of the *volt*.[57]

A similar role is played by the measurement of the *quantum Hall effect*.[58] Generally, *the Hall effect* is used to investigate conduction in materials. A typical quantity that is measured is the so-called *Hall resistance*, R_H. The quantum Hall effect occurs

[56] We do not stop here to explain in detail this effect, which is a very rich and complex phenomenon with several implications in the field of matter physics. For the purposes of this discussion we limit to summarize it in a few lines. For further information it is interesting to read the original *Nobel Lecture* by B.D. Josephson (Josephson 1974) or to consult some monographic texts such as (Barone and Paternò 1982), a classic one in the field.

[57] (Bich 2019).

[58] As before, we do not explain in detail the physics of this effect, which goes beyond the scope of the current discussion. For further information read the original *Nobel Lecture* by K. von Klitzing (von Klitzing 1986) or analogous monographic and *reviews*, such as, for example (von Klitzing 2004).

in some nearly two-dimensional physical systems, at very low temperatures and in the presence of strong magnetic fields. Under these conditions, R_H is the rational multiple of a constant R_K, which is called *von Klitzing's constant*. Then we obtain

$$R_H = \frac{1}{n} R_K = \frac{1}{n} \frac{h}{e^2}$$

with n being either an integer ≥ 1 (*integer quantum Hall effect*) or a ratio (a fraction) between a natural number ≥ 1 and an odd integer (*fractional quantum Hall effect*).[59] Again, the remarkable peculiarity is that, once the quantum Hall effect has been established, the values assumed by the Hall resistance exclusively depend on the e and h fundamental constants.[60] From practical point of view, the constant R_K, which has the dimensions of a *resistance*, directly provides a practical representation of the *ohm*.

The *Josephson's constant* and the *von Klitzing's constant* have got, in the last IS formulation, exact numerical values, due to the exact values assumed by the fundamental constants in the Table 3.2. K_J and R_K are first of all explicitly involved in the effective practical implementation, the so-called *mise en pratique*, of the fundamental standard of the ampere and of all the main measurement units of electromagnetic quantities (ohm Ω, volt V, coulomb C, tesla T, etc.).[61] Secondly, these constants also play an essential role in the *mise en pratique* of the kilogram, whose standard practical implementation, rather articulated, involves mechanical and electromagnetic quantities in the construction of an *electromechanical scale*.[62]

3.3.3 Specificities of the International System

In the previous section we mentioned several properties of the IS, which make it a unique tool in physics and all sciences. Among the various features, it is worth analysing especially the most important one: namely its being *a system*. We had said that a measurement unit is a sample of a certain *physical quantity* which is arbitrarily assumed to have got the value 1. The specific characteristic of the international system is that the measurement units are not defined in a completely arbitrary way, but each of them has got a coherent relationship with the other ones (cf. Fig. 3.2). The logical-rational framework of the system, which has got evidently a mathematical nature, reverberates the relationships among the fundamental quantities. The basic units of measurement can be changed, but there are always relations which, if not identical to those of the IS shown in Fig. 3.2, maintain the same structure. In short,

---

[59] $1/3$, $2/5$, $3/7$, $2/3$, etc.

[60] (von Klitzing 1986).

[61] Appendix 2 of Bureau international des poids et mesures (2019).

[62] This method is the first officially indicated in Appendix 2 of Bureau international des poids et mesures (2019).

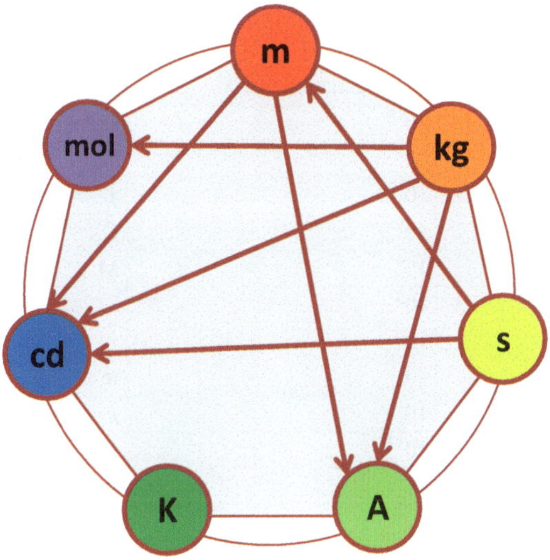

Fig. 3.2 Scheme of the seven fundamental measurement units of the International System. They are second (s), metre (m), kilogram (kg), ampere (A), mole (mol), candela (cd), and kelvin (K). Arrows indicate the correlations among them

the structural aspect of the IS is its most stable feature (cf. quote (2.6)). Among the quantities definable through the IS, the distinction between *fundamental* and *derived* quantities is maintained. The *fundamental* quantities are the seven physical quantities listed in Table 3.1, whose the seven respective corresponding units of measurement are defined. The system provides

the presence of these seven fundamental quantities (Table 3.1), whose measurement units are defined by the seven fundamental standards listed in Table 3.2 (scheme Fig. 3.1);
specific relations among the seven fundamental quantities, so that some of them can be defined in correlations with other ones (scheme Fig. 3.2);
the possibility of defining whatever other physical quantity, and its corresponding unit of measurement, which are products or rations of the seven basic units and of their respective measurement units;
any multiples and submultiples are obtained exclusively by adding appropriate *prefixes* to the IS units[63] (cf. Table 3.3).

These principles are summarized in the following statement

(3.10) The seven constants are chosen in such a way that any unit of the SI can be written either through a defining constant itself or through products or quotients of defining constants.[64]

[63] This choice is specific to modern systems, starting with the eighteenth-century Metric System. As a matter of tact, in the most ancient systems, multiples and submultiples of the units of the same physical quantity had different names, in addition to not being in decimals (cf. above Sect. 3.3.1).

[64] Section 2.2 in Bureau international des poids et mesures (2019).

Table 3.3 Main multiples and submultiples of IS (the complete list can be found in almost all books about general and experimental physics)

Prefix	Exponentiation of 10	Symbol
exa	10^{18}	E
peta	10^{15}	P
tera	10^{12}	T
giga	10^{9}	G
mega	10^{6}	M
kilo	10^{3}	k
	10^{0}	
milli	10^{-3}	m
micro	10^{-6}	μ
nano	10^{-9}	n
pico	10^{-12}	p
femto	10^{-15}	f
atto	10^{-18}	a

So, seven *dimensions* are defined, corresponding to the seven basic physical quantities of IS (Table 3.1). For this reason, the *dimension* of any quantity is expressed by what is called the *dimensional equation*:

$$\dim Q = T^{\alpha} L^{\beta} M^{\gamma} I^{\delta} \Theta^{\epsilon} N^{\zeta} J^{\eta}, \tag{3.2}$$

where the Latin letters represent the physical quantities according to Table 3.1, and α, β, etc. are positive or negative integers, usually small. The *dimension* of any physical quantity *whose measurement provides a numerical value* can be given according to the above relation. In particular it follows that

> (3.11) It is important to emphasize that each physical quantity has only one coherent SI unit[65]

Besides the 7 basic units of measure and the units constructed from powers of them (such as s^2, m^2, m^3, etc.), there are 22 not fundamental units having their own name and symbols (listed in the Appendix A). Finally, it is worth noting that there are some quantities that are not provided through the seven basic quantities of the IS.

> (3.12) There are also some quantities that cannot be described in terms of the seven base quantities of the SI, but have the nature of a count. Examples are a number of molecules, a number of cellular or biomolecular entities (for example copies of a particular nucleic acid sequence), or degeneracy in quantum mechanics. Counting quantities are also quantities with the associated unit one.[66]

[65] Section 2.3.4 in Bureau international des poids et mesures (2019).

[66] Section 2.3.3 in Bureau international des poids et mesures (2019).

Such quantities are called *dimensionless*: by applying the relation (3.2), all the exponents are 0. Particular attention must be paid to these quantities: because of their adimensionality, if it is incorrectly applied, it can lead to contradictory deductions. An indicative example is the following one. It is certainly intuitive to write

$$2\pi \text{ rad s}^{-1} = 1 \text{ cycle s}^{-1}.$$

Nevertheless if, being rad and cycle both adimensional in IS, we mathematically simplified each other, we would obtain

$$2\pi \text{ rad s}^{-1} = 1 \text{ cycle s}^{-1}$$

from which we dimensionally would obtain

$$2\pi \text{ s}^{-1} = 1 \text{ s}^{-1},$$

so

$$2\pi = 1\ (!)$$

which is evidently absurd,[67] due to the fact that, although they are both dimensionless quantities, we are not authorized to do the equivalence between radians and cycles, which are different physical quantities!

This example has been constructed with the aim of highlighting possible ambiguities of the IS, i.e. those aspects that are considered harbingers of misunderstanding and, therefore, loss of precision (which is the main purpose of the whole metrological science). For our analysis, however, this example highlights another, more general and deep feature. The difficulties and the multifaceted characteristics of the IS show that not even the most refined measurement system in the history of mankind is able to work as an automatic mechanism. On the contrary, like all tools used in scientific research, even the conceptual ones, it must be applied with suitable knowledge and awareness of the operations that are being carried out and of the limits and validity areas of the tools used.

However, the presence of quantities which are considered *dimensionless* points out that there is still room for an expansion of the system, in spite of the power and structural complexity of the IS, according to which each quantity can be expressed in relation to the others. There is nothing to prevent the possibility that some of the currently dimensionless quantities may one day be defined as a specific fundamental unit of measurement to which the concept of *dimension* can be correlated: in this sense there are currently debated proposals. For all that, the following should be noted. Let us limit ourselves to the certainly *most important* physical quantities, i.e. those that physical science considers conceptually fundamental: length, mass, time, energy, etc., which have got their well-defined measurement unit within the IS. Let us add the physical quantities measured with dimensionless units, such as angles

[67] This is a deliberately provocative result cited in Sect. 10 of Bich (2019).

or counts: these too, in some way, are taken into account within the IS. Are we sure we have taken everything into account, at least all the most important physical quantities? Are we certain that, at least for the most important quantities, there is *always* a well-defined unit of measurement? We try to answer these questions, at least partially, in Chap. 6 in Sect. 6.3.2.

References

A.B. Arons, *Teaching Introductory Physics* (Wiley, New York, USA, 1997)

A. Barone, G. Paternò, *Physics and Applications of the Josephson Effect* (Wiley, New York, USA, 1982)

G. Basti, *Filosofia della Natura e della Scienza*, vol. 1, 1a edn. (Pontificia Università Lateranense, Roma, Città del Vaticano, 2002)

W. Bich, The third-millennium international system of units. La Rivista del Nuovo Cimento **42**, 49–102 (2019)

U. Bottazzini, E.A. Marchisotto, P. Miller, *Hilbert's Flute: The History of Modern Mathematics* (Spriunger-Verlag, Berlin, Deutschland, 2016)

C.B. Boyer, *A History of Mathematics*, 2nd edn. (Wiley, New York, USA, 1968)

Bureau international des poids et mesures. *Le Système international d'unités (SI)–The International System of Units (SI)*, 9th edn. (BIPM, Sèvres, France, 2019). https://www.bipm.org/utils/common/pdf/si-brochure/SI-Brochure-9.pdf. Available online

Bureau international des poids et mesures, *International vocabulary of metrology-Basic and general concepts and associated terms (VIM)-Vocabulaire international de métrologie-Concepts fondamentaux et généraux et termes associés (VIM)*, 3rd edn. (BIPM, Sèvres, France, 2008)

I.M. Copi, C. Cohen, V. Rodych, *Introduction to Logic*, 15th edn. (Taylor & Francis, New York, USA, 2019)

R. Courant, H. Robbins, *What is Mathematics? An Elementary Approach to Ideas and Methods* (Oxford University Press, Oxford, UK, 1941)

J. de Boer, On the history of quantity calculus and the international system. Metrologia **31**, 405 (1995)

F. Enriques, *Problemi della Scienza* (Zanichelli, Bologna, Italia, 1906)

R.P. Feynman, R. Leighton, M. Sands, *The Feynman Lectures on Physics*, vol. 1, 2nd edn. (Addison-Wesley, Boston, USA, 2005)

R.P. Feynman, R. Leighton, M. Sands, *The Feynman Lectures on Physics*, vol. 2, 2nd edn. (Addison-Wesley, Boston, USA, 2005)

R.P. Feynman, R. Leighton, M. Sands, *The Feynman Lectures on Physics*, vol. 3, 2nd edn. (Addison-Wesley, Boston, USA, 2005)

G. Galilei, *Istoria e dimostrazioni intorno alle macchie solari.* (Accademia dei Lincei, Roma, Italia, 1613)

S.V. Gupta, *Units of Measurement: Past, Present and Future. International System of Units.* (Springer & Metrology Society of India, Heidelberg, Deutschland, 2010)

J. Gyllenbok, *Encyclopaedia of Historical Metrology, Weights, and Measures*, vol. 1 & 2. (Birkhäuser Cham, Basel, Schweiz, 2018)

M. Inguscio, L. Fallani, *Atomic Physics-Precise Measurements & Ultracold Matter* (Oxford University Press, Oxford, UK, 2013)

G. Israel, *La matematica e la realtà* (Carocci, Roma, Italia, 2015)

B.D. Josephson, The discovery of tunnelling supercurrents. Rev. Mod. Phys. **46**, 251 (1974)

A.E. Kennelly, Adoption of the meter-kilogram-mass-second (M.K.S.) absolute system of practical units by the international electrotechnical commission (I.E.C.), bruxelles, june, Proc. Natl. Acad. Sci. U.S.A. **21**(579–583), 1935 (1935)

A.N. Kolmogoroff, *Grundbegriffe der Wahrscheinlichkeitsrechnung* (Springer-Verlag, Berlin, Deutschland, 1933)

A. Levy, *Basic Set Theory* (Springer-Verlag, Berlin, Deutschland, 1979)

G. Lolli, *La guerra dei trent'anni (1900–1930)-Da Hilbert a Gödel* (Edizioni ETS, Pisa, Italia, 2011)

J.C. Maxwell, F. Jenkin, On the elementary relations between electrical measurements. London Edinburgh Dublin Philosop. Mag. J. Sci. **29**(198), 436–460 (1865)

J.C. Maxwell, *A Treatise on Electricity and Magnetism* (Oxford University Press, Oxford, UK, 1873)

D.D. McCarthy, P.K. Seidelmann, *Time: from Earth Rotation to Atomic Physics* (Oxford University Press, Oxford, UK, 2018)

H. Poincaré, *La Science et l'Hypothèse* (Flammarion, Paris, France, 1902)

W. Rautenberg, *A Concise Introduction to Mathematical Logic*, 3rd edn. (Springer, New York, USA, 2009)

W. Rudin, *Real and Complex Analysis* (McGraw-Hill, New York, USA, 1987)

E. Segrè, *From X-rays to Quarks: Modern Physicists and Their Discoveries* (W.H. Freeman & C., New York, USA, 1980)

A. Strumia, *Dalla filosofia della scienza alla filosofia nella scienza*, 1a edn. (ESC-Edizioni Santa Croce, Roma, Italia, 2017)

K. von Klitzing, The quantized Hall effect. Rev. Mod. Phys. **58**, 519 (1986)

K. von Klitzing, 25 years of quantum Hall effect (QHE) a personal view on the discovery, physics and applications of this quantum effect. Séminaire Poincaré **2**, 1–16 (2004)

R.L. Wilder, *Introduction to the Foundations of Mathematics*, 2nd edn. (Dover Publications, Mineola, New York, USA, 2012)

Chapter 4
The Development of the Founding Concept of Physics

When did physics begin? That is not an easy question.
Emilio Segré

This chapter presents some historical elements. However, it does not provide outlines of the history of (experimental) physics in the strict sense; in addition to being a digression, this would require a considerable extension even for a summary exposition. More than the *events* of the historical development of physics, the aim here is to identify the main development lines over time: the historical and cultural development of the founding concept of all physics, specifically exposed in Chaps. 1 and 2, i.e. *the hypothesis according to which natural phenomena can be described and explained through mathematics.*[1] We illustrate not the evolution of *concepts* but the development of the most important *concept* of the discipline.

In this brief exposition we focus on three *fundamental turning points*, with the aim of illuminating the fruitful relationship between mathematics and empirical knowledge that generated physics as well as modern scientific research.

While the *scientific revolution*, conventionally corresponding to the time of Galilei and Newton, necessarily stands out as one of these turning points, there were two other essential periods for the development of the concept of interrelation between mathematics and empirical reality.

The first one was the period that preceded the *scientific revolution*: indeed, it was in the Middle Ages and the Renaissance that the conceptual foundations for a science that was simultaneously based on *empirical findings* and on *conceptual demonstrations* (mathematics) were laid.

The other crucial turning point was the birth of the so-called *modern physics* in the first decades of the twentieth century.

---

[1] See quote 2.1 in Sect. 2.1.

J. Parravicini, *The Foundations of Experimental Physics*,
https://doi.org/10.1007/978-3-031-55452-0_4

For our discussion these three moments are also important from a diachronic perspective, being paradigmatic of different ways in which mathematics and the practice of scientific research can entangle.

4.1 Prodromes of Physical Science

The question on where to locate the roots of modern scientific thought, and especially of modern physics, has no clear answer. Some people place these origins in antiquity, tracing them back to the first Greek philosophers, the so-called "naturalists", and to the knowledge of the so-called ancient "Greek science". To avoid getting into this diatribe, we take the perspective of E. Segrè:

> (4.1) To trace early physics is beyond my competence, and I take advantage of the obvious discontinuities presented by the development of physics as a science to start my story with Galileo.[2]

From his perspective we deduce two criteria:

1. to strictly limit our discussion to modern *physics*, i.e. to the discipline that is methodologically founded on the hypothesis of systematically using mathematics to model and describe phenomenological data;
2. to assume as a recognized *fixed point* the era of the Galilean and Newtonian scientific revolution, a period of turning point and clear discontinuity, and so to evaluate *backwards* the evolution of the concepts.

Indeed, many concepts containing the seeds of modern scientific-mathematical thought are developed in the centuries which immediately preceded the advent of Galilei and Newton, i.e. in the Middle Ages and the Renaissance. In reference to this historical path leading to the natural sciences as we modernly conceive them, among scholars there are two antithetical attitudes (with all the intermediate gradations).

On the one hand, there is the more common opinion that the natural sciences, modernly conceived as methodologically founded on OBSERVATION AND MEASUREMENT and subsequently on EXPERIMENTS, are born *exactly* with Galileo Galilei. Some even go so far as to argue that before him there is nothing outside of an oppressive and slavish Aristotelianism, incapable of dealing with observational data and of separating *theoretical philosophy* from *natural philosophy*, a conception incapable to free oneself from the *auctoritas* principle ("ipse dixit"). In short, an inability to escape from patterns rigidly pursued for millennia, in which the categories of *quantity* and *quality* were not clearly distinct.[3] Hence, the *scientific revolution* would arise in total contrast to the previous culture, as an illumination, a brilliant stroke of love at first sight, completely unthought and unthinkable.

On the other hand, there are others who, aiming to recover the path of knowledge of scholars of Middle Ages and Renaissance, tend to diminish the importance of the

---

[2] E. Segrè, Chap. 1 of Segrè (1984).

[3] Cf. Bellone (2004), Tornaghi and (a cura di) (1997).

Galilean revolution, pointing out that Galilei operates in perfect continuity with the previous centuries: for this reason, the concepts he proposes are nothing more than a sort of synthesis of all the ideas that precede him, which finally have found suitable prominence due to a historical maturity.[4] In short, the *scientific revolution* would have nothing truly revolutionary.

We must clearly recognize that these extreme views are both wrong. On the one hand it is true that Galilei represents a turning point from which experimental physics develops in its modern conception. On the other hand, it is equally true that he does not emerge as a sudden light in a sea of darkness. About the problem of the actual importance of the *scientific revolution*, as well as about that of the Medieval and Renaissance origins of scientific thought, rivers of ink have been spilled.[5] Unfortunately, in such discussions the passions of individual scholars often play a decisive role, with the intention, more or less explicit, more or less conscious, of instrumentally exploiting in a political direction the history of the modern science birth, some exalting the discontinuity factors and maligning the *humus* where it took place (by Enlightenmentists and positivists), others exalting the continuity factors and maligning the importance of revolution (by medievalists and conservatives). On the other hand, in the necessary brevity of this discussion, we try to stick to the facts in remembering both the features of continuity and those of revolution, that are factors which in the birth of modern physical science are inextricably and indissolubly linked.[6]

Therefore, assuming the work by Galileo as a historically decisive moment, we first evaluate what precedes it, in particular, with respect to the problem of the mathematizability of natural phenomena. As a scientist Galilei accepted several intuitions formulated by previous scholars, who anticipate, in some cases even by centuries, some of the methodological steps of the great Pisan. Galilei's most important methodological cornerstones are

1. the *mathematization*[7];
2. the pre-eminence of the *observational investigation* and the consequent *experiments*.[8]

If we have already provided a definition of mathematization,[9] of the second one we give the following:

Definition 4.1 *(Observational investigation)* We call *observational investigation* the process of observation, measurement, and recording, carried out with systematically, accurately, and rationally with the aim of increasing knowledge of a well-defined natural phenomenon.

[4] Cf. e.g. Bellone (2004), Carioti (1977), Tornaghi and (a cura di) (1997).

[5] Cf. e.g. Bellone (2004), Grant (1997), Tornaghi and (a cura di) (1997), Wallace (1981).

[6] To express it with a happily expression widely used in quantum mechanics, one could say that continuity and revolutionality are *entangled* present.

[7] Cf. quote 2.1 in Sect. 2.1.

[8] We highlight here the difference between the *observation* and the *experiment*, which are specified in the following section.

[9] Cf. Definition 2.2.

How could the aforementioned powerful methodological assumptions arise? We need to return to the classification of disciplines according to classical epistemology, as already done in Sect. 1.2. In the Medieval and Renaissance periods, studies and "sciences"[10] are organized according to structured and hierarchical schemes. At the time rather than subdivisions, correlations are highlighted in the human knowledge. As a matter of fact, knowledge is considered as a unitary organism.[11] Briefly, we can recall that, according to the Aristotelian-Thomistic vision which is predominant in the centuries immediately preceding the scientific revolution, "sciences" are divided into two categories[12]:

1. explanatory and deductive sciences;
2. descriptive and inductive sciences.

The former ones employ techniques of pure reasoning and they have got immaterial objects as their investigation field, whose study is carried out by employing demonstrative techniques: among them there are logic, theoretical philosophy, mathematics, metaphysics, etc. The latter ones have got concrete objects as their investigation field and they employ essentially descriptive techniques: among them we find, for example, natural sciences, geology, geography, etc. The former ones are considered foundational of the latter ones, they provide the conceptual tools, for which they are called *scientiæ rectrices*. Although during the Middle Ages and the Renaissance other knowledge classification schemes different from the Aristotelian-Thomist ones are proposed,[13] it can be stated that the cultural background from which Galilei starts assumes these categories. Especially, we highlight the aforementioned two Galilean cornerstones: mathematization is correlated with the deductive disciplines; the empiricism of observational investigation is correlated with the inductive disciplines. Incidentally, we note that this categorization can also be considered valid today:

> (4.2) *Experimentum enim solum certificat in talibus, eo quod de tam particularibus naturis syllogismus haberi non potest.*[14]
>
> Only experiment certifies in such cases, inasmuch as no syllogism can be implied on such peculiar properties of nature.

Our next question is: in the centuries preceding the *scientific revolution*, is it conceivable a discipline like modern physics, which is at the same time explanatory-deductive and descriptive-inductive? The answer is affirmative: there are the *scientiæ mediæ*, such as astronomy, known for millennia, or geometric optics. These *scientiæ* are at the same time deductive and inductive: they straddle both mathematics (abstract and immaterial, but demonstrative) and nature (empirical and subject to description). In

[10] In the sense of "disciplines".

[11] Grant (1997), Strumia (2017).

[12] Basti (2002), Strumia (2017).

[13] In particular, it is worth remembering the approach of the *Chartres school* (Wallace 1981).

[14] Albertus Magnus (1867).

short, there are disciplines where tangible phenomena provide measurable matter, whose procedures and concepts are nevertheless recognized as mathematical.

The case of astronomy is the most immediate: for centuries it has been developing mathematical models that are as capable as possible of describing the observed *phenomena* and, more importantly, are also *predictive*. In short, the obtained models make it possible to arrive at mathematical relationships that give the possibility of predicting the phenomena. In this we see the co-presence of at least four important elements:

1. the study of natural phenomena,
2. the recording of data arising from observations,
3. the development of geometrical-mathematical models, and
4. the development of predictive mathematical relationships.

All this leads to something very close to what are modernly called *physical laws*[15]: the aforementioned elements can be found in modern Galilean physical science. In several passages of his writings, St. Thomas Aquinas himself, the most important and iconic medieval philosopher, repeatedly recognizes the disciplines constructed in this way: the *scientiæ mediæ*. In particular, he clearly identifies the first of the problems of physics in the problem of the mathematical description of motions:

(4.3) *[C]onsiderare motus non pertinet ad mathematicum, sed tamen principia mathematica ad motum applicari possunt. Et ideo secundum hoc, quod principia quantitatis ad motum applicantur, naturalis considerat de divisione et continuitate motus [...] Et in scientiis mediis inter mathematicam et naturalem tractatur de mensuris motuum, sicut in scientia de sphæra mota et in astrologia.*[16]

Studying motion does not belong to the mathematician, but nevertheless mathematical principles can be applied to motion. And therefore according to the fact that the principles of quantity are applied to motion, it is natural to consider the division and continuity of motion [...] And in the intermediate sciences, between mathematics and the natural ones, the measurements of motion are discussed, as in the science that studies the motion of celestial spheres and in astronomy.

Therefore, if the idea of building a discipline based on observations and structured on mathematical models was widely circulating, what of this had actually been developed? The history of pre-Galilean natural sciences[17] is a very vast topic, so we limit to cite a few specific authors.

During the Middle Ages there had been several *schools* which, within the philosophical studies of the knowledge of the time, had developed original concepts regarding the study of natural phenomena. The main ones are the following[18]:

The *School of Chartres* in France (eleventh–twelfth century). It is among the first ones to recognize an autonomous value of the investigation of natural phenomena.

---

[15] Cf. Definition 2.5.

[16] Thomas Aquinas, pars 3 q. 5 a. 3 ad 5 in (Tommaso d'Aquino).

[17] We use the adjective "pre-Galilean" as a synonym for "before the *scientific revolution*".

[18] Clagett (1959), Grant (1997), Wallace (1981).

The *School of Oxford* in England (twelfth–thirteenth century). Its most important representatives are *Robert Grosseteste* (1175–1253) and *Roger Bacon* (1214–1292). The former largely works on geometric optics. The latter highlights the importance of empirical *experiences* in the study of nature in addition to accomplish studies on colours; some English scholars explicitly indicate him as a precursor of the scientific method.

The *School of Paris* in France (twelfth–fourteenth century). Its main representatives are Blessed *Hugh of Saint Victor* (1096–1141), *Jean Buridan* (1295–1361), and *Nicolas Oresme* (1323–1382). Regarding the first scientist, it is worth remembering his proposal for a classification of knowledge which significantly differed from the Aristotelian one. In an intellectual context focused on theology and theoretical philosophy, whom he is dealing with, he strongly defends the dignity of studies having natural phenomena as their object while dealing with these subjects. Buridan is a proponent of the *impetus theory*: an explanation of motion that is clearly different from Aristotelianism. He also develops concepts close to what is modernly called *momentum*. To demonstrate this concept, he studies the motion of boats on the Seine in *controlled conditions*, so carrying out modern *experiments*.[19] Nicolas Oresme is mainly a mathematician and a forerunner of the concept of analytical geometry in introducing a preliminary form of what we today call the "Cartesian plane". Furthermore, he proposes the first concepts of what is today called "vector", with its representation in a plane through position arrows. In addition, he provides the law of uniformly accelerated motion and also suggests the principle of relativity of motions.[20]

Therefore, a fruitful environment for the construction of the physics discipline as we now know had been preparing for centuries in a slow but consistent way. The ideas for developing a mathematized science of natural phenomena, on the other hand, were not linked to single representatives, nor did they circulate in limited more or less visionary contexts, instead they spread throughout Europe. As mentioned previously, astronomy and optics were disciplines carried out and studied everywhere, disciplines that became test beds where the main ingredients of future Galilean science are employed.

In the decades immediately preceding the time of Galilei we see the affirmation of the experimental method in various contexts. Here too, it is worth mentioning some important scientists[21]:

Two generations before Galilei there are the Flemish *Andrea Vesalius* (1514–1564) and the Italian *Gabriele Falloppio* (1523–1592), Professors at the University of Padua. Both found the modern anatomical science based on a systematic and extensive work on cadaver dissection. Vesalius' modern approach in applying the

[19] Clagett (1959), Zupko (2018).

[20] Clagett (1959), Kirschner (2021).

[21] Part One, Chap. 1 of Bellone (2004).

experimental method is not limited to the choice of performing autopsies, but he arrives to carefully plan the experimental conditions in which to accomplish the dissections, in order to obtain the best and most controlled conditions. His main work *De humani corporis fabrica* is considered the first modern treatise on anatomy: the search for expository clarity made him commission the best anatomical drawings among all the illustrations of contemporary treatises.

One generation before Galileo, the observations of the Dane *Tycho Brahe* (1546–1601), astronomer at the Copenhagen Observatory, played a crucial role. He carried out a remarkable work in the systematicity and accuracy of the observations, collecting an enormous quantity of data extremely reliable for the time, providing the first suggestions for overcoming the *Ptolemaic model*.[22]

Contemporary with Galilei there is the English physician *William Harvey* (1578–1657). He studies at the University of Cambridge and specializes at the University of Padua, where he has got the Fallopian's disciples as teachers. Harvey carries out careful experimental anatomical work based on which he understands and describes the cornerstones of blood circulation.

The picture emerging from this brief excursus is the following: during the years immediately preceding the birth of *Galilean physics*, the main concepts are actually present, i.e. the possibility of building knowledge of natural phenomena through *observational data*, and there is also the idea, and, in some cases, the practice of formally organizing them according to a mathematical language. Moreover, these concepts circulate widely among the most important cultural institutions in Europe at the time.

Accordingly, we can affirm that, contrary to what is claimed by the Enlightenment and positivists, in the centuries that preceded Galilei and Newton, scientific research is carried out with many of the current methodological elements; this scientific research is able to provide many data recognized as valuable for the evolution of knowledge, some of which are still considered valid.[23]

Consequently, the following question arises: what is missing for the emergence of modern physics? What great methodological innovations do Galilei and Newton propose to be considered the architects of a true revolution in knowledge?

---

[22] It is known that Tycho Brahe develops his own original cosmological model, which has got the earth at the centre of the cosmos, the sun and moon revolving around it, and all the other planets around the sun.

[23] Grant (1997), Part I Chap. 3 and 4 of Duhem (2016), Chap. 1 of Prigogine and Stengers (1986, 1999), etc.

4.2 Galilei and Newton: Physics Arrives Before Mathematics

4.2.1 The Birth of Mechanics

Despite all the conceptual and methodological prodromes above mentioned, in the vast majority of cases in the last decades of the sixteenth century the *naturalis philosophia*[24] is treated according to two basic criteria[25]:

1. in a definitely *qualitative* way;
2. as part of a theoretical-philosophical system (Aristotelian).

The main methodological innovation by Galilei is to *free the investigation* from the aforementioned two criteria, generating an effective *discontinuity* in the investigation of τὰ φυσικά[26] and triggering a tumultuous development in the study of natural phenomena.

Until the time of Galilei, scholars interested in natural phenomena share the opinion that the growth of knowledge is essentially a problem of *philosophical scheme*. In other words, it is assumed the necessity for identifying an appropriate (general, theoretical) *philosophy* that makes it possible to carry out an explanation of the multifaceted natural world with convincing detail. On the other hand, Galilei understands that the key to obtain significant progresses in the study of natural phenomena is not *changing* the philosophical scheme, but treating such phenomena *separately* from the philosophical system, through SPECIFIC CONCEPTS AND METHODS,[27] the former *mathematical*, the latter *observational and experimental*.

> (4.4) The times immediately preceding Galileo are full of precursors. Astronomers, navigators, artists, and technologists raised practical and theoretical questions that are pregnant of the future [...] The capital methodological discoveries were made by Galileo when he recognized the power of the combination of experiments with mathematics.[28]

For this reason, the use of experimentation embraced by Galilei is an aspect that was already widely used in several disciplines (as we have seen). On the other hand, Galilei wants to decouple the investigation of natural phenomena from their explanation of *ontological nature*, moving on to the use of quantitative categories: this allows him to use mathematics as *basic explanatory and deductive science*, so that it acts as a *scientia rectrix*. The entire construction of knowledge developed up to then, i.e. the *philosophia* of time, had the aim of explaining the *essence* of things.[29] To advance in the knowledge of natural phenomena, Galilei *programmatically* renounces this purpose, as he declares in the quote 3.4 (Sect. 3.2). Consequently, for the Pisan

[24] Cf. quote 1.6 in Sect. 1.2.

[25] Tornaghi and (a cura di) (1997), Strumia (2017).

[26] Cf. Sect. 1.2.

[27] Tornaghi and (a cura di) (1997).

[28] E. Segrè, Chap. 1 of Segrè (1984).

[29] Part I Chap. 1 of Duhem (2016).

scientist the *restricting the field* of investigation of *philosophia*,[30] by detaching the investigation of *natural philosophy* from the rest of knowledge, is precisely what allows us to develop an experimental method and a suitable mathematical language, specific and fruitful.[31] This choice of limitation can be read as a form of obedience to the reality to be investigated: it is the discovery of a conceptual investigation method more suitable for inspecting natural phenomena than those previously employed. After Galilei this detachment and restriction of the investigation field (also mentioned in the eras preceding him) becomes systematic, leading to the results of modern physics.

A further methodologically innovative aspect is the systematic use of the EXPERIMENT.[32]

Definition 4.2 *(Experiment)* We call *experiment* the pure knowledge investigation of a given natural phenomenon carried out under *controlled conditions*.[33] This control can be obtained by *causing*, i.e. *reproducing*, the phenomenon after removing those factors, normally present when it occurs in fortuitous conditions, which generally influence the phenomenon, preventing a clear evaluation, measurement, and understanding.

The above definition[34] takes into account the fact that, while it is true that all experiments are *observation and measurement*, it is not true that observing and measuring is enough to perform an experiment. As a matter of fact, in the era immediately preceding Galileo the *observational method* was widely used and widespread, especially in medicine and astronomy.[35] Conversely, the *experiment* is something that has not yet been established, although there were harbingers of this concept. Few scholars before Galilei had accomplished the form of investigation *in controlled conditions* for the express purpose of verifying hypotheses.[36] However, this practice was by no means systematically employed and widespread. Indeed, the act of *artificially reconstructing* a natural phenomenon, which is what a modern experiment (generally) consists in, is not something conceptually trivial. Reflecting on the first experiments that Galileo carries out on motion, we could pose the question: *a priori* who assures us that the motion studied indoors in a laboratory, under highly controlled conditions, is *qualitatively the same* as that which occurs outside?

(4.5) *Le dialogue expérimental avec la nature, que la science moderne se découvre capable de mener de façon systématique, ne suppose pas une observation passive, mais une pratique.*

---

30 Cf. e.g. quote 3.1 in Sect. 3.1.

31 Part I of Duhem (2016).

32 Part II, Chap. 4 of Duhem (2016).

33 Cf. e.g. Chap. 1 of Lindsay and Margenau (1957).

34 cf. e.g. Chap. 1 of Fornasini (2008), Montgomery (2001), etc.

35 See previous section.

36 One of these had been the aforementioned Buridan. In his experiments aimed at demonstrating the inadequacy of the Aristotelian motion theory there is an embryonic idea of the *principle of inertia*.

> *Il s'agit de manipuler, de mettre en scène la réalité physique jusqu'á lui conférer une proximité maximale par rapport á une description théorique. Il s'agit de préparer le phénomène étudié, de le purifier, de l'isoler jusqu'á ce qu'il ressemble á une situation idéale, physiquement irréalisable mais intelligible par excellence [...]*[37]

> The experimental dialogue with nature, which modern science discovers itself able of carrying out in a systematic way, does not suppose passive observation, but practice. It is about manipulating and staging physical reality to the point of arriving closest to a theoretical description. It requires preparing the investigated phenomenon, purifying it, isolating it until it resembles an ideal situation, which is physically not realizable but it is par excellence intelligible [...]

It is known that the first methodological innovations are precisely applied to the problem of motion, which had been identified as a crucial issue in the study of natural phenomena since the time of St. Thomas (cf. quote 4.3). The discoveries and concepts enunciated by Galileo on this topic are considered by some to be even more important than his discoveries in astronomy.[38] On kinematics and dynamics Galilei knows some works by Oresme. Although the question of the latter's influence on the Pisan scientist is a subject of debate,[39] it is established that the geometric representation of the speeds proposed by the French scholar was repeatedly employed by Galileo in his works.[40] The motion problem had conceptual issues of two kinds:

problems of geometry;
problems regarding *changes*.[41]

Considering the first ones, Galilei is certainly very skilled: on a conceptual level he has got the apparatus of *Euclidean geometry*, which he knows very well, which has been consolidated, handed down, commented on, and studied in depth for centuries. Galileo himself declares to base his vision of the universe on it, in the statement where he introduces his postulate (quote 2.1 in Sect. 2.1). "Triangles, circles, and other geometric figures" are the protagonists of Euclid's *Elements*, a book that had survived almost 2000 years unscathed, which had never needed to be rediscovered because it had never been forgotten. This especially shows the astonishing continuity between Galileo and what preceded him. At the same time, from a practical point of view, Galilei possesses tools for geometric measurements that are rather advanced for the time or he is able to independently develop new ones.

Conversely, regarding the issue, i.e. changes, the Pisan scientist has got completely unsuitable mathematical tools, as the mathematical objects known as *functions* have not yet been conceived. An interesting example of the inadequacy of the mathematics at Galileo's disposal can be found in the investigation into the motion of bodies dis-

---

[37] I. Prigogine et al., Chap. 1 of Prigogine and Stengers (1986).

[38] Cf. e.g. Chap. 1 of Segrè (1984).

[39] Clagett (1959).

[40] Boyer (1968).

[41] Because of the problem of changes philosophy moved from the discussions of the first *naturalist* philosophers to those of Plato and Aristotle, based on *general principles* (cf. Strumia (2017)).

cussed in the work *De motu*.[42] According to the scheme provided by the geometry of the time, only the relationships between homogeneous quantities could be considered in a study of such kind. Therefore, in the mathematics of his time, the *velocity* v of a body cannot be formally expressed as a function of the ratio between the travelled space s and the time taken to travel it t. For this reason, Galilei reasons in terms of proportions, assuming that two velocities v and $\tilde{v}$, the second one correlated to the space $\tilde{s}$ travelled in a time $\tilde{t}$, are in relation between them according to a relationship which, in modern notation, we would write

$$v/\tilde{v} = (s/\tilde{s}) \cdot (\tilde{t}/t).$$

From this example, it is clear that a suitable language was not yet available to describe the investigated phenomena. In his last work, *Discorsi e dimostrazioni matematiche intorno a due nove scienze* (*Discourses and mathematical demonstrations around two new sciences*, 1638), Galilei summarizes and develops many of the physical and mathematical results he has obtained throughout his life. Among others, the scientist enunciates the law of inertia, the composition of motions (which later would be called "vectorial"), the features of uniformly accelerated motion (with the investigation of the motion of falling bodies), and the first previews of the concept of *momentum conservation*.[43]

It is worth underlining that Galileo dealt with ideal cases, assuming what today we call "motions in the absence of friction". One reason for this is precisely his lack of suitable mathematical tools to deal with more realistic cases, which Newton later did in his treatment of hydrodynamics. Galilei understands that he needs to develop a new mathematics in order to progress the newly born *physical science*. Confirmation of this is the fact that he, having maintained for years the intention of writing a treatise about infinity, discusses this topic in several passages of his last treatise.[44] Noteworthy is the fact that, concerning the investigation of motions, the Pisan considers the *infinitely small* much more important than the *infinitely large*: this shows the intuition that to suitably describe motions there is need for an *infinitely large quid*, which nowadays we call *differential*.

4.2.2 The Birth of Infinitesimal Calculus and Mathematical Physics

Galilei's disciples are mainly interested in the problems left open by their teacher. In an attempt to describe the motions in conditions closer to real ones, it is necessary to take into account the media where they happen: air and water. For this reason, Benedetto Castelli (1578–1643), Evangelista Torricelli (1608–1647), and

---

[42] Cf. Chap. 1 of Bellone (2004).

[43] Bellone (2004), Segrè (1984), Tornaghi and (a cura di) (1997).

[44] Boyer (1968).

Vincenzo Viviani (1622–1703) focus on the study of fluids.[45] Furthermore, Torricelli and Bonaventura Cavalieri (1598–1647), also a disciple of Galileo, start to address the mathematical problem of *infinitesimal calculus*.[46]

In short, we can say that physics came first with the study of the motion phenomena, and then came mathematics, following the stimulus given by physics. Only later a rigorous mathematical formulation arrives. While Cavalieri develops the so-called *Method of indivisibles* for the calculation of areas and volumes, his student Pietro Mengoli (1626–1686), who succeeds him in the professorship in Bologna, develops his work anticipating concepts that would later become the *limits* and the *integrals*. For this reason, Mengoli is the last important Italian mathematician to advance in the Galilean program of developing the infinitesimal calculus. His works inspire John Wallis (1616–1703),[47] who writes the *Arithmetica infinitorum* where he introduces the symbol "∞". It is the most important treatise on infinities and infinitesimals before Newton, which the English scholar read and is inspired by.

The time is ripe for a great work of systematization and recapitulation, both in physics and mathematics. As known, this takes place through the work of Isaac Newton (1642–1727), which, in some way, completes Galileo's program. In his writings, Newton exposes fundamental physical concepts using a precise mathematical formalism. Based on experimental data,[48] he builds his own mechanical theory through an *axiomatic* exposition. In his main work, *Philosophiae Naturalis Principia Mathematica* (1687), Newton assumes *physical quantities* such as time and space, giving them a precise, purely mathematical definition.[49]

> (4.6) *I. Tempus absolutum verum & Mathematicum, in se & natura sua absq[ue] relatione ad externum quodvis, æquabiliter fluit, alioq[ue] nomine dicitur Duratio; relativum apparens & vulgare est sensibilis & externa quævis Durationis per motum mensura, (seu accurata seu inæquabilis) qua vulgus vice veri temporis utitur; ut Hora, Dies, Mensis, Annus. II. Spatium absolutum natura sua absq[ue] relatione ad externum quodvis semper manet similare & immobile; relativum est spatii hujus mensura seu dimensio quælibet mobilis, quæa sensibus nostris per situm suum ad corpora definitur, & a vulgo pro spatio immobili usurpatur [...] III. Locus est pars spatii quam corpus occupat estq[ue] pro ratione spatii vel absolutus vel relativus.*[50]
>
> I. Absolute, true, and mathematical time, of itself, and from its own nature flows equably without regard to anything external, and by another name is called duration: relative, apparent, and common time, is some sensible and external (whether accurate or unequable) measure of duration by the means of motion, which is commonly used instead of true time; such as an hour, a day, a month, a year. II. Absolute space, in its own nature, without regard to anything external, remains always similar and immovable. Relative space is some movable dimension

[45] Bellone (2004), Segrè (1984).

[46] Boyer (1968).

[47] Bottazzini et al. (2016), Boyer (1968).

[48] While regarding astronomical data Newton's sources are more or less known (the *Kepler's laws* at time largely demonstrated, the data of the astronomers John Flamsteed and Edmond Halley, etc.), much less is known about the sources regarding the studies of motion (Chap. 1 of Segrè (1984)).

[49] Cf. Sect. 3.1 and in particular Definition 3.1.

[50] I. Newton, *Definitiones*, in Newton (1687), English translation from Newton (1846).

> or measure of the absolute spaces; which our senses determine by its position to bodies; and which is vulgarly taken for immovable space [...] III. Place is a part of space which a body takes up, and is according to the space, either absolute or relative.

Considering the cultural environment of Newton's era and the times immediately preceding him, it is clear how the above definition can easily be read from a purely philosophical point of view. In fact, two of the most important ancient philosophers, Aristotle and Saint Augustine, were concerned with the definition of time and space. Giving such a precise and stringent definition, Newton looks and refers to them and, at the same time, introduces such a novelty from his predecessors.[51] It is also worth underlining that the aforementioned definitions of space and time are completely abstract and idealized: they are essentially and deeply mathematical. As a matter of fact, the characteristics of *absolute* time and space as defined previously are not observable, i.e. it is not possible to provide any specific experimental procedure to empirically demonstrate them.[52] Indeed, "time" is not provided without an effective measurement procedure such as some "distance between two events", i.e. in a way that is anything but absolute. Nor can we speak about "space" except as "space between (at least) two objects". Therefore, the Newtonian statements of 4.6 cannot be thought as *operational definitions*[53]: they play the role of *pure axioms* of mathematical physics model and they show the great *synthesis and rationalization* of the work. In the same writing, the three *principles of classical dynamics* are stated with great clarity and in the form of *axioms*, a word that is used right in the title.

> (4.7) *Axiomata, sive Leges Motus*
>
> *Lex. I. [statuit] Corpus omne perseverare in statu suo quiescendi vel movendi uniformiter in directum, nisi quatenus a viribus impressis cogitur statum illum mutare.* [...]
>
> *Lex. II. [statuit] Mutationem motus proportionalem esse vi motrici impressæ, & fieri secundum lineam rectam qua vis illa imprimitur.* [...]
>
> *Lex. III. [statuit] Actioni contrariam semper & æqualem esse reactionem: sive corporum duorum actiones in se mutuo semper esse æquales & in partes contrarias dirigi.*[54]
>
> Axioms or Laws of Motion
>
> Law I. [provides that] Every body perseveres in its state of rest, or of uniform motion in a right line, unless it is compelled to change that state by forces impressed thereon. [...]
>
> Law II. [provides that] The alteration of motion is ever proportional to the motive force impressed; and it is made in the direction of the right line in which that force is impressed. [...]
>
> Law III. [provides that] To every action there is always opposed an equal reaction: or the mutual actions of two bodies upon each other are always equal, and directed to contrary parts.

---

[51] Strumia (2017).

[52] Approximately 220 years later physics evolved and the models of Einsteinian relativity overcame effectively these assumptions.

[53] Cf. Definition 3.1.

[54] I. Newton, *Axiomata, sive Leges Motus*, in Newton (1687), English translation from Newton (1846).

Alongside these principles, Newton's theory assumes *Kepler's laws* as starting empirical data. The English physicist's treatise manages to provide a theoretical-mathematical justification of them, by stating the *law of Universal Gravitation* which gives a global explanation of the known celestial mechanics.[55]

However, in his *opus magnum* Newton does not intend to provide a complete and concluded system: on the one hand, he wants to rationalize and systematize the already known knowledge, on the other hand, he aspires to provide a work program towards hydrodynamics and celestial mechanics.[56] In short, from this point of view, the great English physicist completes Galileo's program, suggesting its subsequent development. As already seen, the study of hydrodynamics is a topic largely developed by Galileo's first students for the aforementioned reasons; the detailed study of celestial mechanics is made possible, not only by increasingly reliable experimental data, but also by the new mathematical tools of infinitesimal calculus. The theory developed by Newton is called the *fluxion method.*

(4.8) *Quantitates Mathematicas non ut ex partibus quam minimis constantes sed ut motu continuo descriptas hic considero. Lineædescribuntur ac describendo generantur non per oppositionem partium sed per motum continuum punctorum, superficies per motum linearum, solida per motum superficierum anguli per rotationem laterum, tempora per fluxum continuum, & sic in cæteris. Hæ Geneses in rerum natura locum vere habent, & in motu corporum quotidie cernuntur [...]*

[M]ethodum quærebam determinandi quantitates ex velocitatibus motuum vel incrementorum quibus generantur & has motuum vel incrementorum velocitates nominando fluxiones *& quantitates genitas nominando* fluentes, *incidi paulatim annis 1665 & 1666 in methodum fluxionum qua hic usus sum in quadratura Curvarum [...]*

Ex fluxionibus invenire fluentes problema difficilis est, & solutionis primus gradus æquipollet quadraturæcurvarum; de qua sequentia olim scripsi.[57]

I consider here mathematical quantities, not as consisting of the smallest possible parts, but as generated by a continuous motion. Lines are described and by describing they are generated not by the opposition of parts but by the continuous motion of points, surfaces by the motion of lines, solids by the motion of surfaces, angles by the rotation of sides, times by the continuous flow, and so on. These geneses really have a place in real nature, and are seen every day in the movement of bodies [...]

I was looking for a method of determining quantities from the velocities of motions or increments by which they are generated, and I named these velocities of movements or increments fluxions and the quantities generated I named fluents; I published a small part in the years 1665 and 1666 in the method of fluxions which I used here in the quadrature of curves [...]

It is a difficult problem to find the fluents from fluxions, and the first step in the solution is the quadrature of the curves; about which I once wrote.

Newton here introduces his notations:

(4.9) *Quantitates indeterminatas ut motu perpetuo crescentes vel decrescentes, id est ut fluentes vel defluentes in sequentibus considero, designoque literis z, y, x, v, & earum*

[55] Segrè (1984), Wilson (1989).

[56] Bellone (2004).

[57] I. Newton, *Introductio* in Newton (1723b), da Newton (1723a).

> *fluxiones seu celeritates crescendi noto iisdem literis punctatis $\dot{z}$, $\dot{y}$, $\dot{x}$, $\dot{v}$. Sunt & harum fluxionum fluxiones seu mutationes magis aut minus celeres quas ipsarum z, y, x, v, fluxiones secundat nominare licet & sic designare $\ddot{z}$, $\ddot{y}$, $\ddot{x}$, $\ddot{v}$.*
>
> [...]
>
> *[H]æ quantitates [z, y, x, v] consideri possunt ut fluxiones aliarum quas sic designabo $\acute{z}$, $\acute{y}$, $\acute{x}$, $\acute{v}$.*[58]
>
> I consider indeterminate quantities as perpetually increasing or decreasing, that is, as fluents or defluents; in what follows I denote them by the letters z, y, x, v, and their respective fluxions, or velocities of increasing, by the same dotted letters $\dot{z}$, $\dot{y}$, $\dot{x}$, $\dot{v}$. There are also fluxions of these fluxions, i.e. more or less rapid changes these same quantities, which we may call the second fluxions of z, y, x, v, and thus designate $\ddot{z}$, $\ddot{y}$, $\ddot{x}$, $\ddot{v}$.
>
> [...]
>
> These quantities [z, y, x, v] can be considered as fluxions of other quantities, which I will designate $\acute{z}$, $\acute{y}$, $\acute{x}$, $\acute{v}$.

The concepts of *fluents* and *fluxions* introduced by Newton, with the above-mentioned symbolic notations, had a great impact on the development of his theory and later on the development of mathematics in the United Kingdom. However, infinitesimal calculus is mainly established throughout the notations suggested by the Newton's competitor, Gottfried Leibniz (1646–1716); the latter conceives this mathematical tool at the same time and introduces several currently used concepts and symbols, including

- the *function* $f = f(t)$ such as a modern mathematical object;
- the *infinitesimal differential* of a variable, x, y, t, etc., and respective symbols, dx, dy, dt;
- the symbol of *integral* $\int$.

The use of the differential symbol allows us to indicate the derivative with respect to a variable t with the notation $\frac{d}{dt}$: two centuries later, this will favour the development of the concept of *differential operator* contributing to the birth of *functional analysis*.

4.2.3 The Formalization of Calculus

While on the one hand Newton's work is rightly considered a cornerstone of human scientific, physical, and mathematical thought, on the other hand, as already mentioned, it is far from being a complete system, both from the point of view of physics and from that of mathematics. Here a brief review of the main developments of Newtonian mathematical physics is given.

Newton recognizes that the inverse problem, that of obtaining fluents from fluxions (i.e. primitives from derivatives, to use modern language) was particularly difficult.[59]

---

[58] I. Newton, De Quadratura Curvarum (Newton 1723b), da (Newton 1723a).

[59] Cf. quote 4.8.

During the eighteenth century, many problems of what we modernly call *analysis* and geometry has to be rigorously tackled: issues such as the resolution of what we now call *differential equations* and integrals, the related *boundary conditions*, the domains of validity, the geometric conditions, the changes of coordinates, the presence of mutual interactions between the bodies (and the corresponding nonlinearities), just to list the main ones.

Leonardo Euler (1707–1783) works to unify the mathematical theory of Leibniz (integrals) with that of Newton (fluxions). Newton had almost left out the perturbations to planetary motion[60]: regarding problems of celestial mechanics, Pierre-Simon Laplace (1749–1827) and Joseph-Louis Lagrange (1736–1813)[61] are the first to tackle the problem laying the foundations of the *perturbation theory*. Remaining in the field of mathematical physics, Lagrange and later William R. Hamilton (1805–1865) gave the most advanced and complete mathematical formalization of Newtonian classical mechanics, based on the infinitesimal calculus.

However, it is interesting that, at least until the first decades of the nineteenth century, infinitesimal calculus still requires a rigorous mathematical formalization. There are improvements in *successes of calculus* both in providing a solid theory and in physical laws: these allow the results of experiments and observations to be predicted in detail, but concepts such as "infinity" and "infinitesimal" are still not well defined. A finally complete, formally correct and systematically coherent theory of this branch of mathematics is constructed by Augustin-Louis Cauchy (1789–1857) mainly in his monumental *Cours d'Analyise* of 1821, where the mathematical analysis is re-founded on the notion of *limit*, also extends to the field of complex numbers, and the integration of a function $f(x)$ is redefined within this formalism, definitively adopting the name of "derivative", with the terminology and notation $f'(x)$ used by Lagrange.[62]

The work of the French mathematician gives to mathematical analysis, "i.e. the study of infinite processes"[63] and infinitesimals, a clear formal framework within mathematical science. However, this subject emerges from the needs of experimental physics and begins to increasingly pose new questions in the field of fundamental mathematics. In the second half of the nineteenth century, it gives rise to a great effort by mathematicians (mainly German-speaking and from the Berlin school) who intend to release analysis from geometric meanings and re-establish it on numerical and abstract bases. Such a program is called *arithmetization of analysis*.[64] This study gives rise to numerous new lines of mathematical science, often very far from the initially posed problems. The seed born from the plant of physics through the great

---

[60] Chapter 9 of Bate et al. (1971) and Wilson (1989).

[61] Born in Italy, Turin, with the name of Giuseppe Luigi Lagrangia.

[62] Chapter 4 of Bottazzini et al. (2016).

[63] C.B. Boyer in (1968).

[64] Among others, the mathematicians Johann P. Dirichlet (1805–1859), Leopold Kronecker (1823–1891), Karl Weierstrass (1815–1897), Richard Dedekind (1831–1916), Georg F. Cantor (1845–1918), and Herman A. Schwarz (1843–1921) work on such program (Chap. 12 of Bottazzini et al. (2016)).

intuitions of Galilei and Newton, thrown into the soil of mathematics, borne fruit in a luxuriant manner, leading to a development much larger and richer than one might have initially thought.

4.3 Ricci-Curbastro and Hilbert: Mathematics Arrives Before Physics

It is known that the period between the end of the nineteenth century and the first decades of the twentieth century, when *quantum mechanics* and *Einsteinian relativity* are devised, is considered a time of revolution in physical science, to a some extent comparable to what had happened at the time of Galilei and Newton. As a matter of fact, the discoveries and concepts developed at the time make it necessary to re-discuss the conceptions that had been consolidated and that had determined the *Weltanschauung* that was the heritage of the community of physics scholars. In the period described in Sect. 4.2, physical investigation suggests concepts that take more than two centuries to achieve a complete and convincing mathematical formulation.

However, the period in which the so-called *modern physics* is born is paradigmatic of an almost opposite dynamics. Before this period, mathematics has undergone a tumultuous and very broad development, during which a large number of new branches of the discipline open up. At the junction between the two centuries, most of the necessary mathematical concepts and structures are already available when experimental data and reasoning advance on thermodynamics, electromagnetism, the behaviour of atoms, and the speed of light make the development of radically new theories necessary.

An essential feature of the nineteenth-century mathematics is the tendency towards unification and generalization. For example, the aforementioned process of *arithmetization of analysis*, according to which the fundamental concepts of mathematical analysis are brought back to the field of arithmetic. Similarly, an ancient topic such as the resolution of algebraic equations is addressed by giving life to the branch of abstract algebra.[65] As regards the mathematics related to *modern physics*, it could be said that

> (4.10) Quantum mechanics requires the introduction into physical theory of a vast new domain of pure mathematics -the whole domain connected with non-commutative multiplication. This, coming on top of the introduction of the new geometries by the theory of relativity, indicates a trend which we may expect to continue.[66]

So, it is possible to summarize the two branches of mathematics pointed out by Dirac in the following:

---

[65] Bottazzini et al. (2016), Boyer (1968).

[66] P. Dirac in (1938).

1. the theory of *Hilbert's space* and the correlated *functional analysis* for Quantum Mechanics;
2. the *non-Euclidean differential geometry* for the Einstein's Relativity.[67]

Tracing backwards the parallel development of the above-mentioned two branches, through the contribution of the most important scholars who dealt with them, a turning point is represented by a mathematician who is decisive for the development of both: the German Georg Friedrich Bernhard Riemann (1826–1866). It is believed that Riemann's most important contribution concerns *non-Euclidean geometries*: although in this field he was preceded by the Russian Nicolaj I. Lobačevskij (1792–1856) and the Hungarian János Bolyai (1802–1860), the German mathematician was able to identify the great generality of the topic, correlating it with other investigation fields and allowing it to be integrated within the rest of mathematical science.[68] In 1854 Riemann wrote his dissertation for the qualification to teach in university (*Habilitationschrift*) at the University of Göttingen. This revolutionary text, titled *Über die Hypothesen welche der Geometrie zu Grunde liegen*,[69] is published posthumously only in 1867.[70] Therein Riemann provides a very general conception of geometry as the study of *manifolds*[71] in a *generic space* of *any dimension*. According to Riemann, this *generalized* geometry does not necessarily deal with entities such as points, lines, etc., but with sets of ordered n-tuples that satisfy some rules and relations. Given a non-empty set of elements on which to build a geometric theory, conventionally called *points*, the essential property of any conceivable geometry is the *distance* between two of them. This distance is defined as a function which, satisfying given properties, associates a positive number with two generic points.[72] The rule for finding the distance between two objects identifies a *metric*, through which the elementary, or infinitesimal, distance ds, can be defined in different ways. If, in an n-dimensional space, the coordinates x_j, with $j = 1 \ldots, n$, and x_k, with $k = 1 \ldots, n$, are defined, the metric properties of the space, according to Riemann, are built on the *quadratic form*[73]

$$ds^2 = \sum_{j,k=1}^{n} a_{jk} dx_j dx_k, \tag{4.1}$$

where a_{jk} coefficients are, in most general case, depending on x_j, x_k. The object a_{jk} as a whole is currently called *metric tensor* and it is usually indicated through a

[67] In this paradigmatic exposition we neglect the *Special Relativity*, whose historical development had a different path.

[68] Chapter 24 in Boyer (1968).

[69] *On the hypotheses underlying geometry*.

[70] By Richard Dedekind (1831–1916).

[71] In other words *topological space*, which is locally similar to an Euclidean space (for a more precise definition, see, e.g. Wrede (1972)).

[72] We avoid listing here the characteristics of the *distance* function (e.g. it satisfies the *triangular inequality* , etc.), a topic that concerns *topology*; references can be found in other texts, see, e.g. Rudin (1987), Wrede (1972).

[73] Wrede (1972).

notation such as g_{jk}[74]; for the usual Euclidean geometry, it coincides with an *identity matrix* 3×3. The importance of Riemann's theory is its extreme generality. The "points" considered by him, that are the object of geometry, can be entities of any kind[75]: it is sufficient that some properties are satisfied and that a quadratic form such as (4.1) is defined to allow us to calculate the *distances* between the object of the space.

In light of this, regarding the two branches of mathematics listed above it is possible to say:

1. Through the general Riemannian approach it is possible to conceive the *functional spaces*, where the *points* of the space are functions and the (4.1), determining the metric, is appropriately defined. Here we find the prodromes of what decades later would be called *Hilbert spaces*,[76] pivot of the entire formal construction of quantum mechanics.
2. In a more strictly geometric sense, it is possible to construct geometries, characterized by the metrics specified via the g_{jk} object. Through these constructions the *tensor calculus*[77] is developed at the basis of Einstein's General Relativity.

4.3.1 The Language of General Relativity

The mathematical branch that most directly is related to Riemann's work is the non-Euclidean geometry. This theory develops within the generalization work of nineteenth-century mathematics, building analogies with the parallel development of abstract algebra. The fundamental conceptual pivot is the search for *differential invariants*, which easily fits within the theory of ds-type *algebraic invariants*.

Following the evolution of Riemannian non-Euclidean geometries, it is worth remembering important mathematicians such as Eugenio Beltrami (1835–1900) and Erwin B. Christoffel (1829–1900). In particular, the latter one poses the problem of transforming a quadratic form such as (4.1), through a transformation such as $x_j = \phi_j(x'_1, \dots, x'_n)$, into a form $ds^2 = \sum_{j,k=1}^{n} a'_{jk} dx'_j dx'_k$. By providing the answer to this issue, Christoffel introduces what would be called *Christoffel symbols*, generally indicated with $\Gamma_{jk,r}$ and Γ^r_{jk}, mathematical tools which are the starting point for the theory underlying Einstein's relativity.[78] Between 1887 and 1895 Gregorio Ricci-Curbastro (1853–1925), Professor at the University of Padua, conceives the global construction of what he calls *absolute differential calculus*. The conceptual path

[74] We here avoid distinguishing between *covariant indices* and *contravariant*, also because this distinction is clearly understood later, in the works by Ricci-Curbastro and Levi-Civita

[75] Remember the discussion about the interpretations of mathematical entities in Sect. 2.2.3, and, in particular, Definitions 2.3 and 2.4.

[76] This name is coined by Johan von Neumann (1903–1957), disciple of David Hilbert (1862–1943), in 1929 (von Neumann 1995a).

[77] Name coined by Einstein.

[78] Bottazzini et al. (2016), Wrede (1972).

leading to the construction of this powerful tool is summarized in the foreword to the 1900 essay, written together with his disciple Tullio Levi-Civita (1873–1941), where Ricci systematically explains his theory.

(4.11) *M. Poincaré a écrit que dans les Sciences mathématiques* une bonne notation a la même importance philosophique qu'une bonne classification dans les Sciences naturelles. *Evidemment, et même avec plus de raison, on peut en dire autant des méthodes, car c'est bien de leur choix que depend la possibilité de* forcer *(pour nous servir encore des paroles de l'illustre géomètre français)* une multitude de faits sans aucun lien apparent á se grouper suivant leurs affinités naturelles. [...]

L'algorithme du Calcul différentiel absolu, c'est á dire l'instrument materiel des méthodes, sur lesquelles nous allons entretenir les lecteurs des Mathematische Annalen, se trouve tout entier dans une remarque due á M. Christoffel. Mais les méthodes mêmes et les avantages, qu'ils présentent, ont leur raison d'être et leur source dans les rapports intimes, qui les lient á la notion de variété á n dimensions, que nous devons aux génies de Gauss et de Riemann.[79]

M. Poincaré wrote that in the Mathematical Sciences *a good notation has got the same philosophical importance as a good classification in the Natural Sciences.* Obviously, and even with more reason, one can say as much of the methods, because it is indeed on their choice that depends the possibility of forcing (to use again the words of the illustrious French geometry scholar) a multitude of facts without no apparent link to group together according to their natural affinities. [...]

The algorithm of the absolute differential calculus, that is the material instrument of the methods, on which we are going to discuss the readers of the Mathematische Annalen, is found entirely in a remark due to M. Christoffel. But the very methods and the advantages that they present have their reason for being and their source in the intimate relationships that link them to the notion of n-dimensional variety, which we owe to the genius of Gauss and Riemann.

The theory of *absolute differential calculus*, which Einstein renames *tensor calculus*, provides the tools to develop a differential geometry which is formally independent on the chosen coordinate system. However, for some years Ricci's is not considered of sufficient importance, so that the most important member of Ricci's group, Levi-Civita, after the publication of the aforementioned essay gradually starts to deal with other topics.[80] Nonetheless, the results of the Paduan group circulate in Europe among mathematicians dealing with *Riemannian geometry*. At the beginning of the 1910s Einstein is building the *theory of General Relativity*.[81] By his own admission, he cannot find the right mathematical framework to provide a complete and coherent construction; for this reason he writes to his friend and former classmate, the mathematician Marcel Grossmann (1878–1936), asking him for help.[82] The kernel of the theory requires the equations to be formulated so that their form do not vary under any coordinate transformation. It is Grossmann, a scholar of Riemannian geometries, who introduces Einstein to the method of Ricci and Levi-Civita, a key tool for

[79] G. Ricci-Curbastro and T. Levi-Civita, Préface of Ricci-Curbastro and Levi-Civita (1900).

[80] Bottazzini et al. (2016), Rainich (1928).

[81] It is worth remembering here Newton's explicit declaration of *absoluteness of space and time*, as reported in the quote 4.6 in Sect. 4.2.2, which is completely overcome by Einsteinian relativity.

[82] Chapter 5 of Wrede (1972), Janssen and Renn (2015).

translating Einstein's powerful intuitions and physical hypotheses into mathematical language.

A clear imprint of the aforementioned mathematicians' contributions can be found within the fundamental relationship provided by General Relativity, the *Einstein's field equation*, written as[83]

$$R_{jk} - \frac{1}{2}g_{jk}R = -\frac{8\pi G}{c^4}T_{jk}. \tag{4.2}$$

It can be written in a more compact form by introducing the *Einstein tensor* $\mathcal{G}_{jk} = R_{jk} - \frac{1}{2}g_{jk}R$, so obtaining[84]

$$\mathcal{G}_{jk} = -\frac{8\pi G}{c^4}T_{jk}. \tag{4.3}$$

The various elements in the (4.2) are the following:

R_{jk} is *Ricci's tensor*, which is obtained by a suitable contraction from *Riemann's tensor*. Moreover, it can be written through the above-mentioned *Christoffel's symbols* by means of suitable relationships.[85]
g_{jk} is the aforementioned *metric tensor*.
R is the *scalar curvature*, which is obtained from the above-mentioned Ricci's tensor.
T_{jk} is the *stress–energy tensor* already introduced by Einstein in Special Relativity.
G, *Newton's constant*, and c, *in-vacuum speed of light*, are constant already present in classical physics.

Therefore, the relation (4.2) condenses not only wholly physical quantities but also the various mathematical objects which, at the time of its formalization, are already been fully developed. The astrophysicist George Lemaître (1893–1966) recalls the geometric interpretation of Riemannian geometry at the foundation of General Relativity, connecting it to the elements that compose it.

> (4.12) According to Riemann, curvature is a departure from flatness of space, or spece-time, which is, approximately, flat or Euclidean.
>
> In Euclidean space, the square of the elementary distance of two points is a quadratic form of the differences of the coordinates of the points. When these coordinates are chosen conveniently, the quadratic form has got constant coefficients. […]
>
> Nevertheless, the reduction to a form with constant coefficients cannot be achieved rigorously and, when regions are considered which are not too small, the effect of curvature, or lack of flatness, cannot be completely ignored.[86]

[83] Ohanian and Ruffini (2013).
[84] Cf. e.g. Chap. 5 of Wrede (1972).
[85] Cf. Chap. 5 of Wrede (1972) and Chap. 6 of Ohanian and Ruffini (2013).
[86] George E. Lemaître in (1949).

The process by which the models developed by physical science elegantly made use of mathematical tools already prepared and formally refined clearly shows a fruitful dynamics of exchange between mathematical and physical science, mirroring the one occurring between physics and mathematics discussed in Sect. 4.2.

4.3.2 The Language of Quantum Mechanics

When the development of quantum theory is explained, two independent formalizations are often indicated: the *matrix mechanics* conceived by Werner Heisenberg (1901–1976) with the collaboration of Max Born (1882–1970) and Pascal Jordan (1902–1980), and completed in 1925; the *wave mechanics*, conceived by Erwin Schrödinger (1887–1961) and completed in 1926. Shortly afterwards, Paul Dirac (1902–1984) demonstrates their formal equivalence.[87] In this work there is no need to dwell on the details of these formal structures: it is preferable to refer directly to the three axioms of the theory according to a usual modern exposition[88] in order to highlight the mathematical tools present at the time which quantum mechanics makes use of. Similarly to equation (4.2) for Einsteinian relativity, the three *fundamental postulates of quantum mechanics* allow us to immediately highlight the main involved mathematical structures. They can be stated as follows[89]:

Postulate I (Wave function). Each physical system S is associated to a *Hilbert's space* $\mathscr{H}_S$ su $\mathbb{C}$. A given state of such system is completely described by a vector ψ of such space, which is called *state vector* or *wave function*. Time evolution of ψ is provided by the *Schrödinger equation*

$$i\hbar\frac{d}{dt}\psi(t) = \hat{H}(t)\psi(t), \tag{4.4}$$

where $\hat{H}(t)$ is the self-adjoint operator which is called *Hamiltonian of the system*, $\hbar \equiv h/2\pi$ and h is Planck's constant.

Postulate II (Observables). Each physical observable A is associated to a self-adjoint operator $\hat{A}$ which is defined on the above-mentioned $\mathscr{H}_S$. The value obtained by whatever experimental measurement of A coincides with one among the eigenvalues of the operator $\hat{A}$. So we have

---

[87] Chapter 8 of Segrè (1980).

[88] Despite what the concept of "mathematical postulate" leads us to think, different texts gather the fundamental postulates of Quantum Mechanics in different ways. Most of the essays list three postulates, whereas some texts choose to divide their content into more than three axioms, generally *four*, or they divide the axioms into sub-axioms. However, even when three postulates are given, their content can vary from author to author (see, for example, the exposition in Chap. 9 of Lindsay and Margenau (1957) compared to Chap. 2 of Benenti et al. (2018), or to those in Chap. 4 of Picasso (2016), or to Dirac (1957), etc.). Of course the overall elements remain the same. In our discussion inspiration is taken from one of the most modern treatments of the topic, provided in Chap. 2 of Benenti et al. (2018).

[89] For simplicity, we do not use Dirac notation.

$$\hat{A}\psi_k = a_k\psi_k, \tag{4.5}$$

where ψ_k is an orthonormal basis of eigenvectors of $\hat{A}$. Therefore, the state vector of the system can be written as a linear combination of the ψ_k with suitable coefficients $c_k(t)$:

$$\psi(t) = \sum_k c_k(t)\psi_k. \tag{4.6}$$

Postulate III (Wave function collapse).[90] Let be a system described by ψ, whose observable A is measured with a numerical value a_n. Then, immediately after this measurement the state of the system is given by

$$\phi \propto \hat{P}_n\psi, \tag{4.7}$$

where $\hat{P}_n$ is the projection operator on the sub-space which corresponds to a_n.

Several specific mathematical objects can be found in the previous statements. As previously mentioned, the key objects of the theory are *functional spaces*, among which the aforementioned *Hilbert's spaces* play the most important role. It has been already stated that the possibility of considering a mathematical space, whose elements are *functions*, derives from the *general and abstract* formulation of the geometric theory, which has been expressed since the work of Riemann. At a conceptual level, in Sect. 2.2.3 it has already been highlighted that it is possible to provide different formally correct *interpretations* of the same mathematical structure, defined by a suitable system of axioms[91] (deductions, lemmas, theorems, etc.), maintaining their validity. Historically, the branch of *functional analysis*, which deals with functional spaces and their elements, is born from the study carried out, starting from the 1930s, on differential and integral equations. This is the obvious development of the mathematical analysis formalization accomplished by Cauchy, Weierstrass, etc. and also deals the most important equations of what today we call *mathematical physics*. This trend is developed by important mathematicians, including the French Jacques Sturm (1803–1855), Joseph Liouville (1809–1882) and Henry Poincaré (1854–1912), the German Rudolph Lipschitz (1832–1903), the Swede Ivar Fredholm (1866–1927), the Italians Salvatore Pincherle (1853–1936) and Vito Volterra (1860–1940).[92] The publication of the notes entitled *Sopra le funzioni che dipendono da altre funzioni* (*On functions that depend on other functions*) by the latter one in 1887 is considered

[90] Such postulate is also called "wave packet reduction" or "state after measurement" or "von Neumann postulate" (Picasso 2016).

[91] Cf. Definition 2.3.

[92] The French mathematician J. Hadamard explicitly states that Pincherle and Volterra had founded a new branch of mathematics (Chap. 19 of Bottazzini et al. (2016)).

by some to be the birth date of functional analysis.[93] Among the notions introduced before the end of the nineteenth century, there are

- the *orthogonality between functions*

$$\int \phi_j(\xi)\phi_k(\xi)d\xi = 0 \quad \text{for } j \neq k$$

- the concept of *operator* and *functional*, with a notation such as

$$F[x_a^b(t)]$$

- the concept of *function space*, with the definition of a coherent coordinate system.

These different contributions converge in three important works at the beginning of the new century: the *measure and integration theory* (1902) by Henri Lebesgue (1875–1941), the *spectral theory* (1906) by Maurice Fréchet (1878–1973), and the definition of *metric spaces* (1906) by David Hilbert. A few years later, Hilbert provides the definitive systematization of the functional spaces theory. Starting from Fredholm's work about differential equations, the German mathematician grasps the unitary significance of such topic for the development of the whole mathematics. As a part of the great work of generalization that had been the dominant note of the just ended century, he applies the categories and theorems of algebra and finds the theory of quadratic forms as a decisive element, which can be directly correlated with the Riemannian ones of (4.1). Hilbert collects his results in this field in the great treatise *Methoden der mathematischen Physik* ("Methods of mathematical physics"), whose first volume is published in 1924 with the help of his disciple Richard Courant (1888–1972). Thanks to the use of series expansions, a topic studied from the dawn of mathematical analysis and again during the 19th century (see the works of Joseph Fourier (1768–1830), Friedrich W. Bessel (1784–1846), etc.), Hilbert is able to consider functions as elements of spaces of infinite dimensions.

(4.13) *Das Problem der Reihenentwicklung willkürlicher Funktionen*

Viele der im vorigen Kapitel behandelten Zusammenhänge [Die Algebra der linearen Transformationen und quadratischen Fromen] finden eine weitgehende Analogie, wenn man statt der Vektoren im n-dimensionalen Raume Funktionen von einer oder mehreren Veranderlichen betrachtet, die in einem vorgegebenespsciten Grundgebiet G definiert sind. So kann man zu der Tatsache, daßsich im Raume von n Dimensionen jeder Vektor linear durch n unabhängige, sonst aber beliebig gewählte Vektoren ausdrücken läßt, das Problem in Parallele setzen, eine mehr oder weniger willkürlich angenommene Funktion im Grundgebiete G als line are Kombination aus vorgegebenen Funktionen darzustellen. (Die Menge dieser vorgegebenen Funktionen muß, wie sich im folgenden als selbstverstandlich erweisen wird, unendlich sein.) Wir sprechen dann von dem Problem der Reihenentwicklung des willkürlich angenommenen Funktion nach dem vorgeschriebenen Funktionensystem. *Im vorliegenden*

[93] Chapter 19 of Bottazzini et al. (2016).

Kapitel soll diese bei den Aufgaben der mathematischen Physik in der mannigfachsten Form auftretende Fragestellung unter allgemeinen Gesichtspunkten behandelt werden. [94]

Series Expansions of Arbitrary Functions

Many of the relations discussed in Chapter I [The Algebra of Linear Transformations and Quadratic Forms] are closely analogous to theorems in which functions of one or more variables, defined in some given fundamental domain G, take the place of vectors in n-dimensional space. The problem of expressing a vector in n-dimensional space as a linear combination of n independent but otherwise arbitrary vectors is analogous to the problem of representing a more or less arbitrary function in the domain G as a linear combination of members of some given set of functions. (It becomes obvious that the number of functions in this set must be infinite.) This is known as the *problem of series expansion of an arbitrary function in terms of a given set of functions*. In the present chapter this problem, which appears in various forms throughout mathematical physics, is treated from a general point of view.

Courant and Hilbert immediately introduce the concept of *orthogonality between functions* so defining the *inner product between two functions* $f(x)$ and $g(x)$

$$(f, g) = \int f\, g\, dx$$

which satisfies the Cauchy–Schwarz inequality $(f, g)^2 \leq (f, f)(g, g)$. By this it is possible to construct *normalized functions* $\phi_1(x), \phi_2(x), \ldots$ so obtaining an *orthonormal system* which satisfies relations similar to the purely geometric ones[95]

$$(\phi_n, \phi_m) = \delta_{n,m}.$$

Therefore, the discussion advance to define the main mathematical entities of what would be called *Hilbert spaces*, which is typically indicated with $\mathscr{H}$. In particular, the analogies with vector spaces are stressed, also highlighting that they still held even when dealing with an infinite-dimensional space. This analogy turned out to be so well matching that, half a century later, Dirac spoke of it as something almost obvious, with the laconism typical of his character:

(4.14) Hilbert space is just a Euclidean space with an infinite number of dimensions and which is made precise by a convergence condition imposed on the coordinates of the vectors in it.[96]

Similar to what happened with Ricci's essay on absolute differential calculus, this work by Hilbert is also greeted with substantial indifference by the majority of physicists who dealt with quantum phenomena. However, when 2 years later, in 1926, the work on wave mechanics is published, which requires an in-depth study of the solutions of partial derivative differential equations, they realize that the content of the *Methoden* is a solid mathematical structure and already completed, immediately

[94] Chapter 2 of Courant and Hilbert (1924), English translation from Courant and Hilbert (1989).

[95] Courant and Hilbert (1924), Kolmogorov and Fomin (1999).

[96] P. Dirac, Introduction in Dirac (1974).

available to provide the *Grundlagen* of the new physical theory. In this sense it is notable that the subsequent specific development of *quantum formalism* is also due to two other disciples of Hilbert, Hermann Weyl (1885–1955) and Johan von Neumann. The definitive modern systematization of quantum mechanics is consequently provided in 1930 by Dirac in the book *The Principles of Quantum Mechanics*[97] and in 1932 by von Neumann in *Mathematische Grundlagen der Quantenmechanik* ("Mathematical foundations of quantum mechanics").

References

Albertus Magnus. Liber Sxtus De Vegetabilibus - Tractatus I - Capitulum I, in *De Vegetabilibus Liibri VII* (Typis et impensis G. Reimeri, Berolini, 1867)

G. Basti, *Filosofia della Natura e della Scienza*, vol. 1, 1a edn. (Pontificia Università Lateranense, Roma, Città del Vaticano, 2002)

R.R. Bate, D.D. Mueller, J.E. White, *Fundamentals of Astrodynamics* (Dover, New York, USA, 1971)

E. Bellone, *Caos e Armonia - Storia della Fisica* (UTET, Torino, Italia, 2004)

G. Benenti, G. Casati, D. Rossini, G. Strini, *Principles of Quantum Computation and Information - A Comprehensive Textbook* (World Scientific, Singapore, 2018)

U. Bottazzini, E.A. Marchisotto, P. Miller, *Hilbert's Flute: The History of Modern Mathematics* (Spriunger, Berlin, Deutschland, 2016)

C.B. Boyer, *A History of Mathematics*, 2nd edn. (Wiley, New York, USA, 1968)

S. Carioti, Nicole Oresme precursore di Galileo e di Descartes? Rivista Critica di Storia della Filosofia **32**, 11–23 (1977)

M. Clagett, *The Science of Mechanics in the Middle Ages* (University of Wisconsin Press, Madison, USA, 1959)

R. Courant, D. Hilbert, *Methoden der mathematischen Physik*, vol. 1 (Springer, Berlin, Deutschland, 1924)

R. Courant, D. Hilbert, *Methods of Mathematical Physics*, vol. 1 (Wiley-VCH, Weinheim, Deutschland, 1989)

P.A.M. Dirac, The relation between mathematics and physics. Proc. R. Soc. (Edinb.) **59**, 122–129 (1938–39)

P.M.A. Dirac, *The Principles of Quantum Mechanics*, 4th edn. (Oxford University Press, Oxford, UK, 1957)

P.M.A. Dirac, *Spinors in Hilbert Space* (Plenum Press, New York, USA, 1974)

P. Duhem, *La théorie physique. Son objet, sa structure*, 2nd edn. (ENS Éditions, Lyon, France, 2016). http://books.openedition.org/enseditions/6077

P. Fornasini, *The Uncertainty in Physical Measurements. An Introduction to Data Analysis in the Physics Laboratory* (Springer, Cham, Schweiz, 2008)

E. Grant, *La scienza nel Medioevo* (Il Mulino, 1997)

M. Janssen, J. Renn, History: Einstein was no lone genius. Nature **527**(7578), 298–300 (2015)

S. Kirschner, Nicole Oresme, in E.N. Zalta (ed.) *The Stanford Encyclopedia of Philosophy*. Metaphysics Research Lab, Stanford University, Fall 2021 edition (2021)

A.N. Kolmogorov, S.V. Fomin, *Elements of the Theory of Functions and Functional Analysis* (Dover, New York, USA, 1999)

G.E. Lemaître, The cosmological constant, in *Albert Einstein Philosopher-Scientist* (MJF Books, New York, USA, 1949)

[97] Dirac (1957).

R.B. Lindsay, H. Margenau, *Foundations of Physics*, 2nd edn. (Dover, New York, USA, 1957)
D.C. Montgomery, *Design and Analysis of Experiments*, 5th edn. (Wiley, New York, USA, 2001)
I. Newton, *Analysis per quantitatum series, fluxiones, ac differentias* (Sumptibus Societatis, Amsterdam, Paesi Bassi, 1723a). http://digitale.beic.it/BEIC:RD01:39bei_digitool1358078. Available online, BEIC - Biblioteca Europea di Informazione e Cultura
I. Newton, De quadratura curvarum. Newton (1723b). http://digitale.beic.it/BEIC:RD01:39bei_digitool1358078. Disponibile online, BEIC - Biblioteca Europea di Informazione e Cultura
I. Newton, *Philosophiae naturalis principia mathematica* (London, UK, 1687)
I. Newton, *The Mathematical Principles of Natural Philosophy*. Daniel Adee. Translated by A. Motte
H. Ohanian, R. Ruffini, *Gravitation and Spacetime*, 3rd edn. (Cambridge University Press, Cambridge, UK, 2013)
L. Picasso, *Lectures in Quantum Mechanics* (Springer, Cham, Schweiz, 2016)
I. Prigogine, I. Stengers, *La nouvelle alliance - Metamorphose de la science*, 2nd edn. (Gallimard, Paris, France, 1986)
I. Prigogine, I. Stengers, *La nuova alleanza - Metamorfosi della scienza*, 3rd edn. (Einaudi, Torino, Italia, 1999)
G. Rainich, Levi-Civita on tensor calculus. Bull. Am. Math. Soc. **34**, 775–777 (1928)
G. Ricci-Curbastro, T. Levi-Civita, Méthodes de calcul différentiel absolu et leurs applications. Mathematische Annalen **54** (1900)
W. Rudin, *Real and Complex Analysis* (McGraw-Hill, New York, USA, 1987)
E. Segrè, *From X-rays to Quarks: Modern Physicists and Their Discoveries* (W.H. Freeman & C., New York, USA, 1980)
E. Segrè, *From Falling Bodies to Radio Waves: Classical Physicists and Their Discoveries* (W.H. Freeman & C., New York, USA, 1984)
A. Strumia, *Dalla filosofia della scienza alla filosofia nella scienza*, 1a edn. (ESC - Edizioni Santa Croce, Roma, Italia, 2017)
Tommaso d'Aquino. *Super Boetium de Trinitate*. Corpus Tomisticum. https://www.corpusthomisticum.org/cbt.html. Disponibile online, © Fundación Tomás de Aquino
F. Tornaghi, G. Mangiarotti (a cura di), *Galileo Galilei mito e raltá - Itinerario antologico* (CE.Se.D. Edizioni, Milano, Italia, 1997)
J. von Neumann, Method in the physical sciences, in *The Neumann Compendium*, World Scientific Series in 20th Century Mathematics, vol. 1, ed. by F. Bródy, T. Vamós (World Scientific, Singapore, 1995), p. 627
W.A. Wallace, *Prelude to Galileo: Essays on Medieval and Sixteenth-Century Sources of Galileo's Thought* (D. Reidel Publishing Company, Dordrecht, Holland; Boston, USA, London, UK, 1981)
C. Wilson, The Newtonian achievement in astronomy, in *The General History of Astronomy*, vol. 2A, ed. by R. Taton, C. Wilson (Cambridge University Press, UK, 1989)
R.C. Wrede, *Introduction to Vector and Tensor Analysis* (Dover, New York, USA, 1972)
J. Zupko, John Buridan, in E.N. Zalta (ed.) *The Stanford Encyclopedia of Philosophy*. Metaphysics Research Lab, Stanford University, Fall 2018 edition (2018)

Part II
Structure of Scientific Knowledge

This part wants to provide some answers to the question of what kind of knowledge is obtained by means of investigation carried out through the scientific method in general, and especially through concepts and approaches of physics. Among the questions from which the discussion starts there are:

What kind of knowledge is scientific knowledge?
What is scientific knowledge based on?
What are the theories of physics based on?
In what sense and under what conditions can we say that they are true or not?
What does it mean that scientific knowledge is true?
Is physics true? When and in what sense can it be said?
How can we verify the correctness of schemes, concepts, theories of physics?
Does the knowledge provided by the physical science match the actual state of the natural world?
Could scientific disciplines all be reduced to physics, at least in principle?
Is there a way that non-scientific knowledge could be reduced to scientific knowledge, at least in principle?

Chapter 5
The Methodologies of Knowledge

Io stimo più il trovar un vero, benché di cosa leggiera, che 'l disputar lungamente delle massime questioni senza conseguir veritá nissuna.

Galileo Galilei

5.1 Epistemology, that Is, the Study of Knowledge Methods

This is probably the chapter that (almost) no scientist wants to read. As a matter of fact, scientists "carry out scientific research"; expressing the concept with a brachylogy, they "do science", even though in their work they rarely question themselves on what it means to "do science". In university curricula there is no course teaching the *theory of the scientific method*: thus, what "doing science" means is learned by "doing science". It is through practice that physicists learn what "science" is; what "physics" is; how scientific research is carried out; and, consequently, what "scientific knowledge" is and what isn't. In this discussion, however, our specific purpose is to engage in reflecting precisely upon those aspects that are generally taken for granted. So far the fundamental concepts of physics have been evaluated and the main mechanisms through which they have been validated have been reviewed. The guiding criterion of these assumptions is that they have been demonstrated to be able to *increase knowledge* of natural phenomena. Therefore, after reaching this point in our discussion, we cannot avoid the question of what the proper methods of scientific knowledge are. An overview, as agile as possible, of the main approaches and contributions regarding the study of knowledge is thus given. To start with, a definition of the topic of this chapter is provided:

J. Parravicini, *The Foundations of Experimental Physics*,
https://doi.org/10.1007/978-3-031-55452-0_5

Definition 5.1 (*Epistemology*) With *epistemology* we mean the logical and philosophical discipline which studies the methods of "scientific knowledge" and, in particular, the bases for which it can be actually said scientific and sure.[1]

The words come from the Greek ἐπιστήμη, "knowledge", "cognition", but also "awareness". Some scholars underline that *epistemology* is distinct from *gnoseology*,[2] a broader discipline that studies the process of knowledge in its generality. However, this distinction is irrelevant for the synthetic discussion given here. Specifically, with the expressions *study of knowledge*, *theory of knowledge*, *study of knowledge methods*, the study of the cognitive processes of the human mind is omitted (which belongs to other disciplines, such as, for example, *theory of mind*).

The birth and development of modern natural sciences have shown how the *scientific method* introduced by Galilei and Newton is able to lead to a great increase in knowledge of natural phenomena. For this reason, nowadays any deliberation on knowledge and the methods connected to them, even performed in more philosophical than scientific fields, cannot ignore the problem of *science* and the *scientific method*. Indeed, the term to describe such an approach, *philosophy of science*, has in some contexts become a synonym for *theory of knowledge* and related methods. Here the main positions that have been proposed to analyse and classify Galilean scientific knowledge are addressed. What, then, is this form of knowledge? Is it *true* knowledge? And in what sense is it *true*?

5.2 Scientific Explanations as the Result of Method

5.2.1 Positivism and Verificationism

An important proposal about the nature of scientific knowledge is called *verificationism*. This approach has its origins in the considerations arisen in the cultural-philosophical school of *positivism*: originated in France in the first half of the nineteenth century, such a school had a huge and widespread influence on culture. Along with advocating a strongly materialist conception of reality, it trusted in the *scientific method* that bordered on an optimism with fideistic aspects. This movement develops throughout the century with various ramifications.

It is the most recent of these ramifications, the neo-positivism *neo-positivism* or *logical positivism* that flourishes at the begin of twentieth century[3] that proposed verificationism. In the intentions of its proponents, neo-positivism aims at adhering as closely as possible to the methods of logical-mathematical-scientific investigation.

---

[1] From Basti (2002), Sect. 4.1.

[2] Always from Greek, γνῶσις, "knowledge", and λογία, "discipline, reasoning".

[3] It is joined by scholars such as, e.g. the physicists Ernst Mach (1838–1916), Moritz Schlick (1882–1936), Richard von Mises (1883–1953), the mathematician Hans Hahn (1879–1934), the philosopher and logician Rudolph Carnap (1891–1970).

This approach leaves out the romantic optimism of the old French positivism in order to focus on the study of cognitive processes linked to language and to the effective construction of scientific knowledge.[4] The purpose is to identify what are satisfactory scientific criteria allowing clear and precise knowledge in all fields. Indeed, according to neo-positivism, knowledge arising from the scientific method is considered the only way of reaching steady knowledge. They believe that only through the systematic use of the Galilean scientific method, i.e. ultimately, the method of experimental physics, it is possible to establish the veridicity of any assertion (proposition). For this reason, every other discipline has to follow this criterion at least in principle. Such a conception implies a more or less declared program, according to which all disciplines have to be programmatically addressed in order to become both *empirically verifiable* and *mathematizable*, at least in their last instance.[5]

Therefore, a key issue in this process is to discriminate between what is *science* and what is not, that is, between *scientific propositions* and *non-scientific propositions*: only the former ones are considered *meaningful*, while the latter ones are classified simply as *meaningless*. This issue is known as the *demarcation problem* (*Abgrenzungsproblem* or *Demarkationsproblem*) and the principle assumed to discriminate *scientific propositions* from *non-scientific propositions* is called emphAbgrenzungskriterium, i.e. *demarcation criterion*. In the context of logical positivism, this criterion is that of *verificationism* (*Verifikationismus*).

> (5.1) [A]n alleged proposition (or sentence) is genuine if, and only if, it is a truth function of, or reducible to, elementary (or atomic) propositions expressing observations or perceptions.
>
> In other words, it is meaningful if, and only if, it is so related to some observation sentences that its truth follows from the truth of these observation sentences.[6]

Thus, the verificationist criterion states that only what is susceptible to a direct perceptual or observational verification is scientific.[7] Everything else is not only not scientific, but is even not *authentic*, that is, meaningless.[8] Conceptually, they follow this line of reasoning:

All the meaningful knowledge is only the *scientific* one.
All scientific knowledge is, or will be, at least in principle, traced back to *physics*.
All the (so-called) knowledge that cannot be traced back to physics is simply *meaningless* and it is not real knowledge.

---

[4] Chapter 2 of Chibbaro et al. (2014).

[5] Chibbaro et al. (2014), Strumia (2017).

[6] K. Popper, Chap. 11 of Popper (1962).

[7] We note that such a view in the empirical natural sciences has a counterpart in mathematics; accordingly only statements that are rigorously proven belong to mathematics. This attitude is particularly present in the intuitionist and formalist schools, which are presented in Sect. 7.3.1. Such a vision of mathematics, however, although driven by a need for clarity and rigour and, therefore, apparently reasonable, is radically called into question by the results on the incompleteness of Gödel (Sect. 7.3.2).

[8] Some authors distinguish a *weak verificationism* from a *strong verificationism*, but this is irrelevant for our discussion.

Since, etymologically, what is *beyond physics* is called *metaphysics* (μετὰ τὰ φυσικά), it follows that the not considered *scientific* statements are *metaphysical* statements, thus coinciding with *nonsense* statements or *pseudo-propositions*.

Superficially, the criterion stated in the quote (5.1) may seem reasonable. In truth, it immediately shows deep inconsistencies.

1. The first problem, in fact, lies in its contradictory nature: the effectiveness of the *verificationist criterion* (5.1) is not empirically demonstrable through the criterion that it itself enunciates; therefore, it is among the *metaphysical* propositions that one should avoid.
2. The second problem is deeper.

   > (5.2) This criterion excludes from the realm of meaning all scientific theories (or "laws of nature"); for these are no more reducible to observation reports than so-called metaphysical pseudo-propositions.[9]

   Consequently, the arising question falls into the one discussed in Sect. 2.1: no empirical observation, no experiment will ever be able to *show models*, *laws*, abstract *concepts*.[10]

   > (5.3) The point is that all physical theories say much more than we can test. Whether this 'more' belongs legitimately to physics, or whether it should be eliminated from the theory as a 'metaphysical element' is not always easy to say. [...] reference to Mach, Poincaré, and Einstein was unfortunate, since Mach, more especially, looked forward to the final elimination of atomism which he (with many other positivists) considered to be a metaphysical element of physics. (He eliminated too much.) Poincaré tried to interpret physical theories as implicit definitions [...]; and Einstein has for a long time been a believer in metaphysics, operating freely with the concept of the 'physically real'.[11]

The verificationist criterion turns out to be a claim, which, if it is assumed consistently, it will actually lead to the rejection of almost all scientific knowledge as a meaningless construction. The epistemological conception apparently closest to the empirical data reaches positions of scepticism, arriving to reject concepts such as that of the *atom*, as it cannot be immediately experienced through the senses. According to verificationism, statements such as Newton's postulates on space and time in the quote 4.6 (Sect. 4.2.2) or the postulates of Quantum Mechanics—which need objects such as Schrödinger *wave function* (Eq. (4.4)) or the operators on functional spaces (Eq. (4.5) and (4.7)), objects far from being a matter of sensitive experience[12]—would be labelled as non-scientific, *metaphysical*, and even meaningless. It is the actual development of scientific knowledge, and of physical science in particular, which

---

[9] Karl Popper, Chap. 11 of Popper (1962).

[10] Cf. the discussion on *Galileo's postulate*, e.g. the quote 2.1 in Sect. 2.1.

[11] K. Popper, Chap. 11 of Popper (1962)

[12] Part II Chap. 6 of Duhem (2016).

shows how such a suffocating conception risks leading the enterprise of knowledge into a dead end.

5.2.2 *Falsificationism*

While verificationism is shown to be inadequate, a perspective that appears more convincing is the one complementary to it, namely, *falsificationism* (*Falsifikationismus*). It starts from the verificationist position in raising the "problem of demarcation" as the central issue. Karl Popper (1902–1994), that is considered the main epistemologist of the twentieth century, expresses the question in the following terms:

> (5.4) *Die Aufgabe, ein solches Kriterium zu finden, durch das wir die empirische Wissenschaft gegenüber Mathematik und Logik, aber auch gegenüber "metaphysischen" Systemen abgrenzen konnen, bezeichnen wir als* Abgrenzungsproblem.[13]
>
> I call the problem of demarcation the problem of finding a criterion which would enable us to distinguish between the empirical sciences, on the one hand, and mathematics, logic as well as "metaphysical" systems on the other.

Discriminating between what is scientific and what is not means precisely understanding what can be considered an empirical scientific investigation. Such investigation has the advantage of exploiting the experiment as a final criterion: considered *highly reliable*, that is, to use Feynmann's words,[14] you have got a judge capable of establishing the *truth* of the devised explanations. Therefore, from this perspective the *Abgrenzungsproblem* is essential to establish the effectiveness of knowledge. Precisely because the Galilean *scientific method* is found to be so effective in the investigation of natural phenomena, the solution to the problem can be reached not so much with abstract considerations, but rather through the analysis of this way forward.

> (5.5) This [drawing a line of demarcation between scientific and not-scientific statements] is a problem which has agitated many philosophers [...], although I have never found a very explicit formulation of it. The most widely accepted view was that science was characterized by its observational basis, or by its inductive method, while pseudo-sciences and metaphysics were characterized by their speculative method or [...] by the fact that they operated with 'mental anticipations' -something very similar to hypotheses. This view I have never been able to accept. The modern theories of physics, especially Einstein's theory [...], were highly speculative and abstract, and very far removed from what might be called their 'observational basis'. All attempts to show that they were more or less directly 'based on observations' were unconvincing. The same was true even of Newton's theory.[15]

In itself each theory is a reasoned and structured explanation of natural phenomena, expressed in a language that, in physics, is largely overlapping with the mathemati-

---

[13] K. Popper, Chap. 1 of Popper (1935), English translation from Popper (2005).

[14] Quote 1.4 in Sect. 1.1.

[15] K. Popper, Chap. 11 of Popper (1962).

cal one although it does not entirely coincide with it.[16] Since it is an abstract mental construction, which exploits strictly mathematical entities, there is no possibility of empirically verifying it, as mentioned in the previous section. So, what is the solution? Falsificationism is proposed as a *complementary* to verificationism: scientific knowledge is not what can be empirically verified, but what can be *empirically falsified*!

(5.6) *Theorien sind somit niemals empirisch verifizierbar. Wollen wir den positivistischen Fehler, die naturwissenschaftlich-theoretischen Systeme durch das Abgrenzungskriterium auszuschließen, vermeiden, so miissen wir dieses so wahlen, daß auch Sätze, die nicht verifizierbar sind, als empirisch anerkalmt werden können.*

Nun wollen wir aber doch nur ein solches System als empirisch anerkennen, das einer Nachprüfung *durch die "Erfahrung" fähig ist. Diese Überlegung legt den Gedanken nahe, als Abgrenzungskriterium nicht die Verifizierbarkeit, sondern die* Falaifizierbarkeit *des Systems vorzuschlagen; mit anderen Worten: Wir fordern zwar nicht, daß das System auf empirisch-methodischem Wege endgültig positiv ausgezeichnet werden kann, aber wir fordern, daß es die logische Form des Systems ermöglicht, dieses auf dem Wege der methodischen Nachprüfung negativ auszuzeichnen:* Ein empirisch-wissenschaftliches System muß an der Erfahrung scheitern können.[17]

Theories are, therefore, never empirically verifiable. If we wish to avoid the positivist's mistake of eliminating, by our criterion of demarcation, the theoretical systems of natural science, then we must choose a criterion which allows us to admit to the domain of empirical science even statements which cannot be verified.

But I shall certainly admit a system as empirical or scientific only if it is capable of being tested by experience. These considerations suggest that not the verifiability but the falsifiability of a system is to be taken as a criterion of demarcation. In other words: I shall not require of a scientific system to be capable of being singled out once and for all in a positive sense, but its logical form to be such that it can be singled out, by means of empirical tests, in a negative sense: it must be possible for an empirical scientific system to be refuted by experience.

Summarizing, the main concepts introduced in this epistemological scheme are

1. the DEMARCATION PROBLEM ("Abgrenzungsproblem");
2. the DEMARCATION CRITERION ("Abgrenzungskriterium");
3. the FALSIFIABILITY.

Thus, according to the falsificationist approach scientific investigation does not advance so much additively as subtractively: it does not "add further knowledge", but rather it "removes further ignorance". From this point of view, one can never know that a given explanation is true, but rather that it is not false.

It follows that research advances through a process of continuous construction and refinement.

1. – There are natural phenomena to be explained.
   – Observation of them produces empirical (experimental) data.

[16] Part II of Duhem (2016).

[17] K. Popper, cap. 1 di Popper (1935), English translation from Popper (2005).

2. – On the basis of them a theory is established.
   – This theory produces a model, usually expressed in a mathematical language.[18]
   – The above-mentioned construction matches with data.
3. – The theory is considered valid ("true").
   – The experimental observations go on.
4. – At some point the data seem to contradict it: this opens up a *problem*.
   – *Falsification*: this problem is not solvable with the instruments provided by the theory.
5. – It is mandatory to abandon the starting theory: a new theory is built up.
   – The new theory is able to match both the old and the new data.
   – The new theory is considered valid ("true").
6. This process is repeated.

According to this conceptual pattern, a theory is considered valid ("true") only as long as the empirical data do not contradict it. Nevertheless, strictly speaking, it is not true: it is simply "not false", or rather "not yet falsified". This is because we are sure that, with further refinement of the observational data, sooner or later problems will arise so that the theory is falsified. Then it will be necessary to assume a new theory that can explain the new data in a never-ending process of continuous subtraction of ignorance.

Although this point of view may seem to belittle both scientific research and the possibilities of human knowledge, emptying the understanding of phenomena, Popper's purpose is, on the contrary, "to save" knowledge, phenomena, as well as the effectiveness of an increasingly powerful and refined scientific method, a scientific method which is capable to be adapted to the new problems continuously shown by the investigation of nature.

> (5.7) Every time we proceed to explain some conjectural law or theory by a new conjectural theory of a higher degree of universality, we are discovering more about the world, trying to penetrate deeper into its secrets. And every time we succeed in falsifying a theory of this kind, we make a new important discovery. For these falsifications are most important. They teach us the unexpected; and they reassure us that, although our theories are made by ourselves, although they are our own inventions, they are none the less genuine assertions about the world; for they can clash with something we never made.[19]

The declared goal of the falsificationist approach is actually twofold:

1. affirming that scientific investigation produces *true*, reliable, and non-illusory knowledge;
2. recognizing that scientific investigation produces, however, a knowledge that is always approximate and never complete.

---

[18] If the mathematical language is exploited, we will be within the borders of the Galilean paradigm; the falsificationist perspective, especially the Popper's one, does not exclude that other scientific theories can be developed producing models that are expressed through other languages (Strumia 2017).

[19] K. Popper, Chap. 5 of Popper (1972).

Here, falsificationism is reflected by the definition of "science" of Einstein at the beginning of this discussion, quote 1.1 (Sect. 1.1): there he recalls that science represents an "attempt", i.e. something intrinsically never definitive, and therefore always subject to be called into question. It is indicative in this that such perspective abandons the verificationist claim of branding as "meaningless" all statements not susceptible to a direct empirical verification: more realistically, it wants to discern the knowledge that uses the scientific method and the one that uses other methods. On the one hand, Popper's approach proposes a scientific research that is perpetual tension, as it is always susceptible to be called into question. On the other hand, Popper aims at defusing possible attitudes of scepticism, which give in to the temptation to relegate scientific explanations to a sort of "mere conventions"; in this way, he proposes to keep alive the perception that the enterprise of scientific knowledge is increasingly approaching a *kernel of truth*.[20]

This epistemological position has been both accepted and criticized, since it still remains a schematic model that does not exhaust the human experience of knowledge, or even of scientific knowledge. Our analysis does not digress into discussing its strengths and weaknesses in detail, as these insights are the subject of several specialist treatises. However, falsificationist reflection has posed a powerful and realistic perspective from which to evaluate scientific research, a perspective that is important to keep in mind when evaluating the scientific method.

5.2.3 Research Programs

There are at least two aspects of the falsificationist approach described here, which certainly grasps important features of the scientific method, needing to be critically developed:

1. The effective adherence of the falsificationism model to the actual progress of scientific research.
2. The underlining of a *pars destruens*, the falsification, rather than a *pars construens*: the latter one is what, in general, enlivens scientists.

These items were addressed by introducing the concept of *methodology of research programs*, mainly thanks to the work of Imre Lakatos (1922–1974). When we evaluate in detail the typical way in which the scientific method proceeds, it does not develop by addressing a single phenomenon or a single problem, to which the chain of explanations and falsifications previously mentioned may be applied. On the contrary, scientific research advances through more or less large *research programs*. The latter one indicates that a scientific explanation has its own hierarchical *structure*. Indeed, a research program is characterized by an indispensable and specific part which belongs to the program and by a corollary of *auxiliary hypotheses*.

[20] Cf. also Zoglauer (2023) and "perspectival realism".

(5.8) All scientific research programmes may be characterized by their '*hard core*'. The negative heuristic of the programme forbids us to direct the *modus tollens* at this 'hard core'. Instead, we must use our ingenuity to articulate or even invent 'auxiliary hypotheses', which form a *protective belt* around this core, and we must redirect the *modus tollens* to *these*. It is this protective belt of auxiliary hypotheses which has to bear the brunt of tests and get adjusted and re-adjusted, or even completely replaced, to defend the thus-hardened core [of the program].[21]

The emphasis on *research programs* allows us to underline, more than the falsificationist aspect, which however is recognized as an important element: it especially highlights the construction of the confirmation apparatus of a theory, a process which is called "positive heuristic".

(5.9) Research programmes, besides their negative heuristic, are also characterized by their positive heuristic. [...]

The positive heuristic of the programme saves the scientist from becoming confused by the ocean of anomalies. The positive heuristic sets out a programme which lists a chain of ever more complicated models simulating reality: the scientist's attention is riveted on building his models following instructions which are laid down in the positive part of his programme. [...]

Which problems scientists working in powerful research programmes rationally choose, is determined by the positive heuristic of the programme rather than by psychologically worrying (or technologically urgent) anomalies. The anomalies are listed but shoved aside in the hope that they will turn, in due course, into corroborations of the programme.[22]

The schematic rigidity of falsificationism in the strict sense is corrected by a fruitful practice distinguishing the different elements within the research program. An excellent example of this approach is represented by *Newton's research program*, which has been discussed (Sect. 4.2):

(5.10) The classical example of a successful research programme is Newton's gravitational theory: possibly the most successful research programme ever. When it was first produced, it was submerged in an ocean of 'anomalies' (or, if you wish, 'counterexamples'), and opposed by the observational theories supporting these anomalies. But Newtonians turned, with brilliant tenacity and ingenuity, one counterinstance after another into corroborating instances [...] In Newton's programme the negative heuristic bids us to divert the *modus tollens* from Newton's three laws of dynamics and his law of gravitation. This 'core' is 'irrefutable' by the methodological decision of its proponents: anomalies must lead to changes only in the, 'protective' belt of auxiliary, 'observational' hypotheses and initial conditions.[23]

The actual method of scientific research is much more complex than it appears at first sight, since schemes and interpretations can only be modified in specific circumstances. What is the criterion establishing these circumstances? The answer can only be eminently *pragmatic* and *operational* (see, for instance, the problem of what are the "right" questions to examine nature, Sect. 1.1). Acceptable circumstances are those where there is progress in understanding, or rather in knowledge of the observed natural phenomena.

---

[21] Imre Lakatos, Chap. 1 of Lakatos (1978).

[22] Imre Lakatos, Chap. 1 of Lakatos (1978).

[23] Imre Lakatos, Chap. 1 of Lakatos (1978).

> (5.11) We '*accept*' problem shifts as 'scientific' only if they are at least theoretically progressive; if they are not, we '*reject*' them as 'pseudoscientific'. [...]
>
> *A given fact is explained scientifically only if a new fact is also explained with it.* [...]
>
> [N]o experiment, experimental report, observation statement or well-corroborated low-level falsifying hypothesis alone can lead to falsification. There is no falsification before the emergence of a better theory.[24]

We note that the introduction of the concepts of:

1. RESEARCH PROGRAM,
2. POSITIVE HEURISTIC,
3. PROTECTIVE BELT of a theory

allows us to highlight the *pars costruens* of the scientific enterprise without rejecting the falsificationist paradigm. From this point of view, this vision is closer to what enlivens the work of scientists: indeed, they are not especially driven by the desire to reduce ignorance but, on the contrary, by that of increasing knowledge. Consequently, the positive heuristic of a research program allows us to develop this approach.

5.3 Scientific Explanations as the Result of Circumstances

The problem of scientific knowledge, which has been exposed so far by explaining in broad terms the ideas of the main scholars who have dealt with the topic, provides idealizations and general schemes of how scientific research actually works. This systematization is based on the assumption that the development of such a form of knowledge follows completely rational methodological criteria. For this reason, the analysis is carried out with diachronic criteria which in principle are constantly valid over time. However, if we analyse the actual advances of "science", there are elements that appear not to be, or to be only partially, rational. This may be unpleasant for those who believe that scientific research is a somehow "pure" and "aseptic" activity, in which we use a "supreme reason" in a "distilled" form. Conversely, it is namely in these elements that are not immediately rational that one can see the imprint of the *profoundly human activity* in which *everything* of the human nature is involved.[25] It is congenial to become aware of these elements in order to realize that they involve possibilities alongside risks for scientific knowledge.

[24] Imre Lakatos, Chap. 1 of Lakatos (1978).

[25] Cf. *Preface*.

5.3.1 *Partially Rational Elements: The Affirmation of Scientific Theories*

In evaluating the evolution of scientific knowledge, there are elements that are closely related to the historical circumstances when these advances happened. For this reason, it is interesting to ask the following question:

In this *method* of knowledge, has the human history of "science", with its limits, contradictions, accelerations, and slowdowns, got anything to teach?

Assuming a similar perspective means to evaluate the scientific method in a synchronic way. From this, a significant description of scientific research is given by the *theory of scientific revolutions*, mainly due to Thomas Kuhn (1922–1996). He investigates the process of how scientific ideas and explanations have *effectively* established throughout history. The resulting analysis introduces some original concepts:

1. the notion of PARADIGM;
2. the distinction between NORMAL SCIENCE and SCIENTIFIC REVOLUTION;
3. the notion of PUZZLE.

According to Kuhn's picture, scientific knowledge historically advances in periods. In a given period, there are specific conceptual schemes for the explanation of natural phenomena: these are universally accepted and represent the *paradigms*.

> (5.12) [T]he term 'paradigm' is used in two different senses. On the one hand, it stands for the entire constellation of beliefs, values, techniques, and so on shared by the members of a given community. On the other, it denotes one sort of element in that constellation, the concrete puzzle-solutions which, employed as models or examples, can replace explicit rules as a basis for the solution of the remaining puzzles of normal science.[26]

Periods in which a given *paradigm* is well consolidated and considered satisfactory is called *normal science*.

> (5.13) '[N]ormal science' means research firmly based upon one or more past scientific achievements achievements that some particular scientific community acknowledges for a time as supplying the foundation for its further practice. Today such achievements are recounted, though seldom in their original form, by science textbooks, elementary and advanced.[27]

During periods of "normal science" the activity of scientists is to perform "normal research", in other words to answer questions while remaining within the framework of the universally accepted paradigm.

> (5.14) Bringing a normal research problem to a conclusion is achieving the anticipated in a new way, and it requires the solution of all sorts of complex instrumental, conceptual, and mathematical puzzles.[28]

---

[26] Thomas Kuhn, Postscript-1969 of Kuhn (1996).

[27] Thomas Kuhn, Chap. 1 of Kuhn (1996).

[28] Thomas Kuhn, Chap. 1 of Kuhn (1996).

In short, in such periods scholars use standard and well-known schemes and concepts: in doing so, they manage to fit everything into the paradigm framework. This activity is called "puzzle solving" as it does not lead to unexpected results, hence it does not much increase knowledge. The situation is similar to solving a school problem: everybody knows that the solution exists and that it can be achieved through consolidated and tested methodologies. Even if the problem had not already been previously solved, no one would doubt the existence of a solution within the available conceptual tools. Therefore, the resolution leads to a modest advancement in knowledge.

Nevertheless, this does not last forever. In the work of investigating and explaining natural phenomena, sooner or later unexpected events called *anomalies* emerge.

> (5.15) Discovery comences with the awareness of anomaly, i.e., with the recognition that nature has somehow violated the paradigm-induced expectations that govern normal science. [...] [I]t closes only when the paradigm theory has been adjusted so that the anomalous has become the expected.[29]

Initially there is an attempt to bring the anomalies back within the known paradigm. If this cannot be accomplished, usually a group of scholars proposes a different *paradigm*, partly or wholly incompatible with the current one. Such scholars are opposed by supporters of "normal science"; however, if they manage to advance their position, at some point their *new paradigm* become predominant: this is the moment of the *scientific revolution*.

> (5.16) What are scientific revolutions, and what is their function in scientific development? [...] [S]cientific revolutions are here taken to be those non-cumulative developmental episodes in which an older paradigm is replaced in whole or in part by an incompatible new one.[30]

According to such a perspective, the periods of "scientific revolution", contrary to those of "normal science", are those in which knowledge has got an effective growth. For this reason, in some way periods of crisis are the most fruitful. On the other hand, the reasons why one paradigm arrives to undermine another one, i.e. the reasons why a scientific revolution is successful, are complex: they are result of a process where comparison with observational and experimental data plays a role, despite not being the only key element.

> (5.17) [T]the act of judgement that leads scientists to reject a previously accepted theory is always based upon more than a comparison of that theory with the world. The decision to reject one paradigm is always simultaneously the decision to accept another, and the judgment leading to that decision involves the comparison of both paradigms with nature *and* with each other.[31]

Finally, this process is destined to be repeated, as summarized in the diagram of Fig. 5.1. The so far accomplished analysis does not want to catch a rational structure

[29] Thomas Kuhn, Chap. 6 of Kuhn (1996).

[30] Thomas Kuhn, Chap. 9 of Kuhn (1996).

[31] Thomas Kuhn, Chap. 8 of Kuhn (1996).

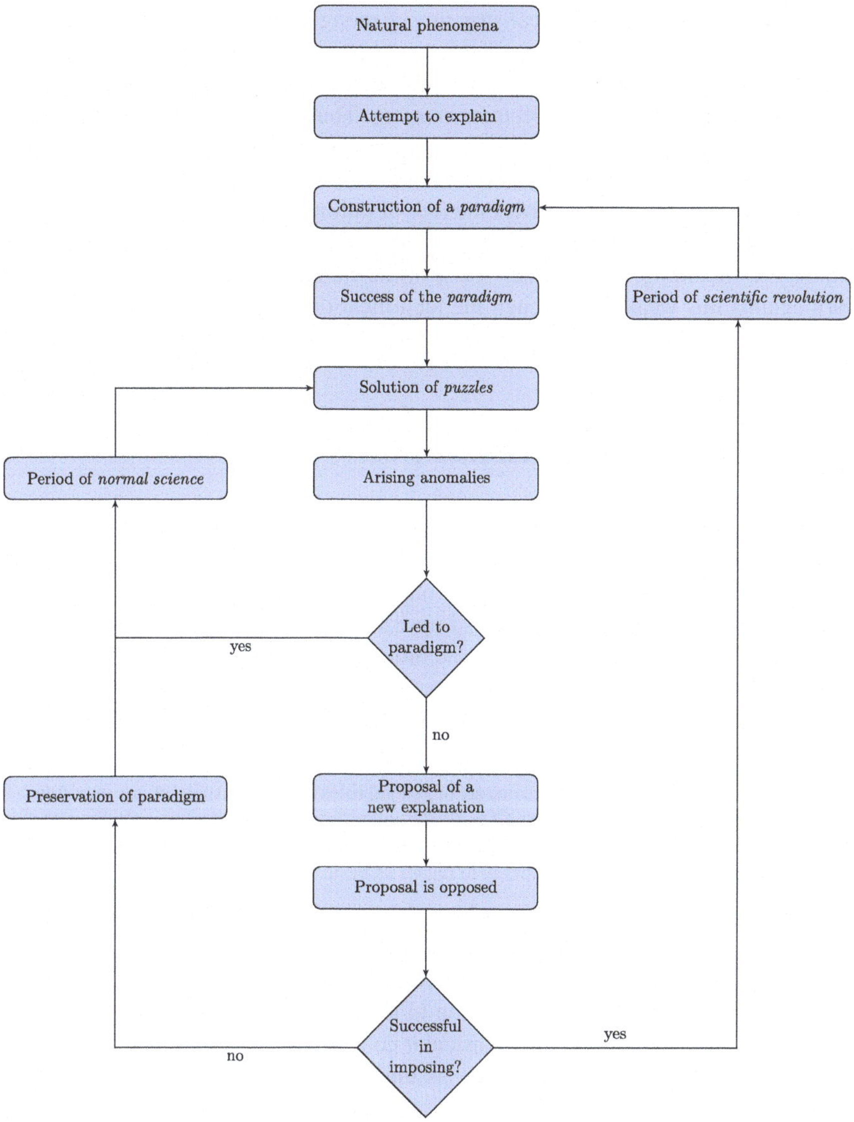

Fig. 5.1 The processes of scientific revolutions according to T. Kuhn

of scientific methodology. On the contrary, it seems to suggest that the criteria by which a *paradigm* imposes on another may be opaque: it may be the result of choices that are not entirely rational, thus overshadowing an attitude of scepticism.

Neglecting this latter feature, our study concentrates on pointing out the merits of such analysis. Kuhn highlights that in the construction of scientific explanations elements that are poorly controllable are also involved, in addition to rational methodology and epistemological structure. Among them, the historical processes and philosophical conceptions of a given period occupy an important place; perhaps the most important factor, however, is the presence of a given community of scholars that has got its own shared *Weltanschauung*. Thus, the problem of *consensus* is introduced: in order to get a form for scientific explanations, there is the need for some *consensus* by a community of individuals. This aspect is a *critical factor* to be aware of, since potentially it carries the risk of polluting the scientific investigation with a clash of opinions, transforming the validation of research into a problem of *majority*. Concluding, an important merit of the study of scientific revolutions is to warn against this weakness through the identification of some mechanisms related to it.

5.3.2 Completely Irrational Elements?

So far our study has exposed explanation schemes, according to strictly rational criteria belonging to the scientific method (Sect. 5.2). However, this method is influenced by extraneous factors, such as historical ones, in such a way to suggest the possibility that the very notion of *method* is imprecise, if not inadequate. Therefore, it is possible to accomplish a much deeper and more radical critique of the idea that there are general methodological norms in research, placing due emphasis on the limits of scientific knowledge. This line is followed by Paul Feyerabend (1924–1994) in some essays written in a highly provocative and, to a certain extent, even ironic style.

> (5.18) My intention is not to replace one set of general rules by another such set: my intention is, rather, to convince the reader that *all methodologies, even the most obvious ones, have got their limits*. The best way to show this is to demonstrate the limits and even the irrationality of some rules which she, or he, is likely to regard as basic.[32]

While following the actual development of scientific knowledge, further elements extraneous to a rational method are identified and give rise to the so-called *epistemological anarchy*. The analysis starts from the study of the historical processes described above and takes the discussion to its extreme consequences. Feyerabend finds that, in our modern societies, particular consideration is given to what is called "science", which however is generally assumed *a priori* and not suitably demonstrated.

> (5.19) *I don't take the excellence of science for granted* [...], *I try to show wherein it consists and how greatly it differs from the naive standards of excellence proposed by rationalists.* [...]
>
> But science is not sacrosanct.[33]

If we observe the actual historical development of scientific knowledge, we do not find well-defined processes, as Kuhn believes, but a great complexity or even chaos.

[32] Paul Feyerabend, Chap. 2 of Feyerabend (1993).

[33] Paul Feyerabend, Part 1 Chap. 1 of Feyerabend (1978).

Indeed, far from constituting an obstacle, this chaos was often the driving force to the development of scientific knowledge.

> (5.20) [T]he history of science will be as complex, chaotic, full of mistakes, and entertaining as the ideas it contains [...]. Conversely, a little brainwashing will go a long way in making the history of science duller, simpler, more uniform, more 'objective' and more easily accessible to treatment by strict and unchangeable rules.
>
> Scientific education as we know it today has precisely this aim. It simplifies 'science' by simplifying its participants: first, a domain of research is defined. The domain is separated from the rest of history (physics, for example, is separated from metaphysics and from theology) and given a 'logic' of its own. [...]
>
> [D]id scientists ever remain within the boundaries of the traditions they defined [...]? [...] [T]o these questions my answer will be a firm and resounding NO.[34]

According to this vision, any claim to identify *pathways of method*, following rationalist schemes such as verificationism and falsificationism, is a form of "domestication" of science, which does not take into account the real involved processes. In light of actual scientific practice, does it make sense to say that a *method* exists? Do or did scientists really follow a rational *method*?

> (5.21) [W]herever we look, whatever examples we consider, we see that the principles of critical rationalism (take falsifications seriously; increase content; avoid *ad hoc* hypotheses; 'be honest' .whatever *that* means; and so on) and, *a fortiori*, the principles of logical empiricism (be precise; base your theories on measurements; avoid vague and unstable ideas; and so on), though practised in special areas, give an inadequate account of the past development of science as a whole and are liable to hinder it in the future. They give an inadequate account of science because science is much more 'sloppy' and 'irrational' than its methodological image. [...] Without 'chaos', no knowledge. Without a frequent dismissal of reason, no progress. [...] We have to conclude, then, that *even within* science reason cannot and should not be allowed to be comprehensive and that it must often be overruled, or eliminated, in favour of other agencies. There is not a single rule that remains valid under all circumstances and not a single agency to which appeal can always be made.[35]

We achieve the conclusion that scientific knowledge is by no means more sure or more rigorous than all other forms of knowledge; on the contrary, the more it does not follow any rational method, the more it is vital. The criterion is that there is no criterion, knowledge develops without following any rules: this is *epistemological anarchism*. If this is the case, why then is "science" held in such high regard in the modern world? Feyerabend answers that this happens because it is allied with power, to whom it has become a prop. Thus, scientific research enters the systems of modern societies in the same way as institutional and economic structures, exchanges of goods, financial transactions, struggles for power, etc.

> (5.22) [S]cience has now become part of the basic fabric of democracy [...] In these circumstances, is it surprising that science reigns supreme and is the only ideology known to achieve worthwhile results? It reigns supreme because some *past successes* have led to institutional measures [...] that prevent a comeback of the rivals. Briefly, but not incorrectly:

---

[34] Paul Feyerabend, Introduction of Feyerabend (1993).

[35] Paul Feyerabend, Chap. 15 of Feyerabend (1993).

today science prevails not because of its comparative merits, but because the show has been rigged in its favour.[36]

It follows there is no reason to affirm that the scientific method produces *true* knowledge: it has simply been able to gain enough consensus in society.

Such a position of marked anarchy is evidently affected by a certain relativistic cultural climate spread in Western societies in the last decades of the twentieth century.[37] This vision is rightly considered extreme and, as mentioned, Feyerabend himself states that his intent is deliberately provocative, starting from the title of his most important essay, "Against the Method". Without entering into the predictable controversies that these ideas raised, it is worth noting some interesting items.

The value of Feyerabend's work is to point out that scientific knowledge and its method should not be accepted as indisputable, intangible, and immutable facts, as stated in the quote 5.19. Scientific knowledge and scientific cognizance must not be treated with sacred criteria, but as human constructions subject to continuous verification and discussion. Furthermore, it highlights the risk that, by becoming part of a social *system*, scientific enterprise can assert itself not through a conviction based on data and rational explanations, but through connivance and mutual utility with power. We can here see how, if grasped in its form of denunciation and declared provocation, the proposal of *epistemological anarchy* has got the merit of drawing attention to some "raw nerves" of the modern scientific enterprise.[38]

5.4 The Pragmatic Approach of Scientific Practice

Beyond the philosophical side, about which we have just provided a summary overview, scientists present a highly pragmatic attitude towards knowledge. As stated previously, it is possible to learn what "science" is by "doing science"; in the same way, it is possible to learn what the scientific method is and how it is applied, what answers it is able to provide, what advantages and limitations it has got, through practice. Beyond the philosophical conceptions embraced (which can be the most varied), each scientist assumes—at least empirically, pragmatically, operationally—the existence of natural phenomena that are possible to investigate somehow and sets themselves the goal of reaching a stable, reliable, and rational knowledge of those as their ideal purpose. If we consider the method to achieve this purpose, the scientific one has got already its main merit and purpose at its recognized birth: as a matter of fact, it was created to have got a "*more suitable*" instrument to investigate the natural phenomena, i.e. it was developed as a form of obedience to the reality studied (Sect. 4.2.1). In this form of "epistemological adaptation" it is perhaps possible to find the highly empirical peculiarities of scientific investigation: its being both rationally

[36] P. Feyerabend, Part 2 Chap. 9 of Feyerabend (1978).

[37] Chapter 8 of Strumia (2017).

[38] We discuss some of these issues in Sect. 10.2.

constructed and powerfully flexible and adaptive, markedly focused on achieving *knowledge*.

(5.23) Ci sono soltanto due possibili conclusioni: se il risultato conferma le ipotesi, allora hai appena fatto una misura; se il risultato è contrario alle ipotesi, allora hai fatto una scoperta.[39]

There are just two possible conclusions: if your result confirms hypotheses, then you've just done a measurement; if your result is contrary to hypotheses, then you've done a discovery.

Consequently, what can we discover through scientific knowledge? Each of the previous reviews has highlighted an existing aspect. In the methodologies exploited by scientists, knowledge is certainly obtained through verification. At the same time, it is equally true that knowledge advances also subtractively—through processes of continuous proposal, verification, falsification, denial, reconstruction. At the same time, various epistemological structures are constructed to fortify powerful and fruitful explanatory schemes. It should also be recognized that in the affirmation of explanatory theories claiming to be scientific, completely non-scientific elements also play a role: the historical circumstances, the ability to reach consensus and emerge, as well as irrational elements of chance, interest, and force.

(5.24) I should make it absolutely clear that the apparent lack of objectivity is not the fault of Nature herself. There is an objective physical world out there, and physicists correctly regard it as their job to find out its nature and to understand its behaviour. The apparent subjectivity that we see in the strong influences of fashion [...] are simply features of our gropings for this understanding, where social pressures, funding pressures, and (understandable) human weaknesses and limitations play important parts in the somewhat chaotic and often mutually inconsistent pictures that we are presently confronted with.[40]

However, in this process of building up, dismantling and reconstructing scientific knowledge, besides the many perishable elements, there are elements that are nevertheless preserved.

(5.25) *Sans doute, au premier abord, les théories nous semblent fragiles, et l'histoire de la science nous prouve qu'elles sont éphémères: elles ne meurent pas tout entières pourtant, et de chacune d'elles il reste quelque chose. C'est ce quelque chose qu'il faut chercher á démêler, parce que c'est lá, et lá seulement, qu'est la véritable réalité.* [41]

No doubt, at first glance, theories seem fragile to us, and the history of science proves that they are ephemeral: however, they do not die entirely and something of each remains. It is this something that must be sought to unravel, because it is there, and there only, that the true reality is.

Genuine knowledge, *true* knowledge, resides just in this heart of knowledge—surviving the vicissitudes of the *method*, so much influenced by *human* factors. What *true* means in physical science is discussed in Chap. 8. Indeed, the value of the *scientific method* is precisely that it allows to get closer to this heart of "genuine

[39] Enrico Fermi, Presentazione in Fermi (2018).

[40] R. Penrose, Chap. 34 of Penrose (2004).

[41] H. Poincaré, Introduction in Poincaré (1902).

reality", of *true* knowledge, with the awareness that, like all human constructions, it accomplishes this work within rough and repeated approximations.[42] When facing and investigating a natural phenomenon, Poincaré suggests that it is possible to grasp a specific *quid*, a universal characteristic, of the phenomenon: this comprehension represents a *robust* knowledge, because one principle of the object has been grasped.[43]

References

G. Basti, *Filosofia della Natura e della Scienza*, vol. 1, 1a edn. (Pontificia Università Lateranense, Roma, Città del Vaticano, 2002)

S. Chibbaro, L. Rondoni, A. Vulpiani, *Reductionism, Emergence and Levels of Reality. The Importance of Being Borderline* (Springer, Cham, Schweiz, 2014)

P. Duhem, *La théorie physique. Son objet, sa structure*, 2nd edn. (ENS Éditions, Lyon, France, 2016). http://books.openedition.org/enseditions/6077

E. Fermi, in G. Pierazzini, N. Robotti, F. Guerra, M.M. Massai, P. Rossi (eds.) *Se il risultato è contrario all'ipotesi. I manoscritti giovanili di Enrico Fermi, un tesoro dell'Università di Pisa* (Pisa University Press, Pisa, Italia, 2018)

P. Feyerabend, *Science in a Free Society* (New Left Books, London, UK, 1978)

P. Feyerabend, *Against Method: Outline of an Anarchistic Theory of Knowledge*, 3rd edn. (Verso Books, London, UK, 1993)

T. Kuhn, *Conjectures and Refutations: The Growth of Scientific Knowledge*, 3rd edn. (The University of Chicago Press, Chicago & London, USA & UK, 1996)

I. Lakatos, *The Methodology of Scientific Research Programmes* (Cambridge University Press, Cambridge, UK, 1978)

R. Penrose, *The Road to Reality. A Complete Guide to the Laws of the Universe* (Jonathan Cape, London, UK, 2004)

H. Poincaré, *La Science et l'Hypothèse* (Flammarion, Paris, France, 1902)

K. Popper, *Objective Knowledge: An Evolutionary Approach* (Oxford University Press, 1972)

K. Popper, *The Logic of Scientific Discovery* (Taylor & Francis, London & New York, UK & USA, 2005). Author's English translation of Popper 1935

K. Popper, *Logik der Forschung. Zur Erkenntnistheorie der Modernen Naturwissenschaft* (Springer, Wien, Österreich, 1935)

K. Popper, *Conjectures and Refutations: The Growth of Scientific Knowledge* (Basic Books, London & New York, UK & USA, 1962)

A. Strumia, *Dalla filosofia della scienza alla filosofia nella scienza*, 1a edn. (ESC - Edizioni Santa Croce, Roma, Italia, 2017)

T. Zoglauer, Truth relativism, science skepticism and the political consequences, in *Truth and Fake in the Post-Factual Digital Age: Distinctions in the Humanities and IT Sciences* (Springer Fachmedien Wiesbaden, Wiesbaden, Deutschland, 2023)

[42] Cf. e.g. Zoglauer (2023).

[43] This possibility is explored within a classical epistemological Aristotelian-Thomist framework (Strumia 2017; Basti 2002).

Chapter 6
Looking for Solid Bases: Attempts to Explain *Everything*

Grandissima mi par l'inezia di coloro che vorrebbero che Iddio avesse fatto l'universo più proporzionato alla piccola capacitá del loro discorso che all'immensa, anzi infinita, sua potenza.

Galileo Galilei

This text is about foundations and in the next two chapters we go to the foundations of foundations. Our focus is asking what lies at the very heart of physics. In doing so, two directions are proposed.

The first one, dealt in this chapter, follows the object of the discipline: the ambition of physical science to reduce the explanation of whole nature to a few principles, concepts, and fundamental mathematical models. In this effort lies the attempt, or the claim, to bring all the natural sciences back to fundamental physics.

The second direction, dealt in Chap. 7, follows the language of our discipline: the problem of the foundations of mathematics, which is connected to the roots of rational thought and recalls the problem of the relationship between mathematics and reality (which is discussed about in Chap. 2).

6.1 A Hierarchy of Knowledge

As already stated, from the Galilei's time to the twentieth century the development of natural sciences has been astonishing. In particular, thanks to the great Newtonian synthesis physics has been shown to be capable of accurately predicting a huge quantity of phenomena within mechanics. Moreover, during the nineteenth-century thermodynamics and electromagnetism have been added. The combination of these theories, which are the result of the application of Galileo and Newton' experimental-

J. Parravicini, *The Foundations of Experimental Physics*,
https://doi.org/10.1007/978-3-031-55452-0_6

mathematical method, proves able to describe in detail very broad classes of phenomena, as well as to create links between natural events that are apparently very different from each other. This outstanding success leads to a vision of the world where modern science comes to take on religious connotations: it follows the creation of a hierarchy where physics and mathematics are held to occupy the first place of all human knowledge. Historically, this vision is expressed in *positivism* and in *neo-positivism*, which have been presented in Sect. 5.2.1. Nevertheless, this attitude is still widespread to varying degrees both among scientists and in society.

> (6.1) [I]n the will of the Swedish dynamite magnate Alfred Nobel, who established the Nobel prizes, the science prizes are listed with physics first, chemistry second, and physiology and medicine third. [...] Mathematics is omitted from Nobel's will [...]
>
> The hierarchy of subjects can in part be traced to the nineteenth-century French philosopher Auguste Comte, who argued that astronomy was the most fundamental scientific subject, physics the second, and so forth. (He regarded mathematics as a logical tool rather than a science.)[1]

In the light of modern discoveries, the aforementioned hierarchy of knowledge could be considered still valid as a first approximation. Perhaps, we could be tempted to consider it *absolutely* valid: as a matter of fact, our analysis starts with stating that physics is *the most fundamental* of the natural sciences.[2] Following this line of thought, which we call *positivist-reductionist*, an approximate hierarchy of scientific disciplines could be the one shown in the diagram 6.1. Everybody is likely to have seen a similar pattern at one time or another:

1. First and foremost, there is *mathematics*, which is not a natural science, but a discipline characterized by the maximum of abstractness and generality, thanks to which it is possible to order, do hierarchies, and create models.
2. *Physics* is the (natural) science underlying all natural sciences; it is based on mathematics, as it aims at explaining nature through such a language.
3. *Chemistry* is the natural science that studies matter and its transformations; it is based on physics. *Geophysics* and *Astrophysics* employ the methods and languages of physics to study, respectively, the planet Earth and the celestial bodies. *Engineering* uses the tools and knowledge provided by physics and chemistry for application purposes.
4. At a fourth level, disciplines are differentiated between those that study living things and those that study non-living ones. *Biochemistry* belongs to the former ones, employing the specific knowledge of chemistry for the study of living beings. *Mineralogy* belongs to the latter ones, exploiting the knowledge of geophysics and chemistry for the study of materials which are naturally present on the Earth's crust. *Astronomy* uses astrophysics for the detailed study of what exists outside the Earth.
5. *Biology* studies the basic mechanisms of living beings. *Geology* studies in detail the general behaviour of rocks and phenomena that naturally occur on Earth when not directly due to living beings.

---

[1] M. Gell-Mann, Chap. 9 in (Gell-Mann 1994).

[2] Cf. quote 1.6.

6. *Physiology* exploits biology to study how living organisms function. It contributes, together with biology and engineering, to the strongly practical *biomedical engineering* discipline.
7. *Botany*, *medicine*, and *zoology* exploit the knowledge of physiology, respectively, for the study of plants, humans, and animals; on these disciplines *veterinary* medicine is based.

Certainly, this scheme may be extended. In particular, the various disciplines into which medicine is subdivided could be hierarchically added, arriving at the discipline studying the mind at a physiological level, i.e. psychiatry. At this point, a radical materialist would also base psychology on exclusively physiological mental mechanisms, taking the extreme step of reducing all mental activity, and therefore all intellectual and cultural activities, to processes which ultimately go back to biochemistry and then to physics. Therefore, according to this "extreme" perspective, even the so-called "humanities" could fall within a scheme such as the one in Fig. 6.1 and finally be traced back to physics and mathematics. Leaving aside these aspects—more philosophical than scientific, we focus on the organization of the natural sciences. The 6.1 diagram can be read in two ways.

1. With a *subsidiary* meaning. The hierarchy represented therein means that the succeeding discipline necessarily employs some of the knowledge, concepts, and results developed by the preceding discipline. The subsequent disciplines also study a more delimited and particular portion of nature than the previous discipline.[3] Ultimately, the disciplines are *subsidiary* to each other.
2. With a *reductionist* meaning. The hierarchy represented therein means that the succeeding discipline is a more or less necessary approximation of the preceding discipline. By this we mean that the discipline capable of providing an *ultimate* knowledge of nature is physics—with mathematics as its prerequisite. The presence of several disciplines is essentially due to our current ignorance; for this reason, the differentiation between the sciences is destined hopefully to diminish and vanish in physics. In principle, each discipline is *reducible* to physics.

The first interpretation originates from the fact that each particular discipline exploits knowledge of more general disciplines. On the other hand, the second interpretation makes a much more demanding and ambitious claim. The idea that all natural sciences ultimately are, or should be, referable to physics, at least in principle, is an idea that still exists today.

This conception arises from two attitudes. The first is the observation of physic's excellent ability to describe and predict a very large number of natural phenomena. This ability is so broad and profound as to have aroused in many a sort of religious amazement.[4] The second attitude coincides with the approach of the human

[3] Recalling Table 1.1 in Sect. 1.2, there may be a different *subiectum* and a different *obiectum materiale*, which become a specific subset of natural phenomena, a different *obiectum formale quod* and *obiectum formale quo*, i.e. different specific objects of study from different points of view.

[4] Cf. e.g., quote 2.5.

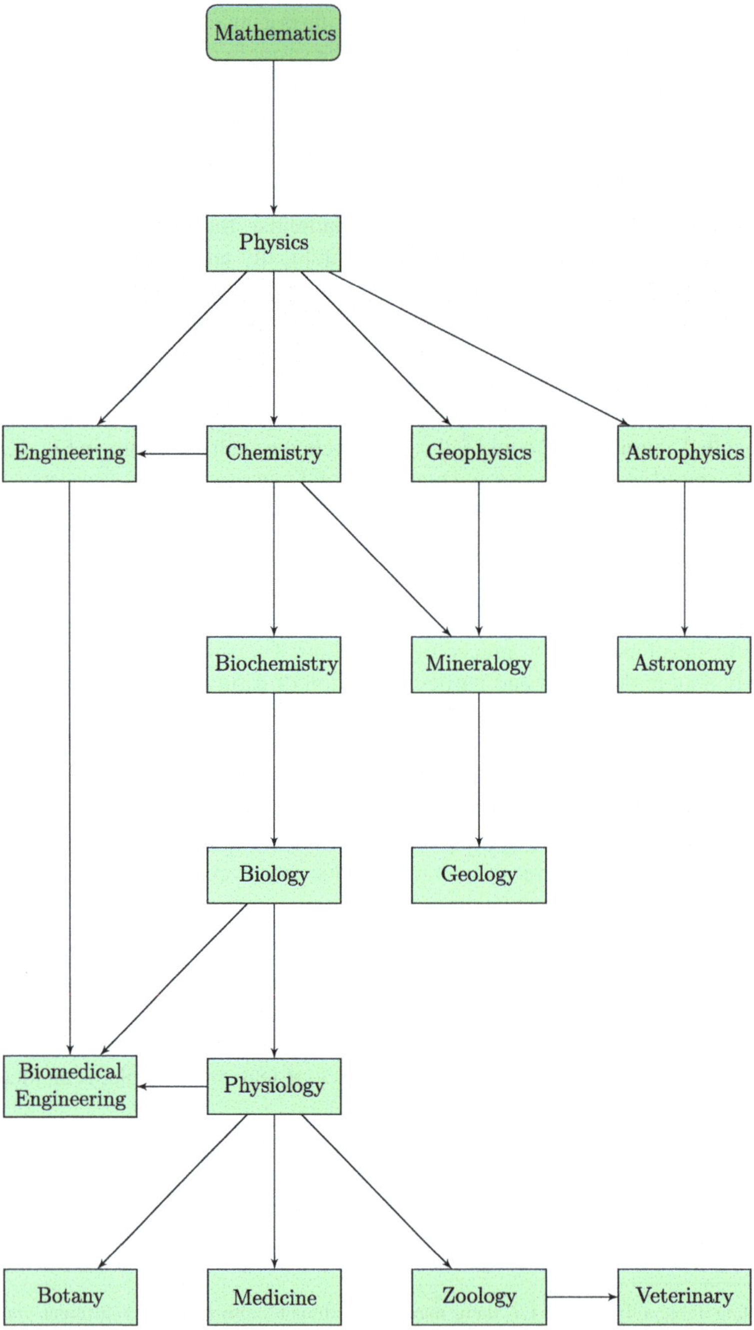

Fig. 6.1 Approximate hierarchy of the main scientific disciplines according to a positivist-reductionist approach

mind to seek the explanation of even complex phenomena in few simple elementary principles.

As a matter of fact, the climax of this cultural position is at the dawn of the twentieth century. An attempt is made to reduce scientific disciplines to their fundamental principles, ideally to physics and mathematics. In particular, this poses the problem of codifying the foundations of scientific disciplines with proper formal clarity.

> (6.2) *Zum Aufbau der Zahlentheorie [...] reichen die Rechnungsgesetze und Regeln für ganze Zahlen aus. In der Statik übernimmt die gleiche Rolle der Satz vom Parallelogramm der Kräfte, in der Mechanik etwa die Lagrangeschen Differentialgleichungen der Bewegung und in der Elektrodynamik die Maxwellschen Gleichungen mit Hinzunahme der Forderung der Starrheit und Ladung des Elektrons. Die Thermodynamik läßt sich vollständig auf den Begriff der Energiefunktion und die Definition von Temperatur und Druck als Ableitungen nach ihren Variabeln, Entropie und Volumen, aufbauen. Im Mittelpunkt der elementaren Strahlungstheorie steht der Kirchhoffsche Satz über die Beziehungen zwischen Emission und Absorption; in der Wahrscheinlichkeitsrechnung ist das Gaußsche Fehlergesetz, in der Gastheorie der Satz von der Enbropie als negativem Logarithmus der Wahrscheinlichkeit des Zustandes [...]*
>
> *Diese grundlegenden Sätze können von einem ersten Standpunkte aus als die* Axiome der einzelnen Wissensgebiete *angesehen werden [...]* [5]
>
> [T]he laws and rules of computation for integers are sufficient for setting up the theory of numbers. The same role is taken over by the parallelogram law of forces in statics, something like Lagrangian differential equations of motion in mechanics, and Maxwell's equations accepting the condition of rigidity and change of electrons in electrodynamics. Thermodynamics is completely built upon the concept of energy function and the definition of temperature and pressure as derivatives from their variables, entropy, and volume. In the midpoint of elementary radiation theory there stands Kirchhoff's law on the relations between emission and absorption; there is the Gaussian law of error in the calculus of probability, the theorem of entropy as negative logarithm of probability of events in the gas theory [...]
>
> Viewed from a primary standpoint, these theorems may be looked upon as the *axioms of the individual fields of knowledge* [...]

The conquest of a few first principles to which all knowledge could be traced back consists in finding axioms from which to deduce all the theories of physics, then all the laws of physics, and ideally from them all the scientific laws. In short, axiomatizing would simply mean knowing, finally knowing with clarity. Given these premises, Hilbert coherently concludes his picture.

> (6.3) *Aber die kritische Prüfung dieser "Beweise" läßt erkennen, daßsie nicht an sich Beweise sind, sondern im Grunde nur die Zurückführung auf gewisse tiefer liegende Sgätze ermöglichen, die nunmehr ihrerseits an Stelle der zu beweisenden Sätze als noue Axiome anzusehen sind. So entstanden die eigentlichen heute sogenannten* Axiome *der Geometrie, der Arithmotik, der Statik, der Mechanik, der Strahlungstheorie oder der Thermodynamik. Diese Axiome bilden eine tiefer liegende Schicht von Axiomen gegenüber derjenigen Axiomschicht, wie sie durch die vorhin genannten zuerst zugrunde gelegten Sätze in den einzelnen Wissensgebieten charakterisiert worden ist. Das Verfhren der axiomatisehen Methode, wie es hierin ausgesprochen liegt, kommt also einer* Tieferlegung der Fundamente *der einzelnen Wissensgebiete gleich, wie eine solche ja bei jedem Gebäude nötig wird in dem Maße, als man dasselbe ausbaut, höher fürt und dennoch für seine Sicherheit bürgen will.*

[5] David Hilbert in Hilbert (1917), English translation from Hilbert (1970).

> *Soll die Theorie eines Wissensgebietes d. h. das sie darstellende Fachwerk der Begriffe ihrem Zwecke, nämlich der Orientierung und Ordnung dienen, so mußes vornehmlich gewissen zwei Anforderungen genügen:* erstens *soll es einen Überblick über die* Abhängigkeit *bzw.* Unabhängigkeit *der Sätze der Theorie und* zweitens *eine Gowähr der* Widerspruchsloigkeit *aller Sätze der Theorie bieten.*[6]

> But the critical examination of these "proofs" makes it discernible that they are not proofs in themselves; rather, they are principally only capable of leading back to certain more deeply lying propositions which in turn are now to be looked upon as new axioms in place of those propositions to be proved. So the proper, currently so-called axioms of geometry, of arithmetic, of statics, of mechanics, of radiation theory or of thermodynamics are originated. These axioms form a more deeply lying layer of axioms opposite to another layer of axioms as they have been characterized by the aforementioned propositions of the first foundation in individual fields of knowledge. Thus the procedure of the axiomatic method, as articulated here, comes up to a deeper-laying of foundations of individual fields of knowledge, similar to the foundations indeed necessary for a building if it were to be enlarged, built higher, and yet vouched for its safety.
>
> If the theory of a field of knowledge, that is, the framework of concepts that represents the theory, is to serve its purpose, namely the orientation and order, it must then satisfy chiefly two fixed demands: it must offer firstly a general view of the dependence or independence of the propositions of the theory and secondly a guarantee of consistency of all propositions of the theory.

In this passage Hilbert suggests the idea that, going up the hierarchy of disciplines, at the root of everything there are wholly mathematical concepts and even there is a form of *distillation* of them, in the form of *axioms*. Axioms represent the maximum possible clarity, capable of encompassing, *in nuce*, all the knowledge of a given discipline: the *axioms of the individual fields of knowledge*. The idea underlying Hilbert's conceptions is resulting from a specific cultural view, as already mentioned. Simplifying, it is possible to summarize such a cultural-epistemological program in the following statement:

> All sciences are, at least in principle, *reducible* to physics, and physics is reducible to its axioms. Physics and its axioms are reducible to mathematics, which is reducible to its axioms. The axioms, both those of physics and the deeper ones of mathematics, must be *consistent*, i.e., immune from contradictions, and *complete*: every proposition of mathematics and physics must be deducible from them through purely logical-formal passages.

> Therefore, once we find the fundamental axioms of physics and mathematics, we build a complete and self-accomplished system of knowledge.

Neither Hilbert, nor the positivists, nor the neo-positivists, affirmed this program in such a categorical way; however, I think that such statements well sum up the underlying cultural idea. Indicative of this position is the sentence that Hilbert wanted engraved on his tomb:

> WIR MÜSSEN WISSEN
> WIR WERDEN WISSEN

---

[6] D. Hilbert in Hilbert (1917), English translation from Hilbert (1970).

that is, "We must know, we will know". During the First International Congress of Mathematics, held in Paris in 1900, Hilbert proposed a list of 23 open problems.[7] They are known as the *23 Hilbert's problems* and they have effectively marked the mathematics of the twentieth century: for decades, solving one of his problems meant gaining a place among the great.[8]

Problem no. 2 dealt with the foundations of mathematics, requiring to *demonstrate the compatibility of the axioms of arithmetic*; problem no. 6 required to *axiomatize all physics*.

Evaluating the general picture as a whole, we can ask ourselves whether, according to this methodology, it was conceptually correct to set these problems, in spite of their being very ambitious. So, how did the problem of foundations in physics and mathematics develop?

6.2 Reductionism: A Useful Step

Recently in physics the issue of the axiomatization of the main theories has been addressed. Rather than finding a set of axioms to which all physics can be traced back, it has been partially possible to axiomatize each of the main theories, such as thermodynamics, probability, quantum mechanics, etc.[9] The general framework where the issue of axiomatization is placed, however, is much broader. It belongs to a very wide methodological and conceptual problem, i.e. the general principle according to which, given a more or less complex problem, it is possible to simplify its formulation and to *reduce* its initial terms. This problem belongs to all sciences, but it affects physics in a specific way, precisely because it is considered *the most fundamental* of the natural sciences.[10]

6.2.1 Features and Achievements

The general approach of reducing to simpler terms a problem, and in particular a problem of knowledge, is wholly intuitive and indeed necessary: in order to deal with an articulated "whole" we start from tackling it and facing its simpler individual parts hoping to be able to provide a step-by-step explanation of the complete system. However, we wonder whether such an intuitive procedure is always correct in physics.

[7] (Hilbert 1900).

[8] (Bottazzini et al. 2016; Bottazzini 2011).

[9] Thanks to the works by C. Caratheodory (1873–1950) for thermodynamics, by Richard E. von Mises (1883–1953) and Andrej N. Kolmogorov (1903–1987) for the theory of probability (see in particular (Kolmogoroff 1933)), by J. von Neumann (1903–1957) for quantum mechanics, by G. Hamel (1877–1954) for mechanics.

[10] Quote 1.2.

This is the problem of *reductionism*: is it always possible to search in a natural phenomenon its "simple" constituents so to explain it as the sum of them?

(6.4) A reductionist believes that a complex system is nothing but the sum of its parts. An account of it can be reduced to accounts of individual constituents. An antireductionist believes that the whole is more than the sum of its parts. There are holistic properties that cannot be described in purely constituent terms.[11]

It is perhaps superfluous to note that elementary particle physicists are generally more inclined towards reductionism, while condensed matter physicists are generally more inclined towards antireductionism.

(6.5) *Réductionisme contre antiréductionisme, ce conflit au cœur de la culture scientifique, né au XIXe siècle, nous divise toujours.*[12]
Reductionism versus anti-reductionism, this conflict at the heart of scientific culture, born in the 19th century, still divides us.

The mathematical property best representing the notion of reductionism is *linearity*. Suppose that we have got a physical system well described by some system of *linear* differential equations such as

$$F\,[\phi] = 0 \tag{6.1}$$

with ϕ is a generic unknown function.[13] We know that the notion of linearity states that, if $\phi_1, \ldots, \phi_i, \ldots \phi_n$, are n functions solutions of Eq. (6.1), then we have

$$\sum_{i=1}^{n} F\,[\phi_i] = F\left[\sum_{i=1}^{n} \phi_i\right] = 0 \tag{6.2a}$$

$$\lambda F[\phi_i] = F[\lambda\phi_i] = 0 \quad \forall i = 1, \ldots, n;\ \lambda \in \mathbb{R}. \tag{6.2b}$$

Intuitively, differential linear equations such as (6.2) specifically represent a condition where the "whole" can be reduced to a "sum of the parts", like $\tilde{\phi} = \sum_{i=1}^{n} \phi_i$. Combining these considerations with the statement of the quote 6.4, we can provide the following definition:

Definition 6.1 (*Reductionism*) Reductionism is the scientific approach where it is believed that any more or less complex system can be explained simply as a pure sum of its more or less fundamental constituents. This attitude applied to physics generally leads to reducing all knowledge to *linear problems*.[14]

On very many physical systems, a reductionist approach works very well. A famous passage by Laplace summarizes an excellent declaration of reductionism and determinism:

---

[11] J. Polkinghorne in Polkinghorne (2002).

[12] I. Prigogine et al., Chap. 3 of Prigogine and Stengers (1986).

[13] We may consider the most general one (vectorial, defined either on $\mathbb{R}^N$ or on $\mathbb{C}^N$, etc.), such as $\underline{\phi} : \mathbb{C}^N \to \mathbb{C}^M$.

[14] Cf. e.g., (Strumia 2021).

(6.6) *Nous devons donc envisager l'état présent de l'Univers comme l'effet de son état antérieur et comme la cause de celui qui va suivre. Une intelligence qui, pour un instant donné, connaîtrait toutes les forces dont la nature est animée, et la situation respective des êtres qui la composent, si d'ailleurs elle était assez vaste pour soumettre ces données á l'Analyse, embrasserait dans la même formule les mouvements des plus grands corps de l'univers et ceux du plus léger atome: rien ne serait incertain pour elle et l'avenir, comme le passé serait présent á ses yeux.*[15]

We must consider the present state of the Universe as the effect of its previous state and the cause of what will follow. If an intelligence which, for a given moment, knew all the forces by which nature is animated, and the respective situation of the beings which compose it, and were also vast enough to analyze these data, she would embrace in the same formula the movements of the largest bodies in the universe and those of the lightest atom: nothing would be uncertain for her and the future, as the past would be present to her eyes.

As a first approximation, the original program of classical mechanics—together with its advanced version of the Lagrangian and Hamiltonian formalism—is reductionist. It analytically and mathematically considers and handles points, which it adds and integrates. The infinitesimal calculus itself has got reductionist components: the symbol $\int$ itself represents nothing more than a sum of several (infinitesimal) components, originated from the letter S. The motion of celestial bodies, the motion of heavy bodies, classical mechanics, the theory of Newtonian fluids, as well as electrostatics are all fields where the reductionist approach provided excellent results. More recently, nuclear and subnuclear physics has revealed the smallest constituents of matter (as known so far) and the interactions between these, leading to the great formalization of the *standard model*.

(6.7) *Die Naturerscheinungen sollen zurückgeführt werden auf Bewegungen von Materien mit unveränderlichen Bewegungskräften, welche nur von den räumlichen Verhältnissen abhängig sind.*[16]

Natural phenomena should be traced back to movements of material particles with invariant kinetic forces, which only depend on their spatial positions.

Philip W. Anderson (1923–2020) attributes the term *intensive* to this kind of physical investigation, since it proceeds from the fundamental laws.[17] Therefore, the approach of investigating matter by looking for its fundamental constituents and their interactions (and aggregations) has led to astonishing progresses. Nevertheless, the reality our senses immediately deals with, which remains the starting point for natural investigation,[18] does not reveal itself as consisting of quarks, leptons, and interaction mediators, but as *complex and articulated* objects. Are these objects just the sum of their basic constituents, or are they something *essentially different*? In the answer to this question, the reductionist and the anti-reductionist visions differ.

---

[15] P.S. Laplace, *De la probabilité* in Laplace (1840).

[16] Hermann von Helmholtz, Chap. 1 of Helmholtz (1847).

[17] (Anderson 1972).

[18] Quote 1.1.

6.2.2 *Limits and Failures*

The problem with the reductionist approach, which is still very popular among physicists, albeit often unconsciously, is that there are several systems which are not *reducible*, i.e. they cannot be explained simply as *the sum of their parts*. Where does this happen? Generally, the reductionist approach cannot be applied in[19]:

1. systems whose constituents have strong interactions with each other, whose mathematical description involves non-negligible *nonlinearities*;
2. systems which have specific *global properties*, i.e. completely undetectable when they are separated into their parts;
3. systems with non-reducible complexity, i.e. whose degree of complexity of their fundamental constituents is the same as that of the system as a whole.

The most important examples of such physical objects are, respectively, *chaotic systems*[20] (including the atmosphere), the *complex systems*[21] (including the living beings), the *systems with self-similarity*[22] (among which the best known and most important ones are objects with fractal symmetry). In physical systems of this kind the reductionist paradigm cannot be applied, except in a very limited way; it is thus necessary to develop conceptual, experimental, and mathematical tools that allow them to be treated with ad hoc techniques.

These approaches have gradually gained more importance within physics over the last half-century. Among others, it is worth mentioning the fundamental work by Philip W. Anderson who, right from the title, marked a turning point in this new anti-reductionist conception of physics: "More is different. Broken symmetry and the nature of the hierarchical structure of science".[23]

> (6.8) The essential idea is that in the so-called $N \to \infty$ limit of large systems (on our own, macroscopic scale) it is not only convenient but essential to realize that matter will undergo mathematically sharp, singular "phase transitions" to states in which the microscopic symmetries, and even the microscopic equations of motion, are in a sense violated.[24]

The motion equations of atoms and particles, classical or quantum, are linear in themselves; when the number of particles becomes very large, the interactions cause *in some sense* to arise new behaviours, new phenomena, completely unpredictable, in which we are mathematically dealing with singular points, thus strongly nonlinear, and in which the notions of symmetry and hierarchy of structures play an essential role. Philip W. Anderson attributes the term *extensive* to this kind of physical investigation: it starts from phenomena, which are tentatively explained in terms of more or less fundamental laws.[25]

---

[19] Cf. e.g., Chap. 10 of Strumia (2017).

[20] Cf. e.g., (Chibbaro et al. 2014).

[21] Cf. e.g., (Ladyman and Wiesner 2020; Thurner et al. 2018).

[22] Cf. e.g., (Strumia 2020).

[23] (Anderson 1972).

[24] Philip W. Anderson in Anderson (1972).

[25] (Anderson 1972).

6.3 Antireductionism: A Broader View

We saw that the reductionist approach is a powerful tool for investigating a complex natural phenomenon by reducing it to simpler elements and to the interactions between them. This procedure can often be carried out, but it represents in fact an approximation.

> (6.9) *Tra le idee qualitative che sono emerse nei nostri tempi sottoline[e]rei l'idea di non linearitá. Per capire la non linearitá occorre ricordare che un problema matematico lineare corrisponde in genere a una situazione [...] in cui c'è proporzionalitá tra cause ed effetti, in cui la somma di diverse cause provoca una somma dei rispettivi effetti. Questo, in genere, succede nella maggior parte dei fenomeni soltanto per tempi abbastanza brevi e per azioni relativamente deboli. Si suol dire che la linearitá dá una descrizione adeguata soltanto in piccolo della maggior parte dei fenomeni. Quando invece si passa allo studio in grande, per esempio si considera l'evoluzione in tempi lunghi di un sistema fisico complesso, vediamo che lo schema lineare non basta più, perché accanto alle conseguenze dirette di una qualsiasi azione che noi possiamo esercitare sul sistema, ci sono una serie di conseguenze indirette che vanno a sommarsi e intrecciarsi in vario modo, per cui anche delle azioni iniziali molto piccole possono dare alla fine delle conseguenze molto grandi, non sempre ad un aumento della intensitá dell'azione corrisponde nel lungo termine un aumento proporzionale del risultato.*
>
> Among the qualitative ideas that have emerged in our time I would emphasize the idea of non-linearity. To understand non-linearity it must be remembered that a linear mathematical problem usually corresponds to a situation [...] where there is a fixed ratio between cause and effect, where the sum of different causes produces a sum of respective effects. This in general happens in the majority of the phenomena only for comparatively short times and for relatively weak interactions. It is usually said that linearity gives an adequate description only on a small scale of the majority of the phenomena. However when you pass to a global large scale, for example if you consider the evolution over a long period of a complex physical system you see that the linear theory is not enough, because besides the direct consequences of whatever action we can exercise on the system there is a series of indirect consequences that add up and intertwine in various ways. For this reason even a small initial action can in the end produce very great consequences and an increase in the intensity of the action does not always correspond in the long term to a proportional increase in the result.[26]

If, as mentioned before, reductionism is well represented by the mathematical concept of *linearity*, the mathematical analysis guarantees that each *sufficiently regular* function in a neighbourhood of a point, i.e. *locally*, can be approximated to a linear function. The same analysis, however, shows us how this approach is only an approximation, a first term: while moving away from the considered point, the approximation becomes less and less accurate and in order to describe the actual behaviour of the function it is necessary to include *nonlinear terms*.

[26] Ennio De Giorgi in De Giorgi (2006a), English translation from De Giorgi (2006b).

6.3.1 Chaos and Complexity

The first physical systems requiring ad hoc tools to be treated, which are different from those employed in the reductionist approach, were chaotic systems, where the *continuous dependence on the initial conditions* ceases. Maxwell yet raises the problem in these terms:

> (6.10) There are certain classes of phenomena [...] in which a small error in the data only introduces a small error in the result. Such are, among others, the larger phenomena of the Solar System, and those in which the more elementary laws in Dynamics contribute the greater part of the result. The course of events in these cases is stable.
>
> There are other classes of phenomena which are more complicated, and in which cases of instability may occur, the number of such cases increasing, in an exceedingly rapid manner, as the number of variables increases. [...] [I]t is to be expected that in phenomena of higher complexity there will be a far greater number of singularities, near which the axiom about like causes producing like effects ceases to be true.[27]

Therefore, to fix the ideas, specularly to the definition 6.1, the definition of an *anti-reductionist* vision can be provided:

Definition 6.2 (*Antireductionism*) Antireductionism is the scientific approach where it is believed that there are systems, more or less complex, which cannot be explained as a pure sum of their more or less fundamental constituents. This attitude applied to physics generally leads to give a large space to *nonlinear problems*.

While with his considerations Maxwell suggests a way for the future, it is Poincaré to indicate the effective presence of features such as extreme sensitivity to initial conditions, the presence of singularities, the emergence of chaoticity in his study of celestial mechanics published at the extreme end of the nineteenth century.[28] Thereby, this represents a crucial turning point in the development of the ideas and methods of physical science. The conceptually decisive aspect is that the systems studied by Poincaré are well described by completely deterministic mathematical relations such as those of classical mechanics (cf. e.g., Eqs. (8.9), (8.10), (8.11)). Therefore, this behaviour does not depend on the fact that the system is described by statistical and/or non-deterministic laws, such as Boltzmann thermodynamic systems, or quantum systems.

In the systems considered by Maxwell and Poincaré, today known as "chaotic systems", small variations of the initial conditions correspond to large variations of the time evolution of the whole system. Thus, if a given system can present two initial conditions, a $t = 0$, $\mathbf{x}(0) = \mathbf{x}_0$ and $\mathbf{x}'(0) = \mathbf{x}'_0$, so they will differ by $\Delta\mathbf{x}(0) = |\mathbf{x}_0 - \mathbf{x}'_0| = \Delta\mathbf{x}_0$. After a time t, if the system has started from the condition $\mathbf{x}_0$, it will be in the condition $\mathbf{x}(t)$, while if it has started from $\mathbf{x}'_0$, it will be in $\mathbf{x}'(t)$, so the difference between them will have become $\Delta\mathbf{x}(t) = |\mathbf{x}(t) - \mathbf{x}'(t)|$. Thanks to Lyapunov's theory of dynamical systems, it is known that the relationship between the two differences is given by[29]:

---

27 J.C. Maxwell in Maxwell (1882).

28 Cf. e.g., Sect. 6 of Chenciner (2015).

29 Chapter 5 of Chibbaro et al. (2014).

$$\Delta \mathbf{x}(t) \approx \Delta \mathbf{x}(0) e^{\lambda t}, \quad \lambda > 0. \tag{6.3}$$

This relationship means that, given two initial conditions that differ by a quantity, even (relatively) small, but finite, the difference between the two configurations will evolve over time, exponentially increasing, with a trend characterized by the constant λ (the *Lyapunov exponent*). This implies that, if the initial conditions of a system are known with a precision within Δx_0 and one intends to predict its evolution with a tolerance of Δx, it will be possible to accomplish this prediction just up to a time tau_p, equal to

$$\tau_p \approx \frac{1}{\lambda} \ln \left(\frac{\Delta x}{\Delta x_0} \right), \tag{6.4}$$

i.e. for times $t > \tau_p$, the incertitude of the prediction will be larger than the requested Δx. These explanations are the basis of the dynamic and chaotic system theory, a field of physics which particularly developed after the Second World War; at the time there was need to develop innovative experimental, mathematical, and conceptual techniques, especially after the fundamental work by Edward Lorenz about meteorology.[30] It is noteworthy that a chaotic dynamic can be found in a large number of physical systems, from meteorology, to celestial bodies, data flows, some social models, living beings, etc.[31]

Compared to dynamic or chaotic systems, a more general concept is that of *complex system*, although it is overlappable to a certain extent. As mentioned in the previous section, a complex system is physical system composed of many strongly interacting parts, so that these interactions cause the presence of non-negligible nonlinearities. As it can be guessed, giving a unequivocal definition of a *complex system* is an open issue.

We here try to provide a strictly "provisional" definition, in the broadest possible sense:

Definition 6.3 (*Complex system (methodological broad definition)*) A complex (physical) system is a system where the *reductionist method* fails to describe its peculiar features and behaviours.

This definition is perhaps too broad; it could be limited with the further conditions that it is made of various components in such a close interaction with each other to produce what are called *emergent behaviours*: specific properties due to the behaviour of the system *as a whole*, whose appearance cannot be predicted a priori starting from the behaviour of the fundamental constituents alone.

To clarify the matter, it is worth to highlight the fundamental elements which can be found in these systems. A *complex system* is always made up of many elements and, in its dynamics, it is described by

---

[30] (Lorenz 1963).

[31] Cf. e.g., (Gell-Mann 1994; Chibbaro et al. 2014; Golan and Harte 2022).

1. *time-changing interactions* among elements;
2. *time-changing states*, which characterize the specific behaviours of the system.

These elements can be formalized as follows:

1. The *interactions* among constituting elements are different and are also of different kind. If, between two generic elements of the system, k and l, there is a specific interaction, which is indicated by indicated ι, we indicate this condition with

$$\mathcal{S}_{kl}^{\iota} = \mathcal{S}_{kl}^{\iota}(t).$$

   These interactions can be mediated through different physical objects (for example, particles of various nature) and they can be either deterministic or stochastic.
2. The *states* of the system are physical quantities that can be scalar, vector, tensor, etc. and they are the result of internal dynamics or external influences; they can also have got a deterministic or stochastic nature. For example, the velocities of a group of planets, the spin of an electron, the internal symmetry of a crystal or of a cluster, etc. can be considered *states* of the system. A state can be indicated in a generic form such as

$$\sigma_k = \sigma_k(t)$$

   highlighting that it evolves as a function of time t.[32]

From the aforementioned ingredients, the evolution of a generic (physical) system will be described by relations such as

$$\frac{d\sigma_l(t)}{dt} \sim \Phi\left[\mathcal{S}_{kl}^{\iota}(t), \sigma_l(t)\right] \tag{6.5a}$$

$$\frac{d\mathcal{S}_{kl}^{\iota}(t)}{dt} \sim \Gamma\left[\mathcal{S}_{kl}^{\mu}(t), \sigma_l(t)\right]. \tag{6.5b}$$

The *first relation*, (6.5a), means that all states of the l element change according to a function Φ, either deterministic or stochastic and usually not-single-valued, which depends both on the $\sigma_l(t)$ current state and on the $\mathcal{S}_{kl}^{\iota}(t)$ mutual interactions. Such expression represents the analytical-reductionist feature of physics: once the form of Φ and the initial conditions $\sigma_l(0) = \sigma_{0l}$ have been specified, we can calculate the evolution of the system and consequently obtain predictions. So, systems describable according to (6.5a) are not complex systems, although they can be *complicated*. It is possible to state that until the discovery of *complexity* physics dealt principally with systems describable with mathematical relations of this kind.

The *second relation* (6.5b), vice versa, describes the evolution of the interactions over time, according to the function Γ, depending on the same elements on which Φ depends, and it can also be deterministic or stochastic. However, here it is highlighted that, contrary to the situations considered in the more traditional (analytic-reductionist) physics, the interactions between the elements of the system

[32] Chapter 1 of Thurner et al. (2018).

evolve over time. Therefore, by combining a situation described by (6.5a) with one described by (6.5b) we obtain the foundations of a mathematical description of a *complex system*.[33]

6.3.2 *Information*

In summary, a non-reductionist approach must be used when dealing with those systems in which the *interaction* of *many* components causes the system as a whole to display phenomena that are not predictable starting just from its elementary constituents. This dynamic is called *emergence* and these phenomena are called *emergent properties*.

Since the concept of *complex system* tends to have a more specific meaning with respect to the characteristics we have described (nonlinearity, chaos, emergency, etc.), in this discussion we prefer to use the expression *Anderson-like systems* as a more general indication. By *Anderson-like systems* we mean both those "large" physical systems, i.e. made up of a sufficiently high number of components such as to display phenomenologies of the kind described in the quote 6.8, and *complex systems* in the strict sense, as described in Sect. 6.3.1. In this sense, an *Anderson-like system* and a system defined in the 6.3 coincide.

In spite of their being more or less complex, the *Anderson-like systems*[34] are *intrinsically different* from their fundamental constituents. In them "the whole" is by no means a "sum of the parts" because "More is different". However, nobody questions that a system is made up of fundamental constituents, reductionistically identified and defined: is there a way to scientifically determine or define what makes a system "different"? What makes a large and/or complex system not reducible to its fundamental constituents alone?

Philosophically, this kind of problem is very ancient and the answer to this questions led to the construction of philosophical systems, among which the most developed one is the Aristotelian-Thomistic *hylemorphic theory*.[35] On the other hand, on a strictly scientific level, it has proved much more difficult to provide a convincing treatment. To address the problem, we can start from the discipline that has got as its object the most complex systems, living beings: in them the key element typical of "large" and "complex" systems emerges.

> (6.11) The laws of biology do depend on the laws of physics and chemistry, but they also depend on a vast amount of additional information about how those accidents turned out.[36]

The key element making an agglomeration something different from its fundamental components is represented by the *information*. The word itself, from "*in-form*",

---

[33] Chapter 1 of Thurner et al. (2018).

[34] (Anderson 1972).

[35] (Strumia 2017).

[36] M. Gell-Mann, Chap. 9 of Gell-Mann (1994).

i.e. "form placed inside", recalls the *hylemorphic theory* of ancient philosophies. According to such a scheme, ontology recognizes two principles: a material *substance* (in Latin *substantia*, i.e. "what lies underneath", in Greek ὕλη, i.e. "material rough, timber") and a *form* (in Latin *forma*, in Greek μορφή).[37]

All this seems intuitive, but in order to make it scientifically based, we need to be able to precisely define *information*, specifying how it can be quantified and measured and how it can be correlated to known physical quantities. To do this, we must refer once again to those "large" and "complex" physical systems (composed of many elements, strongly interacting, strongly nonlinear, etc.) which present the peculiar characteristics we have mentioned. To deal with large and complex systems it is necessary to exploit, in a balanced way, concepts which are

PROBABILISTIC and
COMPUTATIONAL.[38]

Actually, such kind of systems displays a multifaceted set of features rather than a single phenomenon. Therefore, any attempt to measure the related *information* is able to highlight only some aspects of their complexity. For this reason, nowadays it is considered impossible to provide some *general theory of information* capable of measuring complexity in its entirety.[39] To get a global idea, let us evaluate the construction of the two main approaches to the information problem, *classical information theory* and *algorithmic information*—each of them highlighting different peculiarities of the problem.

Historically, systems with evident disorder and chaos are addressed first. Consequently, *probabilistic concepts* are essential tools to be developed.

One of the first problems dealt with concerns the thermodynamic *irreversibility*. Macroscopic systems display this feature, while atoms and molecules, their fundamental constituents which microscopically have got a chaotic behaviour, display reversibility. In his great synthesis Boltzmann introduces the fundamental *probability density function*[40]

$$\varrho = \varrho(\mathbf{r}, \mathbf{v}, t), \tag{6.6}$$

representing the probability of finding a molecule or an atom with a velocity $\mathbf{v}$ in a neighbourhood of $\mathbf{r}$ at a time t. In his *theorem H* Boltzmann demonstrates that for the quantity

$$H(t) = \int \varrho(\mathbf{r}, \mathbf{v}, t) \ln \varrho(\mathbf{r}, \mathbf{v}, t) d\mathbf{r} d\mathbf{v} \tag{6.7}$$

the following inequality holds:

$$\frac{dH(t)}{dt} \leq 0, \tag{6.8}$$

37 Chapter 1 of Strumia (2020).

38 Chapter 1 of Ladyman and Wiesner (2020).

39 Chapter 4 of Lindsay and Margenau (1957).

40 Chapter 4 of Chibbaro et al. (2014).

which shows that a collective and ideally macroscopic quantity, the H function, can identify a specific *time direction* displaying a monotonic behaviour as a function of t. While in classical Newtonian mechanics time plays the role of a mere *parameter*, here it becomes a *fundamental ingredient* of the physical reality, whose behaviour would not be explainable without it.[41] Beyond its own limits, the result of (6.8) is a first and partial attempt to describe specific characteristics of a macroscopic system,[42] which somehow cannot be reduced to the characteristics of its fundamental systems. This agrees with the concept expressed by Anderson in the quote 6.8 according to which in macroscopic systems the microscopic laws are "in a certain sense violated".

Nevertheless, the crucial point of the *theorem H* represents also the decisive point for our discussion: the function $\varrho(\mathbf{x}, \mathbf{v}, t)$, which is a density of probability, *represents a specific physical characteristic of the system.*[43] As reported in quote 8.3, Maxwell has also identified probability not as an accident or a sort of palliative to manage our ignorance of certain natural features, but rather as one of the very foundations of physical discipline. For this reason, the probabilities related to a system are a *specific physical characteristic*: this is found in *specific physical quantities* which, as such, must be measurable.

1. *Classical information theory.* The first *information theory*, called *classical*, starts precisely from probabilistic concepts developed in the 30s and 40s of the twentieth century by Claude Shannon (1916–2001) and Norbert Wiener (1894–1964).The latter clearly recognizes that *information* is a physical quantity in its own right, completely irreducible to the other quantities that previously physics has generally considered as fundamental (such as length, mass, or energy[44,45]). While Shannon's classical information theory is developed to deal with problems involving *signal transmission.*[46] As a matter of fact, the spectra subject to disturbances during communication have got characteristics of chaos that need to be treated within a suitable theory. This theory exploits probabilistic concepts: it considers that a given physical quantity to be measured has got its own probability intrinsically associated with it. Therefore, if the measure depends on a generic (random) variable ξ, which has a probability $p = p(\xi)$ of occurring, the variation of (*classical*) information correlated to the occurrence of the event ξ_i, with $p_i = p(x_i)$, is calculated as

$$\mathcal{I}_i = p(\xi_i) \log \left(\frac{1}{p(\xi_i)} \right) = -p(\xi_i) \log (p(\xi_i)). \tag{6.9}$$

   It is worth pointing out the limitations of the $\mathcal{I}(p)$ function. As it is constructed, it highlights that the occurrence of a *sure* event, for which $p = 1$, brings with it a zero

[41] (Prigogine and Stengers 1986, 1999).

[42] Chapter 4 of Chibbaro et al. (2014).

[43] ibid.

[44] Norbert Wiener, Chap. 5 of Wiener (1948).

[45] Cf. Sect. 3.3.3 and in particular the last paragraph of the section.

[46] Chapter 6 of Thurner et al. (2018), Chap. 2 of Rissanen (2007).

increase of information, $\mathcal{I} = 0$; vice versa, the occurrence of a virtually *impossible* event ($p = 0$) leads to an increase of information $\mathcal{I} = \infty$. Consequently, for a sequence of N events we obtain

$$\mathcal{I}_N = -\sum_{i=1}^{N} p(\xi_i) \log (p(\xi_i)). \tag{6.10}$$

Generalizing this assumption, the theory allows us to obtain the *information* quantity $\mathcal{I}$ correlated with a phenomenon characterized by the above-mentioned ξ and $p(\xi)$, that is,[47]

$$\mathcal{I} = -\int p(\xi) \log (p(\xi)) d\xi. \tag{6.11}$$

Hence, in the various definitions of (6.9) or (6.10) or (6.11), $\mathcal{I}(p)$ is called *classical information* or *Shannon information*. In particular, the (6.11) can be considered the definition of the $\mathcal{I}(p)$ function, which is formally equivalent to the (6.7). This means that the information according to Shannon can be interpreted, up to a multiplicative constant, as *the opposite of entropy*:

$$\mathcal{I} \propto -S, \tag{6.12}$$

so that $\mathcal{I}(p)$ is also called *Shannon entropy*[48] or even *classical information*. Another similarity with thermodynamic entropy is the fact that, according to this theoretical construction, it is necessary to specify the limits of integration in order to calculate the (6.11): thus, it is not possible to calculate absolute information, but a *variation of information* in an interval $\Delta\xi$ of the variable on which $p(\xi)$ depends:

$$\Delta\mathcal{I} = -\int_{\Delta\xi} p(\xi) \log (p(\xi)) d\xi.$$

In the (6.9), (6.10), (6.11), the logarithm base has been deliberately not specified. While mathematically the Napierian logarithm is the preferred choice, in information theory the logarithm is often assumed to have got the base 2 for computational convenience, so as to immediately refer to the binary codes of computers.

Therefore, from (6.10) and (6.11) it is possible to define an actual measurable *physical quantity* and to construct a specific *unit of measurement* of the (classical) information: in the International System this is one of the dimensionless units and

[47] Chapter 3 of Wiener (1948).

[48] In some contexts the term *negentropy* was also used, from *negative entropy*, but this expression can be considered obsolete.

is called *shannon* (Sh),[49] defined as *the information content associated with an event that has probability* $p = 1/2$ *of occurring*.[50]

The classical information theory provides the basis for handling different classes of phenomena, more or less complex: if through the analytical-deductive-explanatory techniques, typical of physics and mathematics,[51] it is not possible to treat these classes, with the classical information theory it is achievable: see, i.e. signal analysis, spectral analysis, protein folding, interconnected systems (networks), nucleic acids (DNA, mRNA...) and their expression, out-of-equilibrium systems, molecular dynamics, etc.[52] Furthermore, since the concepts developed within classical information theory can be naturally extended to strictly quantum systems, they also represent the backbone not only of *quantum information theory*, a research field currently experiencing great growth,[53] but also of other constructions within general frameworks such as *operational probabilistic theory*.[54]

It is also possible to provide a definition of Shannon entropy in a more abstract and general form. An *alphabet* $\mathbb{A}$ is defined as a set of symbols, numerical or symbolic in the strict sense, continuous or discrete, finite or infinite. Symbolic alphabets are discrete, finite, or infinite. Starting from the symbols of the alphabet, a probability $p = p(a), \forall a \in \mathbb{A}$ is assigned. Starting from this, we can start with the construction of (6.9), (6.10), or (6.11) and then define the information related to each set of symbols expressed with $\mathbb{A}$.[55] Exploiting the concept of *alphabet* allows us to expand the theory to many methods of *description* of a given studied physical system. As a matter of fact, from the point of view of knowledge, every natural phenomenon can be described, more or less faithfully and in detail, using some language, which is always expressed through some alphabet.

A similar definition of $\mathcal{I}$ within the classical theory thus suggests the possibility of basing the measurement of information on the concept of *description through an alphabet*. Stressing the *description* problem also raises the question of how to build a *model*, with an appropriate (mathematical?) language and an appropriate alphabet of *Anderson-like systems*.[56]

---

[49] (Rioul 2021).

[50] Consequently, the information content of a single *bit* lies between 0 and 1 Sh. This definition of the shannon is particularly simple when the logarithm of the (6.10) and (6.11) is assumed in base 2. This means that the *bit* is considered a unit of measurement identical to the shannon; however, the difference between the two is that the first one measures the (probabilistic) information *in itself*, while the second measures its implementation on a computer which, in principle, could differ from it.

[51] Cf. Sect. 4.1

[52] (Golan and Harte 2022).

[53] Cf. e.g., (Vedral 2006), and in particular Chap. 5 and following ones.

[54] Cf. e.g., Sect. 1 of D'Ariano and Faggin (2022).

[55] Appendix of Ladyman and Wiesner (2020).

[56] Remember the discussions on the descriptions provided by the logical-mathematical language carried out in Sects. 2.2 and 2.3.

2. *Algorithmic information*. The need to obtain a suitable *descriptive* language highlights the *computational* characteristics linked to complex systems (or rather, *Anderson-like*). The possibilities offered by the development of electronic computers in the second half of the twentieth century make it possible to computationally treat many problems related to systems with a high number of constituents and interactions, previously impossible to address through an analytical-mathematical approach. A similar point of view suggests the introduction of the concept of *algorithmic information*, also called *algorithmic complexity* or *Kolmogorov complexity*,[57] named after Andrej N. Kolmogorov (1903–1987) who proposes it in the 1960s.

The idea starts from considering a given description, called x, of a physical system, expressed in a some language. This description is a sequence of elements of a given alphabet, capable of providing the known details of the system. Therefore, we presume to write a series of instructions y, in a suitable language, for an ideal computer $\mathscr{U}$, such as a *Turing machine*. Following the instructions encoded in the y symbols, this computer is able to produce the detailed description x:

$$\mathscr{U}(y) = x. \tag{6.13}$$

For the sake of simplicity of comparison and implementation on actual calculating machines, the sequences x and y are generally considered as expressed in *binary language*, i.e. in the form of *strings*. An indicative example could be a macroscopic crystalline system: a detailed description of such a system requires providing the three-dimensional coordinates of each of the lattice ions; therefore, the length of x, calculated in the number of symbols used, is at least $\ell(x) \gtrsim 2 \cdot 10^{24}$. On the other hand, it is possible to obtain the detailed description x by compacting it into a series of instructions for $\mathscr{U}$ providing the symmetry and periodicity laws of the lattice. The resulting length $\ell(y)$, thus, whatever the language, the syntax, the alphabet used, is certainly

$$\ell(y) \leq \ell(x).$$

It is therefore clear that a crystalline system, whose description can be compacted into y, has got a relatively uncomplicated structure; on the contrary, a system without internal regularities, such as the distribution of the decimal digits of π which, without the possibility of compacting its description, must be described step by step, it shows a much more "complicated" structure. Such intuitive notions are formalized in the following definition of *Kolmogorov complexity*.

[57] The term *Kolmogorov complexity* is perhaps the most used, but it can be misleading: it is true that this concept quantifies a form of information, but on the other hand it is not universally accepted that a system with a larger quantity of algorithmic information is necessarily more *complex* (see M. Gell-Mann in Part One of Gell-Mann (1994)).

Given a physical or mathematical system $\mathcal{S}$, an ideal calculating machine $\mathcal{U}$ (e.g., a Turing machine) and a binary string of elements x describing $\mathcal{S}$[58]; there is a language that provides y instructions to $\mathcal{U}$. Then, the *algorithmic information content* is defined by

$$K_{\mathcal{U}}(x) := \min_y\{\ell(y) : \mathcal{U}(y) = x\}. \tag{6.14}$$

In other words, it is the *minimum* length of the instruction series through which an ideal machine can produce a detailed description of the system $\mathcal{S}$. Mathematically, it is demonstrated that

$$\mathcal{I} \approx K,$$

i.e. most part of mathematical relationships, mainly inequalities, which hold for the *Shannon entropy* also hold for the *Kolmogorov algorithmic information*, although there are counterexamples due to which it cannot be said that there is strict equivalence.[59] The power of *Kolmogorov information* lies in the generality with which it can formally be defined. On the other hand, it is difficult to provide an *effective* measure of K.

From the definition (6.14), thanks to Charles Bennett (1943), another important concept arises for the measurement of the information contained in systems: the *logical depth*. Given a certain $K_{\mathcal{U}}(x)$, we can see that, in the actual process of passing from y to x, $\mathcal{U}$ takes some time. On the one hand this time is evidently $t = t(y)$, on the other hand it is not necessarily $t \propto \ell(y)$. Indeed, there may be a system encoded by a short y instruction, which nevertheless requires a large number of calculations, implying a considerable time for the actual passage $\mathcal{U}(y) = x$, time not taken into account by the definition (6.14). In that case, behind an *apparently short* statement of y, some form of *complexity* is hidden, which expands into a different dimension: that of *depth*.

Consequently, we define the *logical depth* t_{pl}, at a given level defined by a parameter $\delta \geq 0$, as that time $t = t(y)$ for which

$$t_{\text{pl}} := \min_y\{t(y) : |\ell(y) - K_{\mathcal{U}}(x)| \leq \delta, \mathcal{U}(y) = x\}. \tag{6.15}$$

In other words, this quantity is the (minimum) time taken by a computer machine to produce the detailed description of a system starting from the (minimum) instructions identified through (6.14). Although t_{pl} is extremely difficult to measure in practice, it highlights an aspect of computational information that the Kolmogorov figure of merit does not allow us to grasp.[60] On this matter, in the

---

[58] Chapter 4 of Rissanen (2007).

[59] (Muchnik and Vereshchagin 2006).

[60] Chapter 8 of Gell-Mann (1994), Chap. 4 of Ladyman and Wiesner (2020).

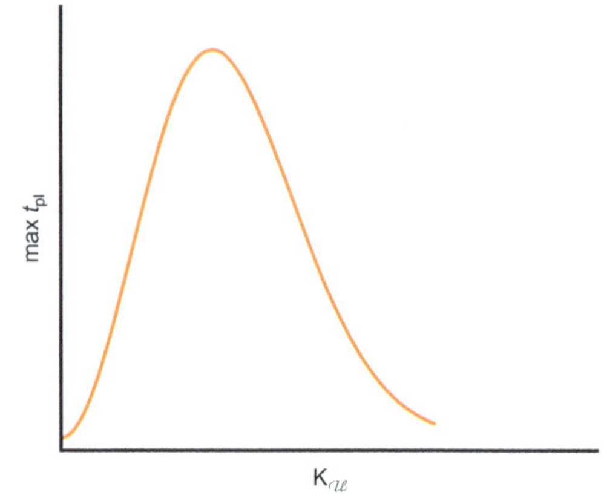

Fig. 6.2 Intuitive plot (according to Gell-Mann (1994)) of the largest possible *logical depth*, max t_{pl} according to (6.15), as a function of the *algorithmic complexity* $K_{\mathcal{U}}$ according to (6.14)

complex systems that can be found (in particular in biological ones), it is worth noting how the algorithmic information and the maximum available logical depth are correlated with each other (Fig. 6.2). Biological systems show a predilection for situations in which it is possible to develop some logical depth, showing a relevant but not particularly high degree of algorithmic information.[61]

The *Shannon entropy*, the *algorithmic information*, and the *logical depth* are only some features of the multifaceted characteristics through which the *information* is displayed. It is remarkable that the three aforementioned concepts are always developed in relation to the technical problems arising in the eras where they are proposed. It is also possible to define other figures of merit: each of them illuminates a further aspect of information and complexity,[62] but none of them, just as no definition of *information* conceived so far, can be considered *exhaustive* of the concept and the phenomena related to it. Therefore, the definitions and theories in this field are to be considered eminently heuristic and operational, until a deeper definition able to grasp the global essence of this quantity[63] is conceived, if it ever will be. On the other hand, for our discussion the most important thing is recognizing that information is a *specific physical quantity* correlated to objective physical systems; it is not a measure subjective to our ability or inability to know, as a paradigm in the strict reductionist sense would assume. Theories are even being developed, inspired by pioneering hypotheses of scholars such as John Wheeler (1911–2008), R. Feynman and David Finkelstein (1929–2016), according to whom the whole physics could be conceived as information processing.[64]

---

[61] Chapter 9 of Gell-Mann (1994).

[62] Chapter 4 of Ladyman and Wiesner (2020).

[63] Chapter 1 of Strumia (2020).

[64] (D'Ariano and Faggin 2022).

(6.12) Information is information, not matter or energy. No materialism which does not admit this can survive at the present day.[65]

Then *information* assumes the same high status of the other physical quantities: it is an element which becomes indispensable even when we limit to provide a description of reality from a strictly materialist perspective.

6.4 Different Levels of Reality

Lastly we draw the conclusions of the so far developed discussion. Just as we need to use different experimental techniques to study different natural phenomena, in the same way considering different (physical) systems requires choosing the most appropriate approach, theory, and discipline. Certainly some elements of one theory or discipline are employed by another and, in the same way, a particularly general discipline, such as physics or chemistry, provides the basis for other disciplines whose investigation fields are more limited, such as geology or medicine. However, it clearly emerges that in physical reality there are classes of systems for which it is necessary to develop own concepts, approaches, theories, and methods; therefore, disciplines (and often even theories within a discipline) maintain their specificity and irreducibility, which make impracticable, or practicable in an extremely limited way, to systematically reduce a particular discipline to a more general one. In other words, the reductionist approach is insufficient and often highly unsuitable to explain several classes of natural phenomena.

Disciplines and theories must often interact with each other. More general disciplines and theories often require the support of less general ones, which they cannot do without. A very important example is the relationship between Quantum Mechanics and Classical Mechanics.

(6.13) Формулировка же основныкх положениий квантовой ме-ханики принципиально невозможна без привлечения механики классической. [...]

Таким образом, квантовая механика занимает очень цвоеобразное положение в пяду физическич теорий - она содержит классическую мекханику как свой предельный сльхай и в тоже время нуждаеця в зтом предельном случае для самого своего обосно-ваня.

However, in principle it is impossible to formulate the basic concepts of quantum mechanics without using classical mechanics. [...]

[65] Norbert Wiener, cap. 5 di (Wiener 1948).

Thus quantum mechanics occupies a very unusual place among physical theories: it contains classical mechanics as a limiting case, yet at the same time it requires this limiting case for its own formulation.[66]

To the fundamental constituents of a physical system, investigated by more general disciplines or theories, additional factors are added, which are more difficult to quantify but no less important or less physically founded. Although not yet exhaustive, a possible interpretation of these factors is to identify them with *information*: this is a peculiar physical quantity, whose features can be identified, treated, and measured.

Hence, if one wanted to follow a strictly reductionist paradigm, for example, reducing chemistry to physics, biology to chemistry, etc., one would need to provide, alongside the fundamental laws and constituents, a quantity of additional information so large as to effectively enter into a new discipline, with new systems, approaches, concepts, and theories. Ultimately we would return to the starting point (cf. e.g., quote 6.11).

(6.14) We can refute the reductionist myth by demonstrating that rules are correct and then challenging very smart people to predict things with them. Their inability to do so is similar to the difficulty the Wizard of Oz has in returning Dorothy to Kansas. He can do in principle, but there are a few pesky technical details to be worked out. [...] The real problem is that Oz is a different universe from Kansas and that getting from one to the other makes no sense. The myth of collective behaviour following from the law is, as practical matter, exactly backward. Law instead follows from collective behaviour [...].[67]

The need to adapt and make disciplines and theories interact with each other, with respect to the considered physical systems, provides an articulated framework, according to which different *levels of reality*[68] can be identified. As a first approximation, these can be classified through the number N of the involved constituents and subsequently through specific parameters.[69] This perspective is well represented by a scheme like the one in Fig. 6.3, which somehow contrasts with the reductionist scheme of Fig. 6.1.

The levels of reality interact with each other in an articulated way, so that it is not always possible to reduce one level to another one and, as in the case of quantum mechanics, sometimes explaining the most fundamental level requires to use concepts belonging to the higher level, although this may perhaps be counterintuitive.

[66] Lev D. Landau & Evgenij M. Lifšic, Chap. 1 of Landau and Lifšic (1989). This is the fourth Russian edition; the English translation of them can be found in Landau and Lifshitz (1965). Sentences also cited by Chibbaro et al. (2014).

[67] Robert Laughlin, Chap. 16 of Laughlin (2005).

[68] (Chibbaro et al. 2014).

[69] Cf. Sect. 8.3 and in particular Eq. (8.12).

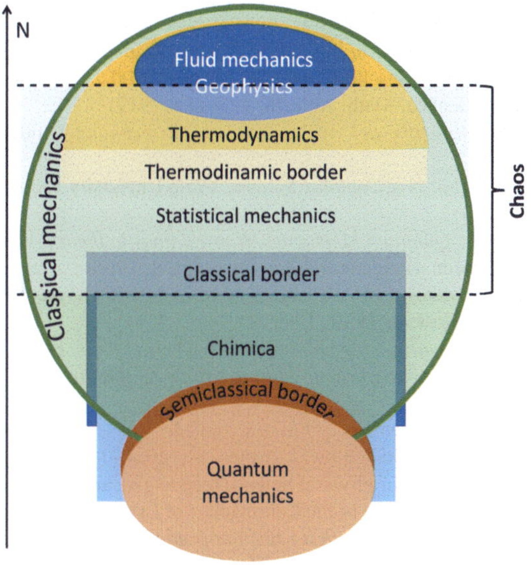

Fig. 6.3 Proposal of a scheme (according to Chibbaro et al. (2014)) with possible interrelations among theories and scientific disciplines and their effective application areas; vertical axis schematically represents the N number of elements of the considered physical system

References

P.W. Anderson, More is different. Science **177**, 393–396 (1972)

U. Bottazzini, Hilbert's problems, in *Mathematical Lives. Protagonists of the Twentieth Century From Hilbert to Wiles*, eds. by C. Bartozzi, R. Betti, A. Guerraggio, R. Lucchetti (Springer-Verlag, 2011)

U. Bottazzini, E.A. Marchisotto, P. Miller, *Hilbert's Flute: The History of Modern Mathematics* (Spriunger-Verlag, Berlin, Deutschland, 2016)

A. Chenciner, Poincaré and the three-body problem, in *Henri Poincaré*. ed. by B. Duplantier, V. Rivasseau. 1912–2012, volume 67 of Progress in Mathematical Physics. (Birkhäuser–Springer, Basel, Switzerland, 2015), pp. 51–150

S. Chibbaro, L. Rondoni, A. Vulpiani, *Reductionism*, Emergence and Levels of Reality, The Importance of Being Borderline (Springer, Cham, Schweiz, 2014)

G.M. D'Ariano, F. Faggin, *Hard problem and free will: An information-theoretical approach, in Artificial Intelligence Versus Natural Intelligence* (Springer, Cham, Schweiz, 2022)

E. De Giorgi, *Lectio magistralis "Il valore sapienziale della matematica" in Ennio De Giorgi Selected Papers* (Springer-Verlag, Berlin, Deutschland, 2006)

E. De Giorgi, *Lectio magistralis: "The sapiential value of mathematics" in Ennio De Giorgi Selected Papers* (Springer-Verlag, Berlin, Deutschland, 2006)

M. Gell-Mann, *The Quark and the Jaguar. Adventures in the Simple and in the Complex* (W.H. Freeman & C., New York, USA, 1994)

A. Golan, J. Harte, Information theory: A foundation for complexity science. Proc. Natl. Acad. Sci. USA **119**, e2119089119 (2022)

H. Helmholtz, *Über die Erhaltung der Kraft-Eine physikalische Abhandlung* (G. Reimer, Berlin, Preußen, 1847)

D. Hilbert, Axiomatic thinking. Philosophia Math. **s1-7**, 1–12 (1970)

D. Hilbert, Mathematische Probleme, Vortrag, gehalten auf dem internationalen Mathematiker-Kongress zu Paris, in *Königliche Gesellschaft der Wissenschaften zu Göttingen (Mathematisch-physikalische*. (Nachrichten, Deutschland, Klasse, 1900), p.1900

D. Hilbert, Axiomatisches denken. Math. Ann. **78**, 405–415 (1917)

A.N. Kolmogoroff, *Grundbegriffe der Wahrscheinlichkeitsrechnung* (Springer-Verlag, Berlin, Deutschland, 1933)

J. Ladyman, K. Wiesner, *What Is a Complex System?* (Yale University Press, New Haven, USA, 2020)

L.D. Landau, E.M. Lifshitz, *Quantum Mechanics. Non-relativistic theory–Volume 3 of Course of Theoretical Physics*, 2nd edn. (Pergamon Press, Oxford, UK, 1965)

L.D. Landau, E.M. Lifšic, *Kvantovaia Mechanika. Nereliativistskaia Teorija–Tom III Teoretičeskaja Fizika*, 2nd edn. (Nauka, Moskva, SSSR, 1989)

P.S. Laplace, *Essai philosophique sur les probabilités* (Bachelier, Paris, France, 1840)

R. Laughlin, *A Different Universe: Reinventing Physics from the Bottom Down* (Basic Books, New York, USA, 2005)

R.B. Lindsay, H. Margenau, *Foundations of Physics*, 2nd edn. (Dover, New York, USA, 1957)

E.N. Lorenz, Deterministic nonperiodic flow. J. Atmos. Sci. **20**, 130–141 (1963)

J.K. Maxwell, I. Does the progress of Physical Science tend to give any advantage to the opinion of Necessity (or Determinism) over that of the Contingency of Events and the Freedom of the Will?, in *The Life of James Clerk Maxwell with a Selection from his Correspondence and Occasional Writings and A Sketch of his Contributions to Science, chapter XIV*. ed. by L. Campbell, W. Garnett (Last Essays at Cambridge. Macmillan & Co., London, UK, 1882)

A. Muchnik, N. Vereshchagin, Shannon entropy vs. kolmogorov complexity, in *CSR 2006*. LNCS 3967. (Springer-Verlag, Berlin, Deutschland, 2006), pp. 281–291

J. Polkinghorne, Reductionism, in *INTERS–Interdisciplinary Encyclopedia of Religion and Science*, eds. by G. Tanzella-Nitti, I. Colagé, A. Strumia (Pontificia Università della Santa Croce, Roma, Italia, 2002). https://doi.org/10.17421/2037-2329-2002-JP-2

I. Prigogine, I. Stengers, *La nouvelle alliance-Metamorphose de la science*, 2nd edn. (Gallimard, Paris, France, 1986)

I. Prigogine, I. Stengers, *La nuova alleanza-Metamorfosi della scienza*, 3rd edn. (Einaudi, Torino, Italia, 1999)

O. Rioul, *This is IT: A Primer on Shannon's Entropy and Information* (Springer International Publishing, Cham, Schweiz, 2021), pp. 49–86

J. Rissanen, *Information and Complexity in Statistical Modeling* (Springer Science Business, New York, USA, 2007)

A. Strumia, Complexity: the role of information in organizing chance. Organisms **5** (2021). https://doi.org/10.13133/2532-5876/17651

A. Strumia, *Dalla filosofia della scienza alla filosofia nella scienza*, 1a edn. (ESC-Edizioni Santa Croce, Roma, Italia, 2017)

A. Strumia, *From Fractals and Cellular Automata to Biology-Information as Order Within Chance* (World Scientific, Singapore, 2020)

S. Thurner, R. Hanel, P. Klimek, *Introduction to the Theory of Complex Systems* (Oxford University Press, Oxford, UK, 2018)

V. Vedral, *Introduction to Quantum Information Science* (Oxford University Press, New York, USA, 2006)

N. Wiener, *Cybernetics or Control and Communication in the Animal and the Machine*, 2nd edn. (The M.I.T. Press, Cambridge, Massachusetts, USA, 1948)

Chapter 7
Looking for Solid Bases: Attempts to Demonstrate *Everything*

There are more things in heaven and earth, Horatio,
Than are dreamt of in your philosophy.

William Shakespeare, Hamlet

The so-called *problem of foundations* (or *foundations crisis*) in mathematics has deep connections with the research on the structure of logical and rational thought. Usually, discussions on these features are not dealt with by physicists, but exclusively by logicians, mathematicians and, sometimes, philosophers. In our opinion, however, the ignorance of this question by natural scientists, physicists as well as other scientists, is a major shortcoming. While mathematics is one of the fundamental pillars of physics, the foundations of mathematics stay among the foundations of physics. As a matter of fact, there is the role played by mathematics, which is the *specific language* of physical science, and, to a certain extent, of the other natural sciences, as discussed in Sect. 6.1. On the other hand, this connection is strongly linked to the problem of the relationship between mathematics and physical reality analysed in Chap. 2. It is therefore possible to find implications also in physics, both at fundamental levels and on application aspects.

7.1 Foundations of Mathematics: The Starting Point

The great process of a scientific knowledge general systematization undertaken during the nineteenth century, which leads to the peak of the reductionist paradigm discussed in Sects. 6.2 and 6.1, reaches its deepest point in dealing with the foundations of what it was considered the *scientia rectrix* of all scientific knowledge (according to the diagram in Fig. 6.1): this is known as the mathematics *foundations problem*. A suitable treatment of this subject would require a larger space than that

J. Parravicini, *The Foundations of Experimental Physics*,
https://doi.org/10.1007/978-3-031-55452-0_7

of our discussion: again we limit to provide the general coordinates in close relation to the overall framework of our analysis.

As stated in Sect. 4.2.3, mainly in the German sphere, after Cauchy nineteenth-century mathematics undertakes the work of founding mathematical analysis, i.e. the study of infinite and infinitesimal processes, on solid exquisitely *arithmetic* bases.[1] The word "arithmetic" comes from the Greek ἀριϑμός, meaning "number"; *arithmeticizing* therefore means tracing concepts, mathematical objects, proofs back to one of the intuitively simplest concepts, that of *natural number*. However, it is required that even arithmetic, in its apparent simplicity, must be founded on formally flawless foundations, following the conceptual line declared by Hilbert in the quote 6.2 and 6.3. A decisive result in this sense is provided by the Italian mathematician Giuseppe Peano (1858–1932), who identifies five fundamental axioms on which it is possible to build all standard arithmetic; from this latter, then, it would be ideally possible to construct all known mathematics.

The five *Peano's axioms* assume the existence of a set $\mathbb{N}$ whose elements are called "(natural) numbers". They are the following ones:[2]

1. **Axiom.** $0 \in \mathbb{N}$.
   0 is a number.
2. **Axiom.** $\exists s : \mathbb{N} \to \mathbb{N}$, i.e. if $n \in \mathbb{N}, m = s(n) \Rightarrow m \in \mathbb{N}$.
   The successor of a number exists and it is a number too.
3. **Axiom.** $\nexists k \in \mathbb{N}$ s.t. $0 = s(k)$.
   0 is not a successor of any number.
4. **Axiom.** $s(n) = k = s(m) \Rightarrow n = m$.
   If two numbers have got the same successor, so they are the same number.
5. **Axiom.** $0 \in \Sigma \wedge (k \in \Sigma \Rightarrow \forall k, k = s(n)) \Rightarrow \forall n \in \Sigma$, that is $\Sigma = \mathbb{N}$
   Induction principle.

Any system built on the aforementioned axioms is called *Peano arithmetic* and is generally denoted by PA. Dealing with arithmetic, an apparently simple and well-established subject with the most advanced mathematical tools raises two fundamental issues:

1. the creation of a specific *theory of numbers*;
2. the role played by fundamental mathematical concepts in the development of logic.

The process of *analysis arithmetification*, as it is called, can be considered to have been essentially accomplished in the last quarter of the nineteenth century. However, the raised issues pose new problems.

On the side of logic, both a method and a symbolism emerge from Peano's works which will prove to be fruitful. In particular, the introduction of symbols such as $\exists$, $\vee$, $\supset$ (see Appendix B), which will later be used by several other logicians, especially Bertrand Russell.

[1] Cf. e.g., Chap. 8 of Wilder (2012).

[2] Wilder (2012), Boyer (1968), Bottazzini et al. (2016).

Once mathematics is reduced to arithmetic, so to the fundamental laws governing the natural numbers, and after defining the first of the numerical sets, $\mathbb{N}$, that is the one satisfying the PA axioms, we reach immediately to the further question of what numbers are.

Do the foundations of mathematics coincide with (natural) numbers? Well, but what are numbers? Georg Cantor (1845–1918) specifically deals this problem.

7.2 A First Response: Set Theories

7.2.1 Naive Set Theory and Antinomies

Cantor recognizes that the act of *counting* lies at the basis of the *natural number* concept. What is counting? It is one-to-one mapping the elements of one set with those of another set. Thus, all sets that can be one-to-one mapped to each other we shall say that they have got the same *number* of elements, in a word they have got the same *cardinality*. A given (natural) *number* $\tilde{n}$ is therefore the *class* of all sets whose elements can be put in *one-to-one correspondence* n times, which is said to have got *cardinality* n. This approach highlights two important aspects:

1. The concept of *set* as key concept of all mathematics.
2. A procedure for *counting* sets with an infinite number of elements, i.e. with infinite cardinality.

Therefore, it would be possible to find the foundations of mathematics in sets, and, following a reasoning such as that by Hilbert,[3] of all scientific knowledge? Undoubtedly a (generic) *collection* of (generic) elements is something more general than a *number*, and yet it also includes numbers.

(7.1) *Unter einer "Menge" verstehen wir jede Zusammenfassung M von bestimmten wohlunterschiedenen Objekten m unsrer Anschauung oder unseres Denkens (welche die "Elemente" von M genannt werden) zu einem Ganzen. In Zeichen drücken wir dies so aus:*[4]

By a "set" we understand any collection into a whole, M, of definite, distinguishable objects m (which are called "elements" of M) of our intuition of thought. In signs we express this thus:

$$M = \{m\}$$

Given this definition,[5] it is necessary to provide a formalization of the thought process referred to by Cantor. This issue seems to receive a satisfactory answer in the

[3] Quote 6.2 and 6.3.

[4] G. Cantor, p. 282 of Cantor (1932).

[5] In German it is used "Menge" to mean "set", this is why the letter M is employed. However, in its daily use, the word has the more generic meaning of "quantity".

so-called *axiom of Comprehension* (not restricted), according to the enunciation by Gottlob Frege (1848–1925).[6]

Axiom of Comprehension[7]
$\forall \mathfrak{P}$ defining a *property*, $\exists M$ s.t. $M = \{m : \mathfrak{P}(m)\}$.
Given any property identified by a proposition, denoted by $\mathfrak{P}$, it is possible to construct a set containing all the elements with that property. This is said to be an *axiom schema*, because it produces a different axiom as $\mathfrak{P}$ varies.

Preparing the definitions of sets according to Cantor and Frege, we obtain what is called the *naive set theory*. Cantor's theory provides important and, for the time, innovative results. Among others, the concept of cardinality allows to "count", and therefore compare, in a formally correct way, infinite sets. If we indicate with $|M|$ the cardinality of a set, Cantor introduces the concept of *countable smallest (infinite) cardinality*

$$|\mathbb{N}| = \aleph_0.$$

In the same theory the *power set* $\mathscr{P}(M)$ is considered, which is defined as the set of all subsets of a given set M. Cantor demonstrates that $|M| < |\mathscr{P}(M)| = 2^{|M|}$, from which the noteworthy results

$$\aleph_0 = |\mathbb{N}| < |\mathscr{P}(\mathbb{N})| = 2^{\aleph_0}$$

and

$$|\mathbb{R}| = \mathfrak{c} = 2^{\aleph_0},$$

i.e. $\mathfrak{c}$ is the *cardinality of continuum*, the set $\mathbb{R}$. Following this line of thought, Cantor formulates the *continuum hypothesis*, stating that

$$\nexists \Psi : \aleph_0 < |\Psi| < \mathfrak{c}. \tag{7.1}$$

The (7.1) is generally indicated with CH: later it would become the first among the 23 *Hilbert's problems*.[8] Naive set theory is therefore able to illuminate several aspects of numerical sets and, consequently, of the foundations of mathematics. The statement of the *Axiom of Comprehension* is also strongly general, making possible to easily consider sets containing elements of any nature. Thanks to it, since any $\mathfrak{P}$ can be in principle assumed as the discriminant of a property shared by elements, it is also possible to construct a direct correspondence between sets and logical propositions. If the object of the logic are the propositions, assuming these as $\mathfrak{P}$ is sufficient to uniquely define a set, according to the aforementioned axiom. This means that set symbolism and logical symbolism tend to overlap (cf. Table B.1).

---

[6] Chapter 2 of Levy (1979), Chap. 1 of Jech (2006).

[7] More precisely, *Axiom Schema of Comprehension (not-restricted)*.

[8] See Sect. 6.1 and (Bottazzini et al. 2016; Hilbert 1900).

Therefore, while set theory plays a specific foundational role in mathematics, the theory of logic is immediately placed next to it, which is developed not only by Frege himself, but also by Alfred N. Whitehead (1861–1947) and Bertrand Russell (1872–1970).[9] The overlap of set theory and mathematical logic highlights many *logical paradoxes* capable of generating *antinomies*.[10] These show how the theory of foundations placed on such bases is by no means a solid ground on which to build the edifice of knowledge. The best known of these is *Russell's antinomy*.[11]

Russell's Antinomy Let $\mathcal{R}$ be the set of *all* sets that are not members of themselves.

$$\mathcal{R} = \{X : X \notin X\} \tag{7.2}$$

It can be easily demonstrated[12] that if $\mathcal{R}$ is a member of itself, then it is not member of itself; if $\mathcal{R}$ is not member of itself, then it is member of itself $\Rightarrow \top$.

Evidently this produces a *contradiction* which, in Frege's words, risks undermining the entire logical-mathematical construction.[13]

What the antinomies highlight is that an interpretation of sets for which the Axiom of Comprehension is always valid is not tenable. In particular, they point out that the correlation between logic and mathematics, made possible precisely through this axiom, is by no means simple and direct, but it requires specific limitations.[14]

7.2.2 Axiomatic Set Theory: Avoiding Antinomies

A first response aimed at overcoming the problem is the construction of an *axiomatic set theory*, thanks to which, through appropriate axiomatic definitions, it is possible, on the one hand, to recover all the previously achieved results, and, on the other habd, to avoid, through appropriate formulations, the arising of antinomies. The best known and most widespread set theory is the *Zermelo–Fraenkel Axiomatic theory*, due to the mathematicians Ernst Zermelo (1871–1953) and Abraham Fraenkel (1891–1965). The objects dealt by the theory are exclusively *sets*: X, Y, Z, etc., which can be elements and/or subsets of other sets. According to the order exposed by Fraenkel himself in Bernays (1958), the axioms are the following:

---

[9] Chapter 9 of Wilder (2012), Chap. 20 of Bottazzini et al. (2016).

[10] See e.g., (Lolli 2011; Jech 2006).

[11] Also called *Russel's paradox*.

[12] Cf. e.g., Chap. 1 of Jech (2006), Chap. 2 of Levy (1979), Chap. 1 of Lolli (2011), etc.

[13] In parallel with Russell, it is possible to find several other antinomies. The most famous ones, in addition to the above, are the *Richard's paradox* and the *Burali-Forti paradox*, the *Grelling paradox*, etc. (Lolli 2011).

[14] Introduction of Part II in Bernays (1958).

1. **Axiom of Extensionality** (given by Frege in 1893).
   $\forall Z, Z \in X \iff Z \in Y \Rightarrow X = Y$.
   Two sets are equal if they have got the same elements.
2. **Axiom of pairing**.
   $\forall X, Y, \exists Z$ s.t. $Z = \{X, Y\}$,
   It is possible to construct a set formed by a couple of elements.
3. **Axiom of Union**.
   $\forall X = \{X_1, X_2, \ldots\}, \exists Y$ s.t. $Y = \bigcup_i X_i$.
   It is possible to construct a set from the union of elements of a set.
4. **Axiom of Power**.[15]
   $\forall X, \exists Y = \mathscr{P}(X)$, with $\mathscr{P}(X)$ the set of all subsets of X.
   For any set it is possible to construct its *power set* $\mathscr{P}$, whose elements are all its subsets.
5. **Axiom of specification**[16] (or *of separation*, *of selection*, *of restricted comprehension*).
   $\forall$ *well-defined* proposition $\mathfrak{P}$ on an element $X \in Z$ (i.e. a *property* of X), then $\mathfrak{P} = \mathfrak{P}(X), \Rightarrow \exists Y$ s.t. $Y = \{U \in Z : \mathfrak{P}(U)\}$.[17]
   Given a set and whatever *well-defined* property on its elements, it is possible to construct a set which contains all the elements with that property, which is a subset of the previous one. This axiom is called *axiom schema* because it produces a different axiom with different $\mathfrak{P}$.

   Axiom of Empty Set From the above-mentioned five axioms it is possible to demonstrate the existence of the *empty set*.[18]
   $\forall X, \exists Y$ s.t. $X \notin Y \Rightarrow Y = \emptyset$.
   There is a set that does not contain any element, and it is indicated with $\emptyset$.
6. **Axiom of choice**.
   $\forall X$ s.t. $\emptyset \notin X$ e $\forall A \subseteq X, \exists f : A \to A$.
   For all non-empty set exists a function f which, for each subset, associates to it one of its elements.
7. **Axiom of Infinity**.
   $\exists W$ s.t. $\emptyset \in W$ and $X \in W \Rightarrow \{X\} \in W$.
   There is a set containing the empty set and for each of its elements, it contains a different one.
8. **Axiom of Replacement**[19] (or *of substitution*).
   $\forall X, \forall \phi : X \Rightarrow Y$ s.t. $Y = \phi(X) = \{\phi(Z) : Z \in X\}$.
   For all function ϕ defined on a set, exists the set of images. This is also a *axiom schema*, because it produces a different axiom for different ϕ.

---

[15] More precisely *Axiom of power set*.

[16] More precisely *Axiom schema*.

[17] We may also write $Y_{\mathfrak{P}} = \{U \in Y | \mathfrak{P}(U)\}$.

[18] Cf. (Bernays 1958).

[19] More precisely *Axiom schema*.

9. **Axiom of Foundation** (or *of Regularity*, *of Restriction*).
   $\forall X \neq \emptyset, \exists Y$ s.t. $Y \in X$ and $\forall Z, Z \in Y \rightarrow Z \notin X$.
   Every non-empty set contains an element disjointed from it.

The Zermelo–Fraenkel theory is denoted by ZF if the *axiom of choice* is omitted (n. 7.2.2, denoted by C), while, if this is maintained, the theory is indicated with ZFC.[20,21] This system has advantages and disadvantages; a few of them are here exposed:

> The formulation of the *axiom of specification* as *axiom of restricted comprehension*, through the expedient of the *subset* formulation, allows us to exclude from the theory "too large" sets, such as that of Russell, as the not-restricted *axiom of comprehension* inevitably led to do. This is a strength, because it avoids most of the antinomies, but it is also a weakness, because it doesn't allow you to deal with some objects, such as, for example, the *universal class* $\bigvee$, i.e. the collection of everything that is present in the universe.[22]
> Another weakness is a certain vagueness about how a "well-defined" $\mathfrak{P}$[23] should be.
> A further limitation lies in the fact that two *axiom schemes* are used instead of simple axioms.

Despite its limitations, ZFC has the advantage of being able to be used as the foundation of almost all standard mathematics, even if it has a compact and relatively simple formulation.

7.2.3 A Deeper Comprehension: Classes, Analogy

Upon closer consideration, an antinomy such as Russell's arises from the fact that the concept of *collection*, on which the concept of *set* is based, can be defined in two different ways. The first way is the one indicated by Cantor in the Definition 7.2.1; it is called *combinatorial collection*. The second one is that Frege's *axiom of not-restricted comprehension*; it is called *logical collection*.[24] We can give the following definition of the first one:

Definition 7.1 (*Combinatorial collection*) A *combinatorial collection*, or *set*, M, is the result of aggregating into a single whole "definite, distinguishable objects of

[20] Chapter 1 of Levy (1979).

[21] The origin of the distinction lies in the fact that C, although having the utmost importance, was the object of bitter disputes among mathematicians: many did not consider its adoption a lawful procedure, due to some technical issues (Bottazzini et al. 2016).

[22] (Bernays 1958; Levy 1979; Jech 2006).

[23] Lolli (2011)

[24] Muller (2001).

our intuition".[25] M is therefore the result of a *combinatorial activity*, as formally codified by the axioms of set theory.

So a *set*, a "*Menge*" according to Cantor (and according to Zermelo, Fraenkel, and von Neumann) is the result of the mind's ability to group together single, distinct objects. Intuitively, a *set* is constructed by grouping distinct objects.

If instead we define a collection through a specific *proposition*, as postulated by the comprehension axiom, we obtain an object that looks like a *menge*, since it is actually a collection of objects, but it is not exactly the same thing, since it is generated by a different mental process. An object obtained in this way is called *class*.

Definition 7.2 (*Logical collection*) A *logical collection*, or *class*, $\mathcal{K}$, is the extension of a proposition, $\mathfrak{Q} = \mathfrak{Q}(X)$, i.e. the integer containing exactly those X objects that satisfy the condition specified by $\mathfrak{Q}$. In symbols $X \in \mathcal{K} \leftrightarrow \mathfrak{Q}(X)$.

Intuitively, a *class*, differently from a set, is constructed by first establishing a *logical criterion*, stated by means of a proposition; subsequently, selecting and grouping the objects which satisfy the established criterion. The construction procedure is therefore very different from that used for a *set*. It is also clear that, according to the above definitions, all sets are classes, but not all classes are sets. Indeed, the difference between *set* and *class* plays a decisive role, since the main root of set antinomies lies in not recognizing it.[26] This distinction is not present in the ZFC, which is limited to avoid objects that may "resemble" a "too big" class such as the $\mathcal{R}$ that the Russell's definition suggests.

The problem of distinguishing between sets and classes is directly faced within an alternative scheme to ZFC, the *von Neumann, Bernays, Gödel axiomatic theory*, generally indicated with NBG. It solves many weaknesses of ZFC (for example, NBG does not need axiom schemes). Furthermore, all theorems and mathematical results demonstrable within the ZFC are also demonstrable in the NBG, so the latter represents a *conservative extension* of the former one. For our discussion, the most important feature is that NBG recognizes the existence of combinatorial collections and logical collections, similar but different from each other.[27]

The NBG theory primarily deals with objects that fall into the category of *class*, in the sense that they are collections of elements. To fix these ideas we write

$\mathfrak{Cls}(X)$
X is a *class* if it is an object satisfying given requirements specified by axioms of NBG (similar to the equivalent of ZFC).[28] Intuitively X is a generic collection of elements[29] which is specified by a proposition, i.e. it satisfies the Definition 7.2.

An *improper class*, or *set*, is defined as

---

[25] From quote 7.1 by Cantor.

[26] Muller (2001).

[27] Muller (2001), Chap. 4 of Mendelson (2015).

[28] Chapter 4 of Mendelson (2015).

[29] Unless it is $\emptyset$.

$\mathfrak{M}(X) \Leftrightarrow \exists Y : X \in Y$
X is a *set*, or *improper class* if there is a *Y* to which *X* belongs.

Finally, a *proper class* is defined as

$\mathfrak{Pr}(X) \Leftrightarrow \neg\mathfrak{M}(X)$
X is a *proper class* if it does not satisfy the previous definition of *improper class*, that is, if *X* does not belong to any class, i.e. $\nexists Y : X \in Y$.

Then, we have that

the *improper classes* coincide with the usual *sets*, i.e. with the objects treated by the ZFC, in the sense that they can contain elements and in turn be elements (of other classes), and they satisfy the Definition 7.1 (i.e. the criteria identified by Cantor in quote 7.1);
the *proper classes*, instead, can contain elements but they cannot be in turn elements.

The distinction between $\mathfrak{M}$ and $\mathfrak{Pr}$ makes NBG more complex than ZFC, which is why it is generally preferred to expose the latter one within treatises on the foundations of mathematics. However, this distinction elegantly solves problems related to Russell's antinomy. The $\mathcal{R}$ object defined by (7.2) is identified by a $\mathfrak{P}$ predicate, so it is a *logical collection*, i.e. a class. In the considered antinomy, the presence of the *self-reference*[30] and the formulation comprising "*all* the sets [of the universe] which...", as constructed, would define $\mathfrak{P}$ as *class own*, for which belonging to something *makes no sense*. There is the remarkable result by von Neumann, who identifies the peculiarity of *proper classes* in their having the same cardinality as the *universal class*[31] $\bigvee$[32]:

$$\mathfrak{Pr}(X) \Leftrightarrow |X| = |\bigvee|.$$

In other words, proper classes are objects so "big" that they cannot be contained in anything, since they can be put in a one-to-one correspondence with all conceivable objects in the universe.

Axiomatic set theories among which ZFC and NBG are the main ones (and ZFC is by far the most widely used) provide powerful tools for developing the vast part of ordinary mathematics[33] and, consequently, they are the basis of the mathematical tools used by physics.

Furthermore, as they are directly connected to mathematical logic, in which they are both a pillar and an essential topic of the discipline,[34] axiomatic set theories allow us to address a large number of both mathematical and logical issues. However, this

[30] See Sect. 7.3, Chap. 7 of Rautenberg (2009), Chap. 11 of Strumia (2017).

[31] Bernays (1958), Levy (1979).

[32] This symbol is introduced by Peano with the word *verum* (Part I of Berto (2009)).

[33] Wilder (2012).

[34] Mendelson (2015).

approach is not the only way to deal with such kind of problems[35]: other theories and other mathematical objects have been proposed for this purpose.[36]

Particularly interesting is the fact that set theory suggests concepts which are close to philosophy and ontology. In fact mathematics, in its search for generality, passes from studying numbers to studying numerical sets, then to sets in general (as generic aggregates of objects) and then to classes (as even more generic aggregates). In short, the various theories developed by mathematics address increasingly broader and more general themes, approaching to knowledge fields that are apparently distant from it.[37] From this point of view, the development of the *proper class* and *improper class* concepts is particularly significant. In classical ontology, which was mentioned in Sects. 1.2 and 4.1, the description of an object can be *equivocal*, *univocal*, and *analogue*.[38] Evidently, scientific knowledge tries to avoid descriptions of the first kind, to focus on those of the second one: an initially unknown object is defined in relation to another known one because it is recognized as sharing the same features. The notion of *analogy*, on the other hand, uses an intermediate process: an initially unknown object is defined in relation to another known one because it is recognized as sharing *some* features.[39] We can see that the definition of classes according to NBG actually uses this category of analogy and surprisingly brings the foundations of mathematics closer to the classical ontology.[40]

7.3 A Broader Response: The Essence of Mathematics

Although the axiomatic set theory, both in the ZFC and the NBG formulations[41] provide a sufficiently large basis from which to deduce the branches of modern mathematics, the problem of having an unassailable foundation for the entire logical-mathematical edifice is still open. In other words, even assuming that axiomatic set theories are able to provide a sufficiently general conceptual basis to deduce *all* the other structures and other results of mathematics, the fundamental question remains open: what is the origin of those same set structures? Or, in even less formal language: from what and of what is mathematics made?

---

[35] See e.g., (Wilder 2012; Lolli 2011; De Giorgi 1997, 1998).

[36] For example, Russell and Whitehead' *theory of types* and the more recent *category theory*.

[37] Basti (2002), Strumia (2020).

[38] Strumia (2017).

[39] Philosophically, we state that "being is given in many ways" or, equivalently, that there is no *univocity of being*.

[40] Strumia (2017).

[41] There are also other axiomatic set theories, but we mention only the two most important ones employed in the construction of modern mathematics.

7.3.1 *Hilbert: The Greatest Attempt at Coherence and Completeness*

When the great debate on the *problem of foundations* reaches its historical apex, it keeps on the horizon the ambitious claim of achieving, one day, a complete and finished structure where to insert all the knowledge of the natural sciences, according to the scheme in Fig. 6.1 and the ideas enunciated by Hilbert (quote 6.2 and quote 6.3). Research in the field is divided into some great streams, among which the main ones give rise to three currents of thought, in strong controversy with each other: the *logicist* approach, the *intuitionist* one, the *formalist* one.

Logicism. Its main exponents are G. Frege, A.N. Whitehead and his pupil, B. Russell. This school finds the foundations of mathematics in *pure logic*. Therefore, according to this logicist conception, those laws of rational thought that have got the name of *logic* are not simply a tool of mathematics, but they are its ultimate origin.

> (7.2) *Es scheint jetzt die Meinung immer mehr Anhänger zu gewinnen, dass die Arithmetik weiter entwickelte Logik ist, dass eine strengere Begründung der arithmetischen Gesetze auf rein logische und nur auf solche zurückführt. Auch ich bin dieser Meinung und gruende darauf die Forderung, dass die arithmetische Zeichensprache zu einer logischen erweitert werden muss.*[42]
>
> Nowadays, the view that arithmetic is a further development of logic as well as the idea that a more rigorous establishment of arithmetical laws reduces them to purely logical laws and to such laws alone are becoming more popular. I also agree with this opinion, and I base upon it the requirement that the symbolic language of arithmetic must be expanded into a logical symbolism.

If mathematics is reducible to arithmetic, mathematical thinking will be ultimately reducible to logic. Following this line of thought, Russel and Whitehad write the ambitious work *Principia Mathematica* (generally indicated with PM), exposed in several volumes published in different years, whose title explicitly refer to the Newton's *opus magnum*.[43] The resulting articulated logical system, still considered a milestone in the field, is generally indicated with the abbreviation PM.

Intuitionism. The main exponent and supporter of this current is the Dutch mathematician Luitzen E.J. Brouwer (1881–1966). The German Hermann Weyl (1885–1955) also plays an important role, adopting intuitionism after being one of the most important pupils of D. Hilbert and supporter of the latter's formalist approach (see below). Brouwer identifies the origins of mathematics exclusively in the construction of human thought: mathematical entities are therefore not discovered, but *constructed*. Therefore, all proofs that do not include an explicit construction process and, in particular, proofs by induction and by contradiction are rejected. Congruently with these assumptions, both the *principle of induction*, assumed by

---

[42] Gottlob Frege in Frege (1891).

[43] *Philosophiæ Naturalis Principia Mathematica*, see Sect. 4.2.2.

Peano to found arithmetic,[44] and the generalized use of the logical principle of the *excluded middle*[45] are rejected in mathematics. Consequently, given a certain mathematical conjecture $\mathfrak{C}$, proving that assuming $\neg\mathfrak{C}$ leads to a contradiction does not imply at all that $\mathfrak{C}$ is true.

It is the generalized use of non-constructive methods where intuitionism finds the source of paradoxes and antinomies. In a perspective of this kind, an object like Russell's "set" $\mathcal{R}$ is not only a not sensible entity, but it is not even a conceivable mathematical object, since there is not a specific defined procedure to build it.

H. Poincaré, who is said to be pre-intuitionist, is also considered close to the intuitionist conception: actually, statements like those in the quote 2.8 suggest a perspective very close to that of Brouwer. Intuitionism proposes an original perspective, for the time, and decisively anti-realist[46] of mathematics. It also undertakes the work of re-demonstrating all mathematical results whose normally accepted proof is of a non-constructive kind (e.g., by contradiction). Since in most cases the obtained proofs confirm the already obtained results, some mathematicians consider the intuitionist perspective a sterile[47] complication.

Formalism. The formalist approach is supported by David Hilbert, along with lots of his pupils, including Johan von Neumann. Hilbert starts from the problem of the antinomies caused by Frege and Cantor' naive set theories: there are different solutions to solve them, but this causes the risk that mathematics moves away from its own rigour.

> (7.3) *Der Heilmittel gegen die Paradoxien warden zu viele empfohlen, die Methoden zur Aufkufklärung waren zu buntscheckig.*[48]
>
> Too many different remedies for the paradoxes were offered, and the methods proposed to clarify them were too variegated.

Directly addressing the issues arisen in the development of the foundational theories, Hilbert believes that the encountered antinomies are caused by treating *infinity* as if it were a mathematical object analogous to finite quantities, i.e. as an *actual infinity*. In order to overcome this problem, and thus to establish mathematics on finally clear and solid foundations, in his fundamental work *Über das Unendliche* (*On infinity*),[49] Hilbert proposes the systematic use of rigorously finitistic methods, i.e. expressing all mathematical entities and mathematical demonstrations in purely *finite* terms. In order to do this, he starts from stating that the too close and, in some way, too simplistic overlap between logic and mathematics (as mentioned at the end of Sect. 7.2.1) represents a serious problem. Consequently, the attempt of founding mathematics on pure logic is stigmatized, because it is pointed out

---

44 See Sect. 7.1.

45 cf. Sect. B.2.1.

46 Or anti-naturalist, see Sect. 2.1 and, e.g., (Sakkurai 2005; Dorato 2005).

47 Bottazzini et al. (2016).

48 David Hilbert in Hilbert (1926).

49 Hilbert (1926).

that mathematical objects do not depend on logic, and that logic itself needs some fundamental concepts on which to lean. According to Hilbert, for example, logical reasoning requires the concept of *sequence*, but this latter is precisely defined only through $\mathbb{N}$, which is a purely mathematical concept.

> (7.4) *Schon Kant hat gelehrt [...] daß die Hathematik über einen unabhängig von aller Logik gesicherten Inhalt verfügt und daher hie mad nimmer allein durch Logik begrindet werden kaun, weshalb auch die Bestrebungen von Frege und Dedekind scheitern mußten. Vielmehr ist als Vorbedingung für die Anwendung logischer Schlüsse und für die Betätigung logischer Operationen schon etwas in der Vorstellung gegeben: gewisse, außer-logische konkrete Objekte, die anschaulich als unmittelbares Erlebnis vor allem Denken da sind.*[50]

> Kant taught [...] that mathematics treats a subject matter which is given independently of logic. Mathematics, therefore, can never be grounded solely on logic. Consequently, Frege's and Dedekind's attempts to ground it were doomed to failure. As a further precondition for using logical deduction and carrying out logical operations, something must be given in conception, i.e., certain extralogical concrete objects which are intuited as directly experienced prior to all thinking.

If it doesn't coincide with logic and yet it has a structure, and a rational structure, then what will mathematics be? Where does it start from? For Hilbert and the formalists, mathematics coincides with its *form* it is the discipline that studies how the very symbols that it uses are configured.

> (7.5) *Und insbesondere in der Mathematik sind Gegenstand unserer Betrachtung die konkreten Zeichen selbst, deren Gestalt unserer Einstellung zufolge unmittelbar deutlieh und wiedererkennbar ist.*[51]

> The subject matter of mathematics is the concrete symbols themselves whose structure is immediately clear and recognizable.

Hilbert himself gives an illuminating example of this approach.

> (7.6) *[U]nd diese* $[a + b = b + a]$ *ist auch gar nicht mehr eine unmittelbare Mitteilung yon etwas Inhaltliehem, sondern ein gewisses formales Gebilde, dessen Verhältnis zu den alten finiten Aussagen*
>
> $$2 + 3 = 3 + 2,$$
> $$5 + 7 = 7 + 5$$
>
> *darin besteht, daß wir in jener Formel für a, b Zahlzeichen 2, 3, 5, 7 einsetzen und dadurch, d. h. durch ein Beweisverfahren–wenn auch ein sehr einfaehes- diese finiten Einzelaussagen gewinnen. So kommen wir zu der Auffassung, daß* a, b, $=$, $+$, *sowie die Gesamtformel*
>
> $$a + b = b + a$$
>
> *an sich niehts bedeuten, so wenig wie die Zahlzeiehen; wohl abet können aus ihr Formeln abgeleitet werden, denen wit eine Bedeutung zuschreiben, und zwar, indem wit sie als Mitteflmgen yon finiten Aussagen auffassen.*[52]

---

[50] ibid.

[51] ibid.

[52] ibid.

[A]nd this latter formula [$a + b = b + a$] is not an immediate communication of something meaningful but is rather a certain formal structure whose relation to the old finitary statements

$$2 + 3 = 3 + 2,$$
$$5 + 7 = 7 + 5$$

consists in the fact that, when a and b are replaced in the formula by the numerical symbols 2, 3, 5, 7, the individual finitary statements are thus obtained, i.e., by a proof procedure, albeit a very simple one. Therefore, we conclude that $a, b, =, +$, as well as the whole formula $a + b = b + a$ mean nothing in themselves, no more than the numerical symbols meant anything.

Then, mathematics has no meaning in itself, except the expression of its own form. It is therefore in its form that mathematics must seek and justify its foundations. According to this conception, all constructions and demonstrations are a pure manipulation of symbols, whose meaning can be, in some way, established *a posteriori* only when a specific interpretation is chosen for them.

(7.7) *Es ist also nötig, die logisehen Operationen und auch die mathematischen Beweise selbst zu formalisieren; dies erfordert eine Umsetzung der logischen Beziehungen in Formeln, so daß wir zu den mathematisehen Zeiehen noch logische Zeichen, etwa*

$$\begin{array}{cccc} \&, & \vee, & \rightarrow, & - \\ \textit{und} & \textit{oder} & \textit{folgt} & \textit{nicht} \end{array}$$

hinzufügen und außer den mathematischen Variablen a, b, c, ...auch logische Variable, nämlich variable Aussagen A, B, C, ...benutzen müssen.[53]

It is therefore necessary to formalize the logical operations and also the mathematical proofs themselves; this requires the logical relationships to be converted into formulas, so that we add logical signs to the mathematical signs, for example

$$\begin{array}{cccc} \&, & \vee, & \rightarrow, & \neg \\ \textit{and} & \textit{or} & \textit{implies} & \textit{not} \end{array}$$

add and besides the mathematical variables a, b, c, ...also logical variables, namely variable statements A, B, C, ...must be used.[54]

In order to achieve a correct enunciation of the mathematics foundations, Hilbert's formalism therefore requires that a theory be developed about mathematical reasoning. Hilbert names this new investigation field METAMATHEMATICS and he begins to develop the project of a broad *Beweistheorie*,[55] i.e. the *theory of proof.*

These approaches, which were introduced in specific historical circumstances, still retain their validity as general categories: as a matter of fact, even today anyone who deals with the foundations of mathematics has a tendency to be more or less close to one of them.

[53] ibid.

[54] In our translation we substituted symbol "-" used by Hilbert with the symbol "¬" used today.

[55] In German "Beweis" means "proof".

Among these three schools, the one that had the greatest development at the time was precisely the formalist one, because of different reasons. Indeed this approach gives rise to very fruitful research lines, which arouse great interest even among mathematicians of non-formalist schools. More specifically, on the one hand, Hilbert reiterates the specificity of mathematical thought with respect to logical thought; on the other hand, it is recognized that it is necessary to rationally investigate not only mathematical entities, but also the specific procedures of mathematical science. Underlying all this, there remains the large project of providing a general framework within arranging the whole edifice of knowledge, an extremely attractive ideal for a large number of scholars, especially at the dawn of the twentieth century, because

> (7.8) *Und wo soll sonst Sicherheit und Wahrheit zu finden sein, wenn sogar das mathematische Denken versagt?*[56]
>
> Where are we to find truth and certitude, if mathematical thinking is defective?

The starting point of the formal systematization is the *axiomatic method*, which is recognized as the guarantee of precision and correctness of modern mathematics, lying at the foundation of all demonstrations.[57] For formalists, mathematical truths coincide with the results provided by demonstrations: for this reason they are *conditional truths*, i.e. their validity depends on the assumptions upstream of the reasoning. Since the mathematical edifice is based on axioms, which for formalists are arbitrarily assumed, within this perspective mathematics is unable to provide anything other than conditional truths.

If the whole mathematical edifice is built on systems of axioms through *demonstrations*, a solid foundation of this edifice will therefore require

1. The BEWEISTHEORIE: the development of a specific meta-mathematical theory that secures the whole demonstration process, i.e. the structure of the mathematical building

   > (7.9) *Fruchtbaren Begriffsbildungen und Schlußweisen wollen wit, [...] sorgfältig nachspüren und sie pflegen, stützen und gebrauehsfähig machen. Aus dem Paradies, das Cantor uns geschaffen, soll uns niemand vertreiben können.*[58]
   >
   > We will carefully investigate fruitful definitions and deductive methods [...], we will nurse them, strengthen them, and make them useful. No one shall drive us out of the paradise that Cantor has created for us.

2. The COMPLETENESS, i.e. the securing of the objectives. Given *any* mathematical statement, i.e. a statement that be *formally* expressible through the symbolic language proper to mathematics, and to metamathematics, it should be possible to demonstrate whether it is true or false using exclusively those axioms identified as fundamental by means of the correct processes and rules of demonstration and inference.

---

[56] D. Hilbert in Hilbert (1926).

[57] Chapter 1 of Wilder (2012), Hilbert (1917).

[58] D. Hilbert in Hilbert (1926).

In other words, mathematics is complete if, given an *axiomatic system* $\tilde{\Sigma}$, which is assumed to be true, about *all* the conceivable statements in the proper form of the discipline (the mathematical language), it is possible to affirm by means of rigorous demonstrative steps whether they are true or false. This makes mathematics, and consequently the other branches of knowledge (Fig. 6.1), a *virtually accomplished* (complete) system.

3. The COHERENCE of the axioms and of the statements derived from them, i.e. securing *foundations*. It is therefore required that, not only the fundamental axioms of mathematics are rigorously *non-contradictory*, but in particular *all* the propositions deriving from such axioms by means of the usual demonstrative processes (systematized thanks to the Beweistheorie).
In other words, mathematics is coherent if the *axiomatic system* $\tilde{\Sigma}$ (on which it is based), a system made up by non-contradictory axioms, is able to provide, through rigorous demonstrative passages, only and exclusively non-contradictory statements. This makes mathematics, and consequently the other branches of knowledge (Fig. 6.1), a virtually *immune from contradictions* system.

Hence it is clear that the three above-listed elements have got a large importance beyond the assumed foundational perspective. As a matter of fact, even if a logicist or an intuitionist perspective were assumed, the systematization of the mathematical edifice arising from the completion of a *Beweistheorie* (that allows for the construction of an axiomatically founded complete and coherent system[59]) would endow mathematics with the maximum conceivable rigour of the human mind, crowning it as queen of all the disciplines, because, for any problem in its field, the existence of a solution would be demonstrated.

(7.10) *[J]edes mathematische Problem einer Lösung fähig ist. Wit sind alle davon überzeugt. Es bildet ja gerade einen Hauptrez bei der Beschäftigung mit einem mathematischen Problem, daß wir in uns den steten Zuruf hören: da ist das Problem, suche die Lösung; du kannst sie dutch reines Denken finden; denn in der Mathematik gibt es kein Ignorabimus. Nun kann zwar meine Beweistheorie nicht allgemein einen Weg angeben, auf dem jedes mathematische Problem sich lösen läßt–einen solchen gibt es auch nicht; aber der Nachweis, daß die Annahme der Lösbarkeit eines jeden mathematischen Problems widersprtmhsfrei ist, fällt durchaus in den Bereich unserer Theorie.*[60]

[E]very mathematical problem is solvable. We are all convinced that it really is so. As a matter of fact one of the principal attractions of tackling a mathematical problem is that we always hear this cry within us: there is the problem, find the answer; you can find it just by thinking, for there is no *ignorabimus* in mathematics. Now my theory of proof cannot supply a general method for solving every mathematical problem–there just is no such a method. Still the proof (that the assumption that every mathematical problem is solvable is a consistent assumption) falls completely within the scope of our theory.

Such a vision of mathematics is very close to the *verificationist* perspective on the experimental sciences, which was discussed in Sect. 5.2.1. We also note that the concept of *finite Weise* (*finitistic method*) is here taken to the extreme: the "finite"

[59] See Sect. 2.2.3.

[60] D. Hilbert in Hilbert (1926).

does not only refer to single objects and methods, but to the entire mathematical discipline. It is not understood only as "not infinite", but as (at least potentially) "wholly complete". If mathematics were not complete and coherent, if there were unsolvable or undecidable problems, this would mean that the building of mathematics cannot be completed, i.e. it cannot be *finished* (*finite*). Consequently, the entire building of human knowledge would always be intrinsically incomplete (see quote 7.8), but Hilbert shares with almost all the mathematicians and scientists of his generation the unflinching conviction that[61] WIR WERDEN WISSEN!

7.3.2 Gödel: The Inherent Incompleteness of Mathematics

We now arrive to discuss one of the most important intellectual achievements of all time. It not only involves mathematics and logic, but it has got repercussions on all human knowledge[62]: Gödel's *incompleteness theorems*.[63]

Much has been written about these theorems. Unfortunately, the great technicality in their demonstrations,[64] the conceptual finesse of their assumptions and their implications has made their correct interpretation very difficult. In particular, there is a widespread idea that Gödel's results demonstrate an extreme version of epistemological relativism, according to which "we cannot be certain of anything, not even mathematics". Nothing further from the Gödel's thought, who, on the contrary, starts from the search for a deeper ideal of (mathematical) truth, which could be, in some way, *absolute*.[65]

In Hilbert's time, logicists, formalists, and intuititionists share the idea that, sooner or later, a complete and coherent formulation of mathematics will be found. Furthermore, on the specific problem of *truth*, almost everyone shares the belief that mathematics is only able to prove *conditional truths*. Logical-mathematical methods are capable of demonstrating true propositions, *provided that* the starting axioms, which are assumed arbitrarily within the non-contradiction, are true. Such a conception of *mathematical truth*, however, is not shared by Kurt Gödel (1906–1978), who believes that it is necessary to seek a mathematical truth *in an absolute sense*.

> (7.11) The metamathematical results I have in mind are all centered around, or, one may even say, are only different aspects of one basic fact, which might be called the incompletability or inexhaustibility of mathematics. This fact is encountered in its simplest form when the axiomatic method is applied, not to some hypothetico-deductive system such as geometry (where the mathematician can assert only the conditional truth of the theorems), but to

[61] P. 120.

[62] Part II of Berto (2009).

[63] In many contexts it is referred to as *Gödel's Incompleteness Theorem*. Indeed, Gödel's proofs must be interpreted in a unitary way, as a single result; however, *two* theorems are involved.

[64] In its original form, reprinted in Gödel (1986), Gödel's work consists of about 22 pages of demonstration, to which is added a 3-page introduction containing a summary of the premises, the procedure and the conclusions.

[65] Kurt Gödel in Gödel (1995a).

mathematics proper, that is, to the body of those mathematical propositions which hold in an absolute sense, without any further hypothesis. There must exist propositions of this kind, because otherwise there could not exist any hypothetical theorems either.[66]

Starting from a certain *axiomatic system*, a logical-mathematical *formal system* is constructed by specifying, in addition to the axioms, a set of symbols of the system, a set of construction rules of correct formulas (a grammar), a set of *inference rules* to validly pass from one formula to another one, etc.[67] Having a well-defined *formal system* allows us to have got a sufficiently rigorous demonstration theory.

Each formal system can be subdivided into *syntax* and *semantics*.[68] The *syntax* deals with establishing the correctness of the procedures *independently* from the truth of the premises. The *truth*, on the other hand, necessarily depends on the *meaning* of the considered objects. This is what *semantics* deals with, necessarily posing the problem of *meaning*, making use of the notion of *truth*. It is worth noting how the distinction between *syntax* and *semantics* today also emerges in physics, and it is often related to information theory.[69]

(7.12) *[La] fisica può solo garantire che il prossimo libro che sarà scritto obbedirà alle leggi dei simboli, però non potrà mai prevedere il contenuto semantico di un libro che non è ancora stato scritto. Ma io penso che la semantica sia la vera ontologia. Quindi la semantica è libera, anche se la sintassi è deterministica.*[70]

Physics can only guarantee that the next book to be written will obey the laws of symbols, but it can never predict the semantic content of a book that has not yet been written. But I think semantics is the true ontology. So the semantics is free, even if the syntax is deterministic.

Regarding the purely abstract articulation of mathematics, as we saw in the previous section, Hilbert's program explicitly aims at constructing a coherent and complete system *regardless of meaning* and, therefore, of any "truth" that is not strictly *conditional*, i.e. dependent on an arbitrary assumption of the starting axiomatic system.

On the contrary, in the quote 7.11 Gödel states that the belief from which he starts is that there exist (logical-mathematical) truths that do not depend on systems of hypotheses and axioms. From this we deduce that no choice of any system of axioms would be sufficient to deduce *all* mathematical propositions that are *true*. In other words, *truth does not coincide with demonstrability*. In a cultural and scientific environment permeated with neo-positivism and the first signs of relativism, this hypothesis is viewed with extreme suspicion.[71] Therefore Gödel undertakes his work

[66] Kurt Gödel in Gödel (1995a).

[67] See e.g., (Mendelson 2015; Enderton 2001; Copi et al. 2019).

[68] Cf. e.g., Chap. 2 of Enderton (2001), Chap. 8 of Copi et al. (2019), etc.

[69] According to these theories, traditional physics, based on usual mathematics, is able to explain only relational, i.e. syntactic, aspects of phenomenal reality; on the other hand, the development of a completely new specific mathematics and physical theory is hoped for in order to explain semantics, i.e. meaning, an aspect inextricably linked to psychic activity and considered *non-reducible* (Faggin 2022; D'Ariano and Faggin 2022).

[70] Federico Faggin in Prefazione of Faggin (2022).

[71] Chapter 9 of Berto (2009).

by assuming a position in open contrast to the prevailing opinion among scholars of the time.

The *incompleteness theorems* are grafted onto the three main logical-mathematical issues we have mentioned so far.

1. the *foundation problem* of mathematics;
2. the *Beweistherorie* and correlated questions about a linguistic-symbolic formalization of the formal structure of mathematics and logic;
3. the *antinomies* problem.

About antinomies, Gödel realizes that the key point is the *problem of self-reference*. In a formulation like the one in (7.2), what really triggers the contradiction is the presence of a *self-reference*.[72] This kind of problem is not at all recent, going back to the famous *liar paradox* of ancient Greek philosophy. It can be stated in roughly: "If the Cretan Epimenides claims that all Cretans lie, is this statement true?" Evidently, it is not possible to answer the question based only on the information provided by the sentence itself.

Such a paradox cannot be solved *within* the logical system where it is formulated, i.e. the truth of the considered statement cannot be deduced through pure (formal) logical steps of the system in which the proposition is formulated. Hilbert's formalist program has the declared goal of *founding* mathematical certainties on *exclusively syntactic* criteria, completely omitting the meaning, i.e. semantics (quote 7.6). Gödel's two theorems exploit the results of the *Beweistheorie*, i.e. the metamathematical syntactical methods typical of Hilbert's formalism. Nonetheless, they demonstrate that the program of demonstrating completeness and consistency of an axiomatic system cannot be accomplished by exclusively exploiting the axioms themselves.

> (7.13) The first of these theorems simply states that, whatever well-defined system of axioms and rules of inference may be chosen, there always exist diophantine problems of the type described which are undecidable by these axioms and rules, provided only that no false propositions of this type are derivable. [...]
>
> The second theorem [...] says that for any well-defined system of axioms and rules, in particular, the proposition stating their consistency [...] is undemonstrable from these axioms and rules, provided these axioms and rules are consistent and suffice to derive a certain portion [...] of the finitistic arithmetic of integers.[73,74]

As already mentioned, the demonstration of these results is particularly long and complex and it largely goes beyond the scope of the present discussion. However, it is worth providing the general idea. Gödel uses a formal system in its simplest form, that of *first order theory*. In this condition, any proposition of the system, such as a theorem, is represented by a formula $\tilde{\varphi}$ which coincides with a sequence of symbols,

---

[72] See e.g., Chap. 7 of Rautenberg (2009).

[73] That is PA axioms or equivalent ones, or analogous extended systems, either mathematical or logical, such as *Principia Mathematica* PM by Whitehead and Russel (Ed.).

[74] K. Gödel, ibid.

constructed according to the above criteria. A *proof* gives a (finite) sequence of m formulas[75]

$$\varphi_1, \varphi_2, \ldots \varphi_m \text{ s.t. } \varphi_m = \tilde{\varphi} \tag{7.5}$$

whose succession in turn obeys the rules assumed in the system.

The basic idea is to build up a mathematical statement, i.e. (from a logical-formal point of view) a proposition $\mathfrak{G}$, formally coinciding with an expression $\varphi_{\mathfrak{G}}$, for which it is not possible to demonstrate neither $\varphi_{\mathfrak{G}}$ nor $\neg\varphi_{\mathfrak{G}}$ (hence, neither $\mathfrak{G}$ nor $\neg\mathfrak{G}$). In other words, given the logical-formal premises, i.e. assuming a certain *formal system*, $\mathfrak{G}$ is an *undecidable* statement. To do this, Gödel translates into "mathematical language" a sentence with an internal self-reference. Gödel ingeniously exploits the properties of natural numbers.

The metamathematical theory, which is made explicit in a *formal system*, makes it possible to translate any mathematical proof into a *finite sequence of symbols*. Therefore it is ideally possible to associate a natural number to each symbol and, thus, also to a generic formula ϕ. To avoid ambiguity, it is necessary to ensure that $\forall\phi$ well-formed of the system is *uniquely* associated with one and only one natural number, and vice versa. For this purpose Gödel exploits the *Fundamental Theorem of Arithmetic*, whose thesis is stated.

Theorem 7.1 (Fundamental of Arithmetic) $\forall n > 1, n \in \mathbb{N}$, *and* $\{p_k\} = \mathbb{P}$ *are prime numbers s.t.* $p_1 < p_2 < \ldots$, *then* $\exists!\{n_k\}$, *with* $n_k \in \mathbb{N}$ *s.t.*

$$n = \prod_{k=1}^{\infty} p_k^{n_k}. \tag{7.6}$$

Moreover, $\exists m$ *s.t.* $n_k = 0, \forall k > m$, *so that the (7.6) can be written as*

$$n = \prod_{k=1}^{m} p_k^{n_k}.$$

In other words, each natural number $n > 1$ *is identified by a unique factorization in prime numbers; such factorization is uniquely determined by the sequence* $\{n_k\}$, *which has got always zero terms beyond the* m*−th term* n_m.

First of all, each symbol of the formal system is associated with a natural number, for example,[76] "$\neg$" is associated with 1, "$\vee$" with 2, "$\supset$" with 3, "$\exists$ " with 4, "$=$" with 5, "0" with 6, etc.; we obtain that $\forall\varphi_\kappa$ is associated with a finite sequence of κ

[75] Of course, φ_1 either coincides with one (or more) axioms of the system, or is in turn the thesis of a previously demonstrated theorem.

[76] Example of coding taken from Chap. 8 of Brewer (2022).

(natural) numbers $n_1, n_2, \dots n_\kappa$. So, the first κ prime numbers are taken and, through some specific rule,[77] we construct what is called *gödelian*[78,79]:

Definition 7.3 (*Gödel number*) Given a formula φ_κ, its *Gödel number* $\mathcal{G}_\kappa$ is defined by

$$\mathcal{G}_\kappa = p_1^{n_1} p_2^{n_2} \dots = \prod_{i=1}^{\kappa} p_i^{n_i}, \quad \forall i \in \mathbb{N},\, p_i \in \mathbb{P}. \tag{7.7}$$

The Fundamental Theorem of Arithmetic 7.1 assures the biuniqueness between Gödel numbers[80] and expressions of the formal system:

$$\forall \varphi_\alpha \exists! \mathcal{G}_\alpha, \ \forall \mathcal{G}_\alpha \exists! \varphi_\alpha. \tag{7.8}$$

By means of this construction, it is possible to transport all metamathematical constructions, including demonstrations, *into* the mathematical system itself and, in particular, the part of mathematics which is defined with the greatest clarity and stability, i.e. the arithmetic of natural numbers! Since proofs can be seen as *a discourse on mathematics* (metamathematics), this implies, through the Gödel's construction, that the proper language of mathematics (the natural numbers) somehow becomes capable of *talk about* himself[81]!

At this point we have got the fundamental tools to prove the *First Incompleteness Theorem*. We consider *all* the expressible mathematical propositions, depending on a variable x, $\{\varphi_i(x)\}$. They are at most a countable quantity and among them all the proofs are included. Following Gödel, they can be arranged as an ordered sequence.[82]

$$\varphi_1(x), \varphi_2(x), \dots, \varphi_n(x), \dots \tag{7.9}$$

Among them, there are also propositions which state that a given formula $\varphi_n(n)$ *is not demonstrable*, which is

$$\varphi_k(n) = \neg \mathrm{Dem}(\varphi_n(n)), \tag{7.10}$$

where $\mathrm{Dem}(\phi)$ means "there is a proof for the formula ϕ".[83] By associating to each of the (7.9) a gödelian by means of the (7.7), we define the set $\mathbb{G} \subseteq \mathbb{N}$ of all Gödel numbers associated with them; among them, there are the numbers associated with

---

[77] Actually Gödel assumes different rules depending on the kind of variables and propositions to be codified (Chap. 8 of Brewer (2022)), but in this informal summary it is irrelevant.

[78] Or also *Gödel number*.

[79] Chapter 6 of Rautenberg (2009).

[80] Of course, this is verified unless there are isomorphisms and if a specific coding that associates a natural number with a symbol is fixed.

[81] Berto (2009).

[82] Gödel (1995b).

[83] Or also "ϕ is a provable formula".

the expression given by (7.10). With appropriate steps, we therefore arrive at the following theorem.[84]

Theorem 7.2 (First Incompleteness Theorem or Gödel's Lemma) *Given a* formal system *where we have got the expressions (7.9), from*

$$\varphi_k(n) = \neg Dem(\varphi_n(n)) \tag{7.11}$$

it is possible to define the class K of natural numbers n s.t.

$$n \in K \equiv \varphi_k(n) = \neg Dem(\varphi_n(n)), \tag{7.12}$$

and, in particular, $K \subset \mathbb{G}$.
In other words, given an expression $\varphi_n(n)$*, it is possible to construct another one* $\varphi_k(n)$ *stating that the first one is not provable, and then to construct a natural number (the* Gödel number*) which uniquely identifies it.*

Here we introduce self-reference. We need to find an expression $\varphi_{\mathfrak{G}}$ saying about itself $\mathfrak{G}$, i.e. "I am not demonstrable". To do this, we construct a *substitution* through a suitable procedure, whereby the expression $\varphi_n(n)$ is replaced by the expression denying its demonstrability.

Theorem 7.3 (Second Incompleteness Theorem or Gödel's Theorem) *A suitable substitution exists s.t. the (7.11) becomes*

$$\varphi_k(k) = \neg Dem(\varphi_k(k)); \tag{7.13}$$

the $\varphi_k(k)$ *is called* gödelian statement $\varphi_{\mathfrak{G}}$. *As a corollary, we can note that the previous* Lemma *(Theorem (7.2)) guarantees that* $\varphi_{\mathfrak{G}}$ *can be associated with a gödelian* $k \in K \subset \mathbb{G}$ *which makes it fit into the system of arithmetic.*
In other words, it is possible to formally construct a mathematical expression which, referring to itself, claims to be unprovable.

From a technical point of view, we note that the two incompleteness theorems need the Fundamental Theorem of Arithmetic; therefore, they apply *exclusively* to the formal systems which are broad enough to allow us to develop an arithmetic where the theorem holds. In particular, theorems (7.2) and (7.3) apply in all axiomatic systems where PA (i.e. traditional arithmetic), or equivalent ones, is assumed; among them, the ZFC system, together with the NBG system, and also the PM logical system by Russel and Whitehead.[85] Since current mathematics always include the PA arithmetic, or a formally equivalent formulation, this implies that all systematizations of it fall under the "axe" of incompleteness theorems. If someone were to propose the expedient of incorporating $\mathfrak{G}$ among the fundamental axioms of the system, postulating its truth, he would not escape incompleteness: as a matter of fact, even

[84] We report it in a formulation very similar to the one described by Gödel himself in Gödel (1995b).
[85] Cf. e.g., (Gödel 1986; Berto 2009; Rautenberg 2009; Mendelson 2015; Lolli 2011).

this further formal system would be subject to the "axe" of Gödel, and so on *ad infinitum*.[86] About the known mathematics formalized with the axiomatic method, therefore, it is not possible to prove

- neither the *coherence*: even assuming that the fundamental axioms are *true*, the presence of undecidable statements shows that it is not possible to demonstrate that the derived system is completely free from contradictions;
- neither the *completeness*: not all statements can be proved to be either *true* or *false*.

Then, the *Hilbert's program* fails: a *sufficiently rich* logical system[87] is unable to *prove* its coherence and completeness. Does Gödel's result, therefore, set limits to knowledge? Is there a way of knowing whether a *Gödel's statement* is true or false? In his original 1931 article, Gödel answered negatively to this question.

> (7.14) *Aus der Bemerkung, daß* $[R(q);q]$ *seine eigene Unbeweisbarkeit behauptet, folgt sofort, daß* $[R(q);q]$ *richtig ist, denn* $[R(q);q]$ *ist ja unbeweisbar (weil unentscheidbar). Der* im System PM *unentscheidbare Satz wurde also durch metamathematische Überlegungen doch entschieden.* [88]
>
> From the remark that $[R(q);q]$ says about itself that it is not provable, it follows at once that $[R(q);q]$ is true, for $[R(q);q]$ is indeed unprovable (being undecidable). Thus, the proposition that is undecidable *in the system PM* still was decided by metamathematical considerations.[89]

We may reason as following. Either φ_G is true or it is false.

1. If φ_G were false, this would mean that it is not true that it is unprovable, so it would be self-contradictory. Therefore we would be dealing with a formal system which, although starting from axioms assumed to be true, produces false statements. Evidently this would make collapse any logical-mathematical building, since in a system of such kind it is possible to prove everything and its opposite.[90] Fortunately, mathematics has got not such features, therefore this possibility is ruled out.
2. If φ_G is true, it is actually what it says; therefore the formal system has produced a statement which is effectively *true*, even if it is not demonstrable.

Mathematical language, strictly rational in nature, is therefore able to express more truths than it is able to demonstrate.[91]

---

[86] The problem of the applicability limits of the Incompleteness Theorems is a very complex and debated issue (see e.g., part II of Berto (2009), or Lolli (2011)). For that belonging to us, it suffices to know that they continue to hold for any countable extension of the initial system.

[87] By this we mean that it is sufficiently expressive to include the PA arithmetic or similar ones. The expression "sufficiently powerful" is often used.

[88] ibid.

[89] Translation from ibid.

[90] This statement corresponds to a logical principle which is classically known as *Scotus' law*, affirming that *ex false sequitur quodlibet*, i.e. "everything follows from the false" (see Part II of Berto (2009)).

[91] Chapter 11 of Strumia (2017).

Among the main problems of mathematics, it is worth mentioning that Gödel's incompleteness is found in the answer to Hilbert's *First Problem*, i.e. the *Continuum Hypothesis* (7.1). CH has indeed been proved *undecidable* in the ZFC system.[92]

Gödel's result, which is considered the greatest conquest of logic in the twentieth century, has got several implications that would take too long to discuss here. Only two of them, which are part of the discussion on the science foundations we are developing, are here reported:

1. As regards the *problem of foundations*, the incompleteness shows that it is not possible to conceive mathematics as an abstractly self-consistent structure, according to the perspective of Hilbert's formalism and, to a certain extent, also of logicism. Gödel believed that his results confirm a Platonist interpretation of mathematics, a perspective that is still much debated.[93] Of course, the idea that it would be possible to construct a mathematical, and therefore epistemological, structure perfectly closed in itself, coherent and self-consistent, according to what discussed in Sect. 6.1 and what proclaimed by Hilbert in the quote 7.10, definitively fades. It is substituted by a conception of mathematics that is not more uncertain or more insecure, as someone naively has believed, but, on the contrary, much richer and more open.
2. This aspect has got implications not only on the foundations, but on the general structure of mathematics, and above all on involving its future advances and on its close correlations with the natural sciences, especially physics.

> (7.15) Kurt Gödel [...] proved that the world of pure mathematics is inexhaustible. No finite set of axioms and rules of inference can ever encompass the whole of mathematics. [...] This discovery of Gödel came at first as an unwelcome shock to many mathematicians. It destroyed once and for all the hope that they could solve the problem of deciding by a systematic procedure the truth or falsehood of any mathematical statement. After the initial shock was over, the mathematicians realized that Gödel's theorem, in denying them the possibility of a universal algorithm to settle all questions, gave them instead a guarantee that mathematics can never die. No matter how far mathematics progresses [...], there will always be, thanks to Gödel, fresh questions to ask and fresh ideas to discover. It is my hope that we may be able to prove the world of physics as inexhaustible as the world of mathematics. [...] Fortunately, the recent successes of particle physics and of cosmology do not exclude the possibility that the world of physics is truly inexhaustible, [...] that Emil Wiechert was right when he said: "The universe is infinite in all directions."[94]

The idea that mathematics is *entirely* reducible to a *finite* set of perfectly complete and consistent axioms represents the mathematical counterpart of reductionism in physics. As in the latter one, axiom systems, together with mathematical logic and metamathematics, provide very powerful tools for investigating, analysing,

---

[92] Gödel already proves that ZFC is consistent with CH, i.e. it is not possible to prove ¬CH starting from the axioms ZFC. Later Paul Cohen (1934–2007) proves the independence of CH on ZFC (Berto 2009).

[93] See the discussion in Sect. 2.1 as well as e.g., Part II of Berto (2009), Brewer (2022), Lolli (2011), etc.

[94] F. Dyson in Chap. 3 of Dyson (1989).

explaining, demonstrating a huge number of entities and abstract structures. However, it is not possible to explain and demonstrate *everything* using these tools only, which turn out to be *intrinsically limited* beyond the advancement and refinement degree that they can, even hypothetically, achieve.
Just as physical reductionism is able to explain *much* but not *everything*, so mathematical reductionism is able to prove *much* but not *everything*. Physics remains an "open" discipline both in the explanations that it provides starting from phenomena (discussed in Chap. 6) and in the language and conceptual structure which give a form to such explanations (discussed in the present chapter).

As far as the specifics of mathematics are concerned, there is therefore no way for which it can be obtained from some mechanical procedure. It is at the very heart of its structure where mathematics shows its wideness and breadth. In this, it shows its *profoundly human* nature: even mathematics shows itself as a discipline irreducible to a closed schematization; on the contrary, it is an ever-expanding and evolving form of knowledge. In this sense, the use of the term *incomplete* is synonymous with *unfinished*. Incompleteness which, far from being a defect, turns out to be a strength in its ability to open even new horizons to human knowledge.

7.4 The Undertaken Direction

After the great debate of the first half of the twentieth century, after the systematization provided by the axiomatic set theories, the development of metamathematics as well as the crucial results by Gödel, in which direction has the investigation into the foundations of mathematics gone? In fact Gödel's results have shown that abstraction taken to the extreme does not make it possible to build up a suitable and self-consistent structure, despite the fact that abstraction is necessary for the development of mathematics. The connections established by the abstract and mental reality of mathematics with reality are therefore coming back.

> (7.16) The assertion, however, that they [all propositions produced by an axiomatic system] are all true could at most be known with empirical certainty, on the basis of a sufficient number of instances or by other inductive inferences.[95]

According to this vision, therefore, mathematics, after reaching an extreme abstraction, can, in some of its declinations, recover some even *empirical* aspects. In this, it is markedly close to both the discipline of physics and the physical reality, which it can illuminate in an innovative way by highlighting the inexhaustibility elements.[96]

> (7.17) Gödel's theorem implies that pure mathematics is inexhaustible. No matter how many problems we solve, there will always be other problems that cannot be solved within the existing rules. [...] [B]ecause of Gödel's theorem, physics is inexhaustible too. The laws of

[95] Kurt Gödel, "Some basic theorems on the foundations of mathematics and their implications" in Gödel (1995a).

[96] Cf. e.g., (Buchanan 2007; Dyson 2004).

physics are a finite set of rules, and include the rules for doing mathematics, so that Gödel's theorem applies to them.[97]

So, the interpretation that many give about gödel's results is that we must speak of a "quasi-empirical origin" of mathematics,[98] for which, even, the relation[99] could be valid

$$\text{Mathematics} \approx \text{Physics?!}$$

We can guess that such an interpretation is by no means shared by the entire scientific community, also because, if taken to the extreme, it would represent a form of reductionism. Of course, in the results of incompleteness there are many facets that might lead to a large number of interpretations.[100] The idea that there are few epistemological elements capable of explaining and demonstrating everything certainly has waned.

Physics cannot be reduced to a finite group of a few entities and interactions, mathematics cannot be reduced to a finite group of a few elements, and disciplines cannot be reduced to physics and mathematics. On the contrary, such two disciplines show an ability to adapt with the precise aim of increasing human knowledge. Physics develops methods, theories, models, and concepts to investigate *irreducible* phenomena. On the other hand, mathematics sprouts a fascinating research line so that categories such as *quality* (generally linked to philosophical fields, certainly not mathematical and non-scientific) could be included in its theories. For example, some sketches of a mathematical theory of qualities developed by Ennio De Giorgi and collaborators are indicative, such as the following one illustrates.

(7.18) *[M]olte conversazioni con studiosi di diverse discipline (Matematica, Fisica, Logica, Informatica, Biologia, Storia, Filosofia, Economia, Teologia, ecc.) [...] hanno fatto emergere l'opportunità di superare il cosiddetto "riduzionismo insiemistico", cioè la tendenza a ridurre tutta la Matematica alle teorie degli insiemi, cercando invece di inserire le teorie matematiche e possibilmente anche altre teorie scientifiche in un quadro più ampio [...] Certamente le teorie degli insiemi proposte da Cantor, Zermelo, Gödel, Bernays, Von Neumann [...] restano tra le espressioni più elevate dello spirito umano [...] tuttavia per comprendere meglio le principali idee della Matematica, della Logica e dell'Informatica sembra opportuno inserirle in un quadro più generale dominato dalle due idee di qualità e relazione. Queste discipline, come del resto le discipline fisiche, chimiche, biologiche, economiche, linguistiche, ecc., considerano infatti oggetti qualitativamente diversi e studiano relazioni tra tali oggetti [...]*

Introduciamo dunque come concetti primitivi l'idea di "qualità" e l'idea di "godere di una data qualità". Conveniamo che dati un oggetto x di qualsiasi natura ed una qualità q, quando scriveremo

$$qx$$

[...]

$Qqual\, x$ *vuol dire che* x *è una qualità*;

$Qrel\, x$ *vuol dire che* x *è una relazione*;

---

97 Freeman Dyson in Dyson (2004).

98 G. Chaitin in Chaitin (1995).

99 Suggested by the mathematician G. Chaitin in Chaitin (1995).

100 Cf. e.g., Part II of Berto (2009) and also Lolli (2011), Penrose (1989, 2004).

Qrelb x vuol dire che x è una relazione binaria;

Qrelt x vuol dire che x è una relazione ternaria;

Qrelq x vuol dire che x è una relazione quaternaria;

Qrun x vuol dire che x è una relazione univoca;

Qrbiun x vuol dire che x è una relazione biunivoca.[101]

> [F]rom many conversations with scholars of different disciplines (mathematics, physics, logic, computer science, biology, history, philosophy, economics, theology, etc.) [...] it seemed appropriate to overcome the so-called "set theoretic reductionism", i.e. the trend to reduce all of mathematics to set theories. On the contrary, we try and include the mathematical theories, and, if possible, also other scientific theories, within a wider framework [...] Surely, the set theories proposed by Cantor, Zermelo, Gödel, Bernays, and Von Neumann [...] stay among the highest expressions of the human mind [...] However, in order to get a better understanding of the main ideas of mathematics, logic and computer science, it seems appropriate to put them in a more general framework, ruled by the two ideas of quality and relation. As a matter of fact, these disciplines, as well as physical, chemical, biological, economical, linguistic disciplines, etc., all consider qualitatively different objects and study relations among these objects [...]
>
> Thus we introduce as primitive notions the idea of "quality" and the idea of "enjoying a given quality". We stipulate that given an object x of any kind and a quality q, when we write
>
> $$qx$$
>
> [...]

Qqual x means that x is a *quality*;

Qrel x means that x is a *relation*;

Qrelb x means that x is a *binary relation*;

Qrelt x means that x is a *ternary relation*;

Qrelq x means that x is a *quaternary relation*;

Qrun x means that x is a *univocal relation*;

Qrbiun x means that x is a *biunique relation*.

Just as it was once possible to weld logic, a discipline belonging to theoretical philosophy, with mathematics, in the same way a program of such kind is being developed, whereby mathematics might one day be able to develop what it could be called *formal ontology*.[102] Mathematics is an excellent basis for this, as it has got deep commonalities with many branches of human learning.

> (7.19) *Notiamo che varie forme di autoreferenza (o di mutua referenza) si incontrano non solo in Matematica, Logica e Informatica, ma anche in molti rami del sapere umano.* [...]
>
> *[R]iteniamo che valga la pena di perseguire questo obiettivo, in quanto né il formalismo, né il riduzionismo insiemistico, come ogni altra forma di riduzionismo, sembrano offrire una prospettiva adeguata per una vera comprensione dei molti problemi che la cultura del nostro tempo deve affrontare.*

---

[101] E. De Giorgi et al. in De Giorgi (1997), translation from De Giorgi et al. (2006).

[102] See e.g., Chap. 11 of Strumia (2017).

Ogni ricerca sugli assiomi fondamentali di Matematica, Logica e Informatica, come pure delle diverse scienze sperimentali, umanistiche, filosofiche comporta fra l'altro il superamento di una visione troppo chiusa delle diverse specializzazioni ed un'idea più ampia del rigore matematico o scientifico.[103]

Notice that various forms of selfreference (or of mutual reference) are met not only in mathematics, logic and computer science, but also in many other branches of human knowledge. [...]

[W]e think that it is worthwhile to pursue this goal, because neither formalism, nor set theoretic reductionism, along with any other form of reductionism, seem to offer a suitable perspective for a real understanding of many problems that the culture of our time has to face.

Any investigation of the fundamental axioms of mathematics, logic and computer science, as well as of various experimental, human and philosophical sciences needs, among other things, to overcome a too restricted vision of the different specialities, and requires a broader idea of mathematical and scientific rigor.

References

G. Basti, *Filosofia della Natura e della Scienza*, vol. 1, 1a edn. (Pontificia Università Lateranense, Roma, Città del Vaticano, 2002)

P. Bernays, *Axiomatic Set Theory (with a historic introductin by Abraham A. Fraenkel).* (North-Holland Publishing Company, Amsterdam, Holland, 1958)

F. Berto, *There's Something About Gödel. The Complete Guide to the Incompleteness Theorem.* (Wiley, UK, 2009)

U. Bottazzini, E.A. Marchisotto, P. Miller, *Hilbert's Flute: The History of Modern Mathematics* (Spriunger-Verlag, Berlin, Deutschland, 2016)

C.B. Boyer, *A History of Mathematics*, 2nd edn. (Wiley, New York, USA, 1968)

W.D. Brewer, *Kurt Gödel* (Schweiz, The Genius of Metamathematics (Springer, Cham, 2022)

M. Buchanan, The incompleteness of physics. Nat. Phys. **3**, 141 (2007)

G. Cantor, *Gesammelte Abhandlungen* (Springer-Verlag, Berlin, Deutschland, 1932)

G. Chaitin, Is incompleteness a serious problem? in *Thinking about Gödel and Turing–Essays on Complexity 1970–2007* (World Scientific, 1995), pp. 299–302

I.M. Copi, C. Cohen, V. Rodych, *Introduction to Logic*, 15th edn. (Taylor & Francis, New York, USA, 2019)

G.M. D'Ariano, F. Faggin, *Hard problem and free will: An information-theoretical approach, in Artificial Intelligence Versus Natural Intelligence* (Springer, Cham, Schweiz, 2022)

E. De Giorgi, Dal superamento del riduzionismo insiemistico alla ricerca di una più ampia e profonda comprensione tra matematici e studiosi di altre discipline scientifiche ed umanistiche. *Atti Accad. Naz Lincei, Cl. Sci. Fis. Mat. Nat.***9** (1998)

E. De Giorgi et al., Verso i sistemi assiomatici del 2000 in matematica, logica e informatica. Nuova civiltà delle macchine **57–60**, 248–259 (1997)

E. De Giorgi, M. Forti, G. Lenzi, *Towards the axiomatic systems of the third millennium in mathematics, logic, and computer science, in Ennio De Giorgi Selected Papers* (Springer-Verlag, Berlin, Deutschland, 2006)

M. Dorato, The laws of nature and the effectiveness of mathematics, in *The Role of Mathematics in Physical Sciences*. ed. by G. Boniolo et al. (Springer, Berlin, Deutschland, 2005)

F. Dyson, The world on a string. New York Rev. **13** (2004)

[103] ibid.

F. Dyson, *Infinite in All Directions* (Harper & Row, New York, USA, 1989)
H.B. Enderton, *A Mathematical Introduction to Logic*, 2nd edn. (Harcourt Academic Press, San Diego, USA, 2001)
F. Faggin, *Irriducibile-La Coscienza, la Vita, i Computer e la Nostra Natura* (Mondadori, Milano, Italia, 2022)
G. Frege, *Function und Begriff* (Verlag von Hermann Pohle, Jena, Deutschland, 1891)
K. Gödel, Some basic theorems on the foundations of mathematics and their implications, in *Collected Works*, vol. III Unpublished essays and lectures. (Oxford University Press, Oxford, UK, 1995), pp. 304–323
K. Gödel, Über formal unentscheidbare Satze der Principia mathematica und verwandter Systeme I, in *Collected Works*. Volume I Publications 1929–1936. (Oxford University Press, Oxford, UK, 1986), pp. 144–195
K. Gödel, über unentscheidbare Sätze, in *Collected Works*, vol. III Unpublished essays and lectures. (Oxford University Press, Oxford, UK, 1995), pp. 30–35
D. Hilbert, *Mathematische Probleme, Vortrag, gehalten auf dem internationalen Mathematiker-Kongress zu Paris, in Königliche Gesellschaft der Wissenschaften zu Göttingen (Mathematisch-physikalische* (Deutschland, Klasse, Nachrichten, 1900), p. 1900
D. Hilbert, Axiomatisches Denken. Math. Ann. **78**, 405–415 (1917)
D. Hilbert, Über das Unendliche. Math. Ann. **95**, 161–190 (1926)
T. Jech, *Set Theory. The Third Millennium Edition, revised and expanded*, 4th edn. (Springer, Berlin, Deutschland, 2006)
A. Levy, *Basic Set Theory* (Springer-Verlag, Berlin, Deutschland, 1979)
G. Lolli, *La guerra dei trent'anni (1900–1930)-Da Hilbert a Gödel* (Edizioni ETS, Pisa, Italia, 2011)
E. Mendelson, *Introduction to Mathematical Logic*, 6th edn. (CRC Press-Taylor & Francis, Boca Raton, USA, 2015)
F.A. Muller, Sets, classes and categories. Br. J. Philos. Sci. **52**, 539–573 (2001)
R. Penrose, *The Emperor's New Mind. Concerning Computers, Minds, and The Laws of Physics.* (Penguin Books, New York, USA, 1989)
R. Penrose, *The Road to Reality (A Complete Guide to the Laws of the Universe* (Jonathan Cape, London, UK, 2004)
W. Rautenberg, *A Concise Introduction to Mathematical Logic*, 3rd edn. (Springer, New York, USA, 2009)
S. Sakkurai, Revisiting the 'unreasonable effectiveness' of mathematics. Curr. Sci. **88**, 415 (2005)
A. Strumia, *Dalla filosofia della scienza alla filosofia nella scienza*, 1a edn. (ESC-Edizioni Santa Croce, Roma, Italia, 2017)
A. Strumia, *From Fractals and Cellular Automata to Biology-Information as Order Within Chance* (World Scientific, Singapore, 2020)
R.L. Wilder, *Introduction to the Foundations of Mathematics*, 2nd edn. (Dover Publications, Mineola, New York, USA, 2012)

Chapter 8
True and False in Physics

Il ne suffit pas qu'une théorie n'affirme pas des rapports faux, il faut qu'elle ne dissimule pas des rapports vrais.
Henri Poincaré

In the last two chapters we focused on rather abstract conceptual aspects. First we provided an overview of some very general schemes classifying the scientific method, and we highlighted their strengths and weaknesses, i.e. ultimately, their specific capabilities (and their weaknesses) to arrive at the *truth* of knowledge. Then we examined the *solidity* of the whole physics building, as the basis of all natural sciences. We discussed the *foundation problem* of the language of physics, i.e. of mathematical logic and later the problem of the *simplification* of the physical world knowledge, i.e. the *reductionist* attempt in physics.

Following this line, we got to see *the foundations' foundations* by discussing the *first principles* of the mathematical language and the *first principles* to which physical science tries to reduce its explanations. This represents the *great framework* within which physics moves, indicating the portions of *truth* to which it can or cannot access.

In this chapter we leave these remote regions and begin a path to some topics closer to the *practical* conduct of scientific research. The issue on the horizon remains the *truth* of the knowledge provided by physics—what kind it is, what degree of reliability it has got, and how it is obtained.

> (8.1) What is truth? How do we form our judgements as to what is true and what is untrue about the world? [...] This seems a difficult question. Our judgements depend upon complicated interconnected combinations of sense-data, reasoning, and guesswork.[1]

Penrose's quotation well summarizes the various elements that contribute to establish what is the truth in physics. We address the question of how much the knowledge

[1] Roger Penrose, Chap. 4 of Penrose (1989).

J. Parravicini, *The Foundations of Experimental Physics*,
https://doi.org/10.1007/978-3-031-55452-0_8

provided by physics is actual knowledge. The issues we face can be summarized in the following questions:

"How can we say that physics *is true*?"
"What does it mean that it is true?"
"Under what conditions can we say that it is true?"

Once more we find the fundamental features determining the truthfulness of the explanations provided by the physical science: on the one hand, the empirical data, the result of measurements, and experiments; on the other hand, the theories, i.e. the structured explanations built on them.

8.1 The Measurement Problem

At the beginning of our discussion (Sect. 1.1) we talk about how the *experiment* can be considered the peculiar criterion for identifying scientific knowledge (quote 1.4 in Sect. 1.1). The experiment is even the last judge of the *truth* of the knowledge obtained through the physics methods. Later, we remark that the reality investigated through observations and experiments is one of the two pillars of physics (Sect. 3.1). So, what is *experimental truth*? And how do we arrive at it? Given the practical and empirical nature of this topic, the first question is answered through answering the second one.

Prerequisite of the experiment, formally described in Definition 4.2 (Sect. 4.2.1), is the *observational investigation*, according to the features listed in Definition 4.1 (Sect. 4.1). Observational investigations, but experiments even more, require turning the inquiry of natural phenomena towards those features that are called *physical quantities*[2] (Definition 3.1). The result of this investigation process is called *measure*, a concept which is inextricably linked to that of *objectivity*. The descriptions of these two concepts are in fact complementary to each other and lead us to give a further definition of the measurement concept, more precise than Definition 3.2 initially provided (Sect. 3.2).

Definition 8.1 (*Measurement (precise definition)*) We call *measurement* that set of *operations* carried out during an experiment, or a simple observational investigation, which, in the verification of a feature of the *considered natural phenomenon*, allows us to record the value of one of its *physical quantity* (according to Definition 3.1).

Definition 8.2 (*Objectivity of a measurement*) A necessary condition for a *measurement* to be considered genuine, i.e. *true*, is that it satisfies the *objectivity condition*, requiring that the result of the measurement operations does not depend on who actually carries out the procedure.

[2] Part II Chap. 4 of Duhem (2016).

It is worth noting that the definition of *objectivity* refers to that of *measurement* in its formulation; the reason lies in the fact that experimental physics considers objectivity in an *operative sense*. Objectivity is provided only referring to a given measurement (or to the measurement outcomes, such the *data* are). Conversely, a measure according to Definition 8.1 is really such only if it has got the features given by Definition 8.2. Ultimately, the statements Definitions 8.1 and 8.2 do not exist alone but together they develop the same concept.

As remarked in Definition 8.1, the physics method requires a measurement process to be carried out within an experiment, so that the expression *experimental measurement* is widespread. In this procedure there are five fundamental characteristics: the first four are exposed together, the last one—the need to provide an *uncertainty* estimation—is exposed later. In order for a *measure* to be *physical* it must have got these "4 + 1" ingredients, discussed in Sects. 8.1.1 and 8.1.2.

8.1.1 Building Blocks

The first four features which can be found in a physical measurement process are the following ones:[3]

1. SELECTION: among all natural phenomena perceived by the experimenter, only a well delimited part of them is taken into consideration; all the others are neglected (in Definition 8.1 we speak about "considered natural phenomenon").
2. PRELIMINARY HYPOTHESES.
3. OPERATIONAL CHARACTER (in Definition 8.1 we speak about "operations").
4. QUANTITATIVE CHARACTER (in Definition 8.1 we speak about "physical quantity").

The first constitutive feature of scientific measurement process states that carrying out an experiment requires paying attention to a well-defined and very *small* part of the overall physical phenomena experienced by our senses and intellect. This selection cannot be ignored and must be accomplished with accuracy. It is a work of mental *abstraction*, whose criteria are not easily analysed or encoded. We can certainly say that it includes *observing the right things in the right way by asking the right questions*—fundamental actions for the scientific knowledge development already mentioned (Sect. 1.1). We have also stated that the concept of *right* can only be defined on the wholly empirical logic of *increasing knowledge* of nature; the same feature can be identified in the process of mental abstraction that leads to measure.

The fact that it is a mind activity (such as abstraction) to begin any measurement highlights the second feature: at the beginning there is the inevitable presence of *preliminary hypotheses* on the phenomenon under investigation, on its behaviour and, consequently, on the operational and conceptual tools that should be employed.

---

[3] Chapter 1 of Lindsay and Margenau (1957).

(8.2) *On dit souvent qu'il faut expérimenter sans idée préconçue. Cela n'est pas possible; non seulement ce serait rendre toute expérience stérile, mais on le voudrait qu'on ne le pourrait pas.*[4]

It is often said that one must experiment without preconceived ideas. This is not possible; not only it would make sterile any experiment, but, although one could aspire to, one cannot achieve it.

All measurements start from some already conceived idea about the investigated natural phenomenon. It is necessary to know *where* and *when* to find it, as well as *what* about it to investigate and *how*. It would make no sense, for example, to carry out a measurement of a molecular bond if you have no idea of what a molecule is; if you do not assume some form of atomic hypothesis; if you do not know what spatial and temporal scales you are dealing with; and, consequently, if you do not prepare specific instruments (equipment) and techniques (conceptual, statistical, etc.) to estimate them. The problem of preliminary hypotheses sometimes requires a rapid evaluation (such as the problem of measuring the focal length of a thin lens) or a planning of decades (such as the inquiry for the Higgs boson): however, in this multiformity, each measurement identically requires the experimenter to have *expectations* and to carry out *preliminary estimates* on the object of the measurement.

Experimental measurement also requires an operational activity directly accomplished by an *experimenter*. Evidently, this noun indicates both an individual and a more or less large and structured team. On the basis of the previously described elements, the experimenter rationally plans and carries out a number of mutually synergistic *practical operations*, aimed at visualizing and recording the selected feature of the investigated natural phenomenon. These operations can be the most varied, both of a purely *conceptual* kind and of an entirely *manual* kind, or a mixture of the two. Purely conceptual operations are generally carried out in the measurement preparation phases, at the time of *planning*, and in the last phase, at the time of *data processing and analysis*. Purely manual operations are often carried out in the first stages of the measurement, when the apparatus needs to be set up and/or calibrated according to the initial planning. Operations of mixed type are accomplished in the central phase of the process, the actual measurement, the heart of the experimental process. The *operational feature* remains throughout the process, as a *reasoned sequence* of activities aimed at a goal,[5] where the experimenter is by no means a mere recorder but plays a highly active role.[6]

While the aforementioned three elements are found in all the natural sciences, the fourth feature belongs only to physics: the *quantitative*, that is, *mathematizable*,[7] character of measurement. The result of this process must be objectively associated with mathematical concepts. Generally, these concepts are understood to be *quantitative in a strict sense*, i.e. numerical; however, a physical measurement can also

---

[4] H. Poincaré, Chap. 9 of Poincaré (1902).

[5] Faggiani (1957).

[6] Chapter 1 of Lindsay and Margenau (1957).

[7] Cf. quote 2.2 in Sect. 2.1.

indicate the presence or absence of symmetries,[8] algebraic structures,[9] set classifications, which are not immediately numerical aspects but which can be considered *quantities in an extended sense*. Therefore, it is constantly verified that any discussion on the physical meaning of a measurement implies an analysis of the concepts of *space* (often also in an abstract sense), of *time* and of *number* (cardinal and ordinal), as well as of the relationships between them that can be found in the recorded data.

The measurement process described above is typically provided in a *highly idealized and concise form*, i.e. by exploiting a mathematical symbolism: the measurement operations and the involved relationships are symbolically represented, so the experiment is described using these symbols.[10] The key assumption defining *physical science* (quote 1.3) is a sort of bet. It justifies the method of assigning mathematical symbols, which represent mathematical entities, to represent entities that are not mathematical but empirical: the *physical quantities*.[11] However, this method is not demonstrated, but it is axiomatically assumed to be true.[12]

8.1.2 Incertitude

Given a specific physical quantity to be measured, the *Bureau Internazional des pois et des mesures* gives the following general norm:

> When a measurement result of a quantity is reported, the estimated value of the measurand (the quantity to be measured), and the uncertainty associated with that value, are necessary.[13]

The keystone of the measurement process in physics is the estimation of the *uncertainty*—concept that can be said iconically:

A measurement is not given without providing the relative uncertainty.

It is possible even better to summarize the concept in a negative form:

If there is no uncertainty, there is no measure.

Suppose we measure a physical quantity and obtain the numerical value X_0 with associated uncertainty δX. The measurement result is X, given by

$$X = X_0 \pm \delta X \text{ measurement units.}$$

---

[8] Important examples are crystalline structures and molecular symmetries.

[9] A notable example is the classification of baryons and mesons into *octets*, as representations of the *algebraic group* SU(3).

[10] Chapter 1 Lindsay and Margenau (1957), Part II Chap. 4 of Duhem (2016).

[11] Definition 3.1.

[12] Chapter 1 of Lindsay and Margenau (1957).

[13] Section 2.1 in Bureau international des poids et mesures (2019).

The presence of uncertainty is *intrinsic* to the very nature of the process and is generated by several elements. These must not be understood as being added together, as if the total uncertainty δX were the algebraic sum of various components; instead, these elements have got a multiple root or they are different points of view.

1. The *first root* of uncertainty is deep and conceptually subtle. The measurement process connects the real physical "world" with the ideal "world" of concepts, the world where the mathematical language is displayed. From this point of view, the presence of an uncertainty is the signal of the inherent gap between the "empirical world" and the abstract mathematical language through which it is known and described. Uncertainty is a signal of this *gap*. We can provide an indicative example, noting that any measurement process supplying a numerical value X_0, even the most sophisticated one, employs a *finite* time. This effectively makes it impossible for the value X_0 to be a number $\in \mathbb{R}$ as it would require an infinite number of digits. Therefore, inside δX there are also the decimal digits which are not known.
2. The *second root* of uncertainty lies in the imperfections of the tools (which we discuss in the following section) and of the used methods: this is certainly the most intuitive and, very often, the most controllable item.[14]
3. The *third root* are the so-called *random uncertainties*. For a detailed treatment of these, please refer to specialist texts on the so-called *error analysis*.[15] For our analysis, the key point lies in the meaning of these uncertainties: they represent the *not complete control*, by the experimenter, of all the variables influencing the considered phenomenon.

Uncertainty is sometimes also called *error*, a very different concept from *mistake*: any measurement necessarily has got an error even if no mistake has been made. In the measurement context, the importance of providing uncertainty is essential in experimental physics. As a general rule it can be stated that it *does not make any sense* to provide a measurement value without an error, it is simply not physics.

There are several reasons for such a categorical assumption, but here two of them are highlighted. It is mandatory to specify, next to the measurement value, the error value, for the purpose of

1. stating the effectiveness, the limits and the consequent confidence degree of one's own measuring apparatus and one's own experimental activity;
2. specifying the limits within which physical theories are required to provide predictions.

The *first item* helps to get the objectivity of the measurement. As a matter of fact, it is an essential element for the results to be checked by other researchers.

The *second item* has got a particular importance in the development of scientific knowledge. An explanation scheme for a given phenomenon, i.e. a theory, *must* provide predictions in agreement with what is measured and with (at least) the measurement uncertainties. A theory that does not explain the data within the experimental

[14] Fornasini (2008), Grabe (2005).

[15] Cf. e.g. Fornasini (2008), Grabe (2005).

uncertainty is surely incorrect (or at least it needs to be amended in some way). On the other hand, in the development of physics, an opposite dynamic has often been found: it is the improvement of experimental precision that has made it necessary to adopt new theories, thus progressing in the understanding of nature. Some examples of these events are given in Sect. 9.4 and in particular in Sect. 9.4.1.

8.1.3 *Measured Quantity*

After discussing the *necessary* features of a physical measurement, we are reaching its heart. Accomplishing a measurement is evidently aimed at obtaining a *specific value* of a physical quantity. Given the aforementioned features, how can one claim that a given numerical value is the *measured quantity*? Therefore, being any measurement intrinsically affected by uncertainty, how does one actually arrive at claiming that X_0 is the result of the measurement process? Certainly, this question on "how" does not involve the specific technical aspect of the measurement, which strictly depends on the contingent conditions of the studied phenomenon—such as the kind of apparatus or the experiment. Instead, the question is understood in a general and conceptual sense.

Let us that all the operations have been performed in a workmanlike manner, with the instrumentation calibrated correctly and in the absence of *systematic errors* (instrumental or methodological), i.e. in the best reasonably obtainable conditions: in order to obtain a reasonable certainty degree about data, we certainly need to repeat the measurement several times. The procedure through which we arrive at affirming that a given X_0 is the numerical value measured is based on two assumptions:

1. an *empirical assessment*;
2. a *logical-mathematical interpretation.*

Here we are briefly discussing.

1. If we measure N times a physical quantity x, we obtain that each obtained numerical value x_i has been measured n_i times, $n_i = n(x_i)$. If we construct a plot of n_i as a function of x_i, we see that the points *line up approximately around the curve* described by the relation

$$n(x) \propto \exp\left[-k^2(x-\mu)^2\right] \tag{8.1}$$

with $k, \mu \in \mathbb{R}$. The curve is the *Gaussian*, whose width σ is shown to be $\sigma = 1/\sqrt{2}k$ and μ the point where it reaches its maximum. Note: we have not stated that the $n_i = n(x_i)$ *follow* the (8.1), but that they *approximately align* around the *plot* of the function described by (8.1). This item is quite subtle: the validity of the aforementioned assumption is optimally and experimentally confirmed, despite the fact that each measurement x_i of the considered quantity shows *deviations* from the prediction by (8.1). Therefore, this law cannot be considered entirely empirical. The effective justification of the aforementioned law, the well-known *normal distribution*, is fully theoretical: it is the result of logical-deductive and mathematical considerations, which have got their fundamental pivot in the *central limit theorem*.[16] About the observation summarized through (8.1), the error theory gives a probabilistic interpretation (introducing the necessary normalization factors), thus producing the well-known *Gaussian distribution*.[17]

2. If we consider the arithmetical average of the values obtained by the measurement

$$\langle x \rangle = \frac{1}{N} \sum_{i=1}^{N} n_i x_i = \sum_{i=1}^{N} x_i \frac{n_i}{N}, \tag{8.2}$$

we can get a probabilistic interpretation of the quantity n_i/N at the *probability of measuring the x_i value of the quantity* under investigation

$$p_i = p(x_i) = \frac{n_i}{N} \tag{8.3}$$

so the (8.2) becomes

$$\langle x \rangle = \sum_{i=1}^{N} x_i \, p(x_i), \tag{8.4}$$

which, after extending to continuous, gives

$$\langle x \rangle = \int_{-\infty}^{+\infty} x \, p(x) \, dx. \tag{8.5}$$

If such a probabilistic interpretation is combined[18] with the observation described by the relationship (8.1), we obtain the probability p of finding a certain specific value $\tilde{x}$, given a set of measurements of some physical quantity.

$$p(\tilde{x}) = \frac{1}{\sigma\sqrt{2\pi}} \exp\left[-\frac{(\tilde{x}-\mu)^2}{2\sigma^2}\right]. \tag{8.6}$$

Through the logical-mathematical and probabilistic development of the *error theory*, we also obtain the demonstration that the μ parameter coincides with the above-mentioned average

$$\langle x \rangle = \mu \tag{8.7}$$

and that, similarly, we can calculate the *variance* D, from which obtaining σ

---

[16] Chapter 4 of Lindsay and Margenau (1957).

[17] Chapter 3 of Grabe (2005), Chap. 4 of Fornasini (2008).

[18] In the present discussion it is not necessary to assume the most advanced interpretation, such as the abstract one provided by Kolmogorov, which conceives probability as a *mathematical measure* Kolmogoroff (1933); it is sufficient to construct the probability as a *frequency estimation*.

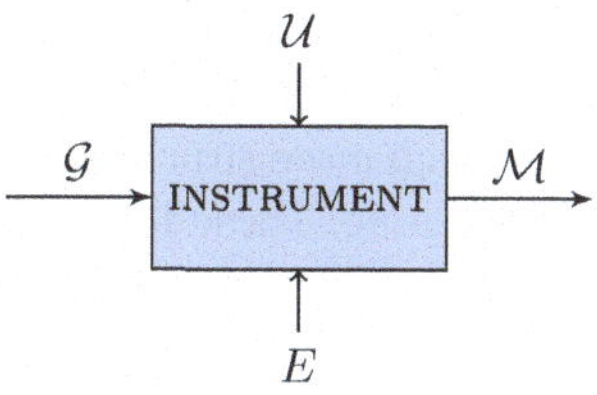

Fig. 8.1 Schematic representation of a generic measuring instrument. $\mathcal{G}$ and $\mathcal{M}$ are the input and output quantities, respectively, $\mathcal{U}$ is the reference unit and E is the energy

$$\sigma^2 = D = \int_{-\infty}^{+\infty} (x - \langle x \rangle)^2 \, p(x) \, dx. \tag{8.8}$$

Although the relation (8.1) is an idealization, the proximity of the values obtained through a measurement to the curve is an established fact, an empirically obtained observation. The relation (8.3) is an assumption resulting from a precise *interpretation* of the measurement data which, applied to the relation (8.2), allows the theory to proceed thanks to the passage (8.4). The remaining discussion, summarized in the equations (8.5), (8.6), (8.7), (8.8), is the result of logical reasoning and pure mathematical abstraction. This theory is the tool employed to arrive at affirming that X_0 is the measure of a given quantity if everything has been carried out in a workmanlike manner, $X_0 = \langle x \rangle$.[19]

The actual value of the measurement[20] is thus the result of both the actual measurement procedure and the logical-mathematical reasoning correlated to it: even the mere measurement data is not a mere recording, but it is given by a logical procedure resulting from reasoned experience.[21]

8.1.4 Measuring Instruments

Any measurement in physics is always performed using a specific *instrument*.[22] It is an object whose fundamental logical structure can be schematized and visualized according to the simple diagram in Fig. 8.1.

The input is a *physical quantity* $\mathcal{G}$. The instrument must always be supplied with *energy* E in order to work. Thanks to its own features and to the inserted energy, the device extracts from $\mathcal{G}$ a numerical value $\Xi_{\mathcal{G}} = \Xi(\mathcal{G})$; by applying appropriate device-specific conversion factors, the tool performs a comparison with a *reference standard* $\mathcal{U}$. The result of this process is a measure $\mathcal{M}$, whose *numeric value* is given $X_{\mathcal{M}} = X(\mathcal{M})$.[23] $\mathcal{G}$, $\mathcal{U}$, and $\mathcal{M}$ are variables transmitting information, while

[19] And in the hypothesis that the only error sources are random uncertainties, δX is given by the *standard deviation of the mean* as $\delta X = \sigma/\sqrt{N}$.

[20] The error theory speaks about "true value".

[21] Galilei would have called it "sensata esperienza", "sensible experience".

[22] Or set of instruments.

[23] Chapter 3 of Fornasini (2008).

E is a physical quantity in the traditional sense of the term. However, this remains a simplified idealization: in the vast majority of cases, the instrument employed has got a much more structured conceptual framework, where there are sub-instruments that integrate. These form a sort of chain: the $\mathcal{M}$ output measurement from one of the elements becomes the $\mathcal{G}$ input quantity of the next one.

1. When the features of a measuring instrument are evaluated, there is an important *first classification*: it is the coarsest one and it is based on the fundamental kind of implemented measurement. As already mentioned, it is possible through operational definitions to determine a *physical quantity*. According to this procedure, one declares what a given object is by explaining *how to measure it* or, at least, *how to identify it* (Sect. 3.2). This last clarification is not accidental: we could specify this concept by saying that:
   The measurement procedure can be of two kinds: (1) a measurement *in a strong sense*, which results in some precise value, generally numerical, of a mathematical variable; (2) a measure *in a weak sense* which might or might not result in a simple identification of the examined physical phenomenon.
   The *strong* measure can also be considered a measure in the proper sense, while the weak one is a sort of *improper* measure since, rather than measuring, we limit ourselves to *identify*. This distinction is reflected in the instrument classification terminology. Actually, it is not always possible to measure a phenomenon, sometimes we need to limit to identify it.

   -scope By the suffix *-scope* (from σκοπιά, observation, inquiry) we indicate an instruments devised to *identify* only ("measurement" in the weak sense) the investigated $\mathcal{G}$ quantity.
   -metre By the suffix *-metre* (from μέτρον, measure) we indicate an instrument devised to *measure* (in the strong sense) the investigated $\mathcal{G}$ quantity.

   The transition from a simple investigation to an actual measurement is a decisive change to develop conceptual schemes based on physical laws and theories, which allow to go beyond a merely descriptive science to a deductive-explanatory one.[24] A very clear example of this step, which has got a great historical importance, is that of the *Volta bottles* in Fig. 8.2a, b. These instruments were in vogue at the end of the eighteenth century for the study of electrostatics, a discipline which, at the time, was considered the *new physics*. They are made up of glass bottles with non-curved walls, inside which a thin metal tongue in the shape of an inverted "Y" is placed; its lower arms are called "leaves" (or "straws") and, when they are in rest position, they are normally parallel under the action of gravity. The Y-shaped element is placed in contact with a small sphere of metallic material (today we would say "conductive"), outside the bottle, and the whole is sealed with sealing wax (today we would say "insulating"). Operation is as follows: after suitably rubbing a bar of insulating material, this is placed in contact with the sphere; this operation distributes the charges on all the conductive

[24] Cf. Definition 2.5; see also the discussion about the nature of the disciplines in Sect. 4.1.

material and, in particular, on the straws. At this point, due to the Coulomb repulsion, the two straws separate, indicating the presence of a force with a non-mechanical nor gravitational nature, a *vis electrica*,[25] as it was called. The flat, non-curved walls of the bottle allow to appreciate the separation between straws while minimizing optical effects. The instrument in Fig. 8.2a is an *electroscope* and it works exactly as described previously. Using modern terms, we can say that this instrument allows to identify the presence of electrostatic charges. In his research on the foundations of electricity conducted at the University of Pavia, Alessandro Volta (1745–1827) extensively uses such a kind of instruments.[26] However, at a certain point new objects (such as the one in Fig. 8.2b) appear among Volta's instruments. On a practical level, the only difference with the first one is the presence of a small label glued to one the bottle's side, on which a graduated scale is drawn. Therefore, apparently the instrument is *practically identical*. As a matter of fact, the presence of the graduated scale makes it possible to *exactly measure*, albeit approximately, the separation between the straws and, consequently, the quantity of *vis electrica*. With the application of the graduated scale, the instrument, from being an electroscope, has become an *electrometer*. The conceptual transition from the Fig. 8.2a instrument to Fig. 8.2b instrument is essential for the development of electromagnetism.[27] Thanks to the possibility of quantitatively measuring the electric charge (as we would say today), Volta arrives to state the *capacitor law*, in modern terms[28]

$$C = \frac{Q}{\Delta V}.$$

It is one of the first applications of the mathematical language to electromagnetism, which was an essential step in getting electromagnetic phenomena out of a purely descriptive treatment and in starting the construction of an organic theory.[29]

The distinction between *-scope* and *-metre* is the first and crudest classification of measuring instruments. A second classification level involves two areas: the function of the fundamental elements composing it and the features displayed by the instrument during operation.

2. A *second classification* of the instruments can be accomplished in relation to their fundamental components. From the practical arrangement point of view, in fact, whatever complexity an instrument presents, it is always possible to identify the following elements.

---

[25] The Latin adjective *electricus* comes from the Greek ἔλεκτρον, which means "amber".

[26] Bellodi (2002), Bellone (2004).

[27] Chapter 4 of Segrè (1984), Bellodi (2002).

[28] Where, of course, Q is the quantity of electric charge, ΔV is the potential difference, i.e. with terminology more similar to that of the time, the *voltage*, C the capacity of the capacitor.

[29] Chapter 5 of Bellone (2004).

Detector Also called *sensor*, it is the part which is sensitive to the physical quantity to be measured.

Transducer This is the part transforming the physical quantity measured by the *detector* into a signal, and then it transmits it.

Display Through *transducer*, the signal from the *detector* is addressed, so it transforms this signal in an indication which can be immediately read by the experimenter.

A fourth part is almost always added to the aforementioned three elements in nowadays instruments of research laboratories.

Recorder, i.r. a *memorization system* allowing to storage, more or less automatically, the reading provided by the *display*.

In order to fix ideas, one tool still widely used in several scientific laboratories (and yet one of the simplest) is the classic column thermometer. The aforementioned elements are

The detector is the bulb of the thermometer, which is filled with a suitable substance.
The capillary is the transducer and one of the components of the display.
The graduated scale, combined with the capillary, represents the display.

The grouping into the above elements plays an important role in the practical implementation of the instruments. As a matter of fact, each of them has to submit to specific criteria, which nevertheless manage to harmonize with the whole of the instrument.

3. A *third classification* of measuring instruments is provided by the *operational requirements*. Keep in mind the previously mentioned general principle: in order to perform a measurement, we need some previous knowledge about the physical quantity being studied. While, on the one hand, the value of the physical quantity to be measured is by definition unknown, it is however equally clear that we need to have got some knowledge of it, at least a summary one. In particular, for the *numerical value* of the quantity $\mathcal{G}$, be $\Xi_0 = \Xi_0(\mathcal{G})$, it is necessary to have got a (more or less) rough estimation of the order of magnitude within which it is expected. Therefore, we can summarize the most important operational requirements of the instruments.[30]

Operating conditions They are the environmental conditions where the instrument is placed and can properly work.

Measurement range It is the range of numerical values Ξ assumed by the investigated physical quantity $\mathcal{G}$, i.e. the range within which the instrument can work according to the designed and tested accuracy degree.

[30] Cf. e.g. Chap. 3 of Fornasini (2008).

(a) Leaves electro*scope*. (b) Leaves electro*metre*

Fig. 8.2 Original instruments by A. Volta. Note in the right instrument the glued label, with the drawn graduated scale. (Courtesy of the Museum for the University of Pavia History ©.)

Transparency In principle the measurement process somehow *perturbs* $\mathcal{G}$: *transparency* is the level of this perturbation by the instrument which, clearly, should be minimized as much as possible.

Linearity range It is the optimal operating regime of an instrument, where the relation $X(\mathcal{M}) \propto \Xi(\mathcal{G})$ is satisfied. Usually, the instrument is kept within this range.

Sensitivity This can be quantified as the ratio $\Delta X/\Delta\Xi$ between the (input) numerical value of $\Xi(\mathcal{G})$ and the variation of the numerical (output) value of $X(\mathcal{M})$. If the instrument response were nonlinear, sensitivity would be defined by the derivative $dX/d\Xi$.

Accuracy This is the instrument resolution, i.e. the smallest variation of $X(\mathcal{M})$, corresponding to a variation of $\Xi(\mathcal{G})$, which the instrument is able to point out.

Responsivity This feature belongs to the (temporal) *dynamics* of the device. It indicates the smallest response time $\Delta\tau$ for the instruments as a consequence of a variation $\mathcal{G}$ during a time Δt. If $\Delta\tilde{t}$ is the typical time variation scale of $\mathcal{G}$, or it is the time scale considered during the measurement, generally it is necessary that $\Delta\tau \ll \Delta\tilde{t}$.

Stability It is the instrument capability to keep its response unchanged over time, i.e. with the same conditions of the quantity to be measured, to provide the same response over time.

Given a physical phenomenon to be measured, the aforementioned classifications of measuring instruments affirm the principle that it is necessary:

1. that there is some *knowledge* about the phenomenon;
2. that some suitable *hypotheses* about it should be conceived.

Without such a preliminary work it is not possible to build reasonable measuring instruments and thus to accomplish measurements and experiments.

8.2 Physical Theories

8.2.1 The Notion of Theory

Previously, in the field of physics the specific case of the most general *scientific laws* has been presented: these represent a series of statements capable of expressing with precision and conciseness the general behaviour of a certain class of phenomena, also providing predictions on what can reasonably be expected under same conditions.

> (8.3) Consider how all events are interconnected [...] Everywhere order reigns, so that when some circumstances have been noted we can foresee that others will also be present. The progress of science consists in observing these interconnections and in showing with a patient ingenuity that the events of this evershifting world are but examples of a few general connections or relations called laws.[31]

As it is explained further in Sect. 9.2 and especially in Sect. 9.3, these remarks are exploited, within a systematic and coherent logical scheme, as a general investigation method to determine causes and consequences. However, the development of a *true* knowledge within a discipline is not limited to construct single laws and single concepts, though useful and explanatory.

> (8.4) *Le savant doit ordonner; on fait la science avec des faits comme une maison avec des pierres; mais une accumulation de faits n'est pas plus une science qu'un tas de pierres n'est une maison.*[32]
>
> A scholar must order; one makes science by facts like a house by stones; but an accumulation of facts is no more a science than a heap of stones is a house.

Scientific knowledge is displayed according to specific articulated *structures* which, at least tentatively, aim at being logically coherent. These structures are *scientific theories*. A scientific theory can be synthetically defined as the logical systematization of the interpretation of a certain class of phenomena. For this reason, it is a coherent scheme, expressed through the language of the discipline, and summarizes the deductions of the aforementioned scientific laws.

---

[31] Alfred N. Whitehead in Chap. 1 of Whitehead (2017).

[32] H. Poincaré, Chap. 9 of Poincaré (1902).

Definition 8.3 (*Scientific theory*) By "scientific theory" we mean a doctrine, or a system of definitions, demonstrations, investigation methods, which is not limited to describe its objects of study, but which seeks to *explain* them according to a *demonstrative procedure* typical of its own investigation method, so that these explanations have a *universal* and *necessary* value.[33]

The aforementioned "universal" and "necessary" value is provided by the fact that whoever performs the same investigation on the same objects, starting from the same assumptions (postulates) and using the same method, is only able to reach, if not the same results, at least the same conclusions.

> (8.5) *Il nome di teoria scientifica spetta dunque ad ogni concetto o sistema di concetti, sorto per induzione da osservazioni preliminari, cui, mediante uno sviluppo deduttivo, vengono subordinati dei fatti supposti, più o meno perfettamente verificati.*[34]
>
> The name of scientific theory belongs to every concept or system of concepts, which arose by induction from preliminary observations, to which are subordinated by means of a deductive development, predicted facts, more or less perfectly verified.

As already stated, the proper language of physics is mathematics and consequently physical theories, as well as and the (physical) *laws* deriving from them, are expressed in mathematical language. Therefore, combining Definition 8.3 with Definition 1.3, it follows:

Definition 8.4 (*Physical theory*) By "physical theory" we mean a "scientific theory" within physical science. Namely, it is a set of concepts and interpretations, based on a system of definitions, demonstrations, and methods: this builds an *explanation* of the phenomena investigated by *physics*, according to *demonstrative procedures* typical of the experimental method and of the mathematical language, which is specific to this discipline, in order to obtain *universal* and *necessary* explanations.[35]

At the end of Sect. 8.1.1 it has been seen how the symbolic-mathematical form assumed by a *physical law* can be interpreted according to the measurement processes as an idealized description of the involved concepts, measurement operations, and relationships, provided with the maximum conciseness and compactness. Such a point of view suggests the possibility that, starting from the obtained relations, which are based on empirical data, a set of further physical laws can be deduced through *purely mathematical* operations.[36] Therefore, when these conditions occur, it is possible to say that there is *physical theory*.[37]

> (8.6) *La matematica non sostituisce l'osservazione, però riesce a potenziarla permettendoci di trarre delle conclusioni molto interessanti da esperimenti più facili e meno costosi di quelli che a prima vista sembrerebbero necessari.*[38]

---

[33] Cf. e.g. Sect. 4.2.1. of Basti (2002); also cf. Duhem (2016).

[34] F. Enriques, Chap. II Sect. 23 in Enriques (1906).

[35] Cf. e.g. Duhem (2016).

[36] This feature is anticipated in Sect. 2.2.2 in the discussion about the analytic-predictive value of the mathematical language.

[37] Chapter 1 of Lindsay and Margenau (1957), Duhem (2016).

[38] Ennio De Giorgi in (2006a), English translation from De Giorgi (2006b).

> Mathematics does not substitute observation but manages to strengthen it allowing us to draw very interesting conclusions from simpler and less expensive experiments than those that at first sight would seem necessary.

In the construction process of it, two key methodologies can be found, whose application allows the use of mathematical structures:

the attention to the quantitative features, and always to the strictly mathematizable elements of the investigated phenomena;
the *simplification* of the phenomena representation.

The first aspect has been already extensively discussed (Sect. 2.1). The second one recognizes that in natural conditions the natural phenomena under investigation are always found with a high complexity degree; therefore, the methodology requires that in such phenomena only those aspects that can be described through physical variables and mathematical relationships are isolated, starting from the simplest ones, to then employ more and more refined structures. Indicative of such an approach is the typical dynamics shown by the development of the most important physical theories.

> (8.7) As usual when a profoundly original theory is born, the equations come first and a clear understanding of their physical meaning comes later.[39]

The construction of a sufficiently large, powerful and well-verified physical theory always represents a conquest acquired along the knowledge path of the discipline. Usually, the construction of a theory takes place in three large steps.[40]

1. *Selection of some concepts* that are more or less intuitive or "primitive"; at least at a first estimate, they seem important for the class of phenomena under investigation.
2. *Identification of physical quantities and construction of the relative symbols*; they are defined in relation to the previous concepts, for which it is possible to perform empirical measurements and experiments according to specific *operations*. This step includes the identification of the corresponding *operational definitions* (cf. Sect. 3.2).
3. *Choice of hypotheses and fundamental relationships* between the symbols, through appropriate logical-mathematical deductions. From these hypotheses and relationships, taken as a foundation, the theory must be able to provide all results, i.e. the *physical laws*, again through logical-mathematical procedures, i.e. a pure manipulation of the used symbols.

As an indicative example, the development of classical mechanics can be singled out, as follows:

1. Basic *primitive concepts* are *space*, *time*, and *movement* (cf. quote 4.6 in Sect. 4.2.2).
2. *Physical quantities* are concepts, such as, e.g., velocity, acceleration, and mass, with the relative definitions of their operational measurement methods.

[39] F. Dyson, Chap. 2 of Dyson (1989).

[40] Chapter 1 of Lindsay and Margenau (1957).

3. *Fundamental hypotheses* are motion laws, such as principles of Newtonian dynamics (cf. quote 4.7 in Sect. 4.2.2).

In the construction of scientific theories, it is noteworthy that the final step of assuming fundamental elements from which to deduce results is well pointed out by Lakatos (quote 5.8 in Sect. 5.2.3) as an important item of the scientific method. The construction of a mature theory reaches such a point that the aforementioned elements are structured and stated with the utmost clarity, so requiring an *axiomatic* formulation (cf. quotes 6.2 in Sect. 6.1 and 6.3 in Sect. 6.1).

Regarding the *postulates* which are assumed during the construction of a physical theory, in principle they could be completely arbitrary, with the only precaution that they obey the rules of mathematical constructions (among which the main one is *non-contradictory*). Actually, the previous experience about the examined phenomena strongly guides the choice of initial ideas and concepts. During the development of a theory, it is also necessary to analyse some more delicate issues.

> (8.8) Whatever the meaning assigned to the term *complete*, the following requirement for a complete theory seems to be a necessary one: *every element of the physical reality must have a counterpart in the physical theory* [...] The elements of the physical reality cannot be determined by *a priori* philosophical considerations, but must be found by an appeal to results of experiments and measurements.[41]

In Chap. 3 it has been said that such elements of physical reality are represented by *physical quantities*: their key features are defined in Sect. 3.2 so that a correlation with data from experience (i.e. experimental) is guaranteed. This correlation with natural reality is a fundamental element for a theory to be *complete*, as Einstein recalls in the quote 8.8, and it is equally fundamental to ensure the objectivity of physical knowledge. However, as already mentioned in Sect. 5.2 (quote 5.3 in Sect. 5.2.1 and quote 5.6 in Sect. 5.2.1), physical theories often introduce many concepts, objects, structures, and schemes which are not directly referable to empirical data, but are necessary for the construction of the whole structure. Congruently, in many discussions it is specified that there are *direct* and *indirect* quantities: the former ones are those that can be directly measured using some tool (i.e. *physical quantities* in the strict sense), while the latter ones can be obtained from measurements of other quantities. This distinction is important since, stating that there must be a *correlation with experimental data* does require that every object of a theory must be *directly measurable*, i.e. a *physical quantity*.

Therefore, from the point of view of theory (a mirror-like to the empirical one provided in Definition 3.1 (Sect. 3.1)), it is possible to classify the objects with which theories operate, and physical laws which can be obtained, into three categories:

1. *Physical quantities in the strict sense* (or proper): Objects of the theory that can be directly correlated with empirical data, because they can be measured using instruments. The most immediate examples are *length*, *gravitational mass*, and *time*.

---

[41] A. Einstein et al. in (1935).

2. *Physical quantities in the broad sense* (or improper): Objects of the theory which, although not directly measurable through instruments, can nonetheless be well defined by means of a description of their measurement process, i.e. by means of the corresponding operational definitions. Notable examples are *phase* and *energy*.
3. *Purely mathematical objects*: The theory needs them, but it is not possible to establish an operational procedure for their measurement. Examples are the infinite-dimensional Hilbert spaces $\mathscr{H}$ or the operators of Eq. (4.5), which are defined on functional spaces, or the same Schrödinger *wave function* $\psi = \psi(\mathbf{x}, t)$ of Eq. (4.4)[42]

Consequently, it is possible, and indeed necessary, that the structure of a physical theory involves mathematical entities which are not physical quantities, i.e. which, not being measurable, cannot have got an operational definition. Regarding this, it is also worth noting that, to build classical mechanics, differential calculus was invented (cf. Sects. 4.2.2 and 4.2.3), whose formal mathematical expression has got the notion of *continuity* as a cornerstone; however, no continuity can ever be experienced: despite the fact that many measures can be recorded, these are always a countable quantity without having the *cardinality of the continuum*.[43]

When a theory reaches its full development—from the initial hypotheses, made explicit through the enunciation of mathematical concepts and relationships[44]—it is possible to deduce the individual *physical laws* related to the class of phenomena considered by the construction. In this sense, it is important to distinguish between the aforementioned laws and the theory's *principles*. As a matter of fact, while the *physical laws* are defined according to Definition 2.5 (Sect. 2.3) and can be *deduced* from the structure of the theory through purely logical-mathematical procedures, the principles can be defined as

Definition 8.5 (*Physical principles of a theory*) *Principles* are the fundamental hypotheses on which a correct verified theory is based; from them it is possible to perform a logical-mathematical deductive process in order to obtain the aforementioned *physical laws* of the phenomena considered by the theory.

It is interesting to note how different principles can produce the same physical laws, so as to result in *equivalent hypotheses* of the same physical theory. The clearest example arises again from classical mechanics. Given N free coordinates of material points, be $k = \{1 \ldots N\}$, thus, $\forall k$ we have

1. Newtonian formulation,[45] following the

---

[42] It assumes values in $\mathbb{C}$ (where, as usual, $\mathbf{x}$ indicates the position and t the time) regarding which we cannot give an operational definition and which we cannot measure. Therefore, it is not a physical quantity, but it is evident that it plays a fundamental role within the theory.

[43] Levy (1979), Wilder (2012).

[44] For mature theories we have stated that they are generally axiomatized.

[45] F_k is the kth component of the force, p_k is the kth component of the momentum $\mathbf{p}$, t is the time.

$$F_k = \frac{d}{dt} p_k; \tag{8.9}$$

2. Lagrangian formulation, following the[46]

$$\frac{\partial \mathcal{L}}{\partial q_k} - \frac{d}{dt}\left(\frac{\partial \mathcal{L}}{\partial \dot{q}_k}\right) = 0; \tag{8.10}$$

3. Hamiltonian formulation, following the[47]

$$\dot{p}_k = -\frac{\partial \mathcal{H}}{\partial q_k} \tag{8.11a}$$

$$\dot{q}_k = \frac{\partial \mathcal{H}}{\partial p_k}. \tag{8.11b}$$

The three aforementioned equations represent three starting hypotheses, three principles, all equally valid, of classical mechanics. Beyond specific nuances, it is remarkable that, starting from them, the same *classical mechanics* theory can be developed and thus the same laws of motion can be deduced, such as, for example, the law of uniform rectilinear motion, that of uniformly accelerated motion, and that of harmonic motion.

8.2.2 Relationship Between Theory and Physical Reality

Previously, we have talked about the conceptual construction of theories. Now, our analysis investigates the question of their relationship with the phenomenological reality. As already stated, in physical science this is investigated through observations, measurements, and experiments. We pose this first question:

Do experiments allow us to *see scientific laws*, or, at least, *physical laws*?

Even among some scientists, it is a widespread opinion that experiments *show* the physical laws, or that they are able to unequivocally demonstrate the truth or falsehood of a scientific theory, law, or principle. In short, we could say that there is the opinion that the laws of physics are *seen* in experiments. As seen before, this opinion is completely incorrect for several reasons (cf. e.g. Sects. 2.3, 3.1, 5.2); furthermore, the relationship between experiments and theory is markedly more complex. Actually,

---

[46] $\mathcal{L} = \mathcal{L}(\mathbf{q}, \dot{\mathbf{q}}, t)$ is the Lagrangian function, q_k is the kth component of the free coordinate $\mathbf{q}$, t is the time.

[47] $\mathcal{H} = \mathcal{H}(\mathbf{q}, \mathbf{p}, t)$ is the Hamiltonian function, p_k is the kth component of the conjugate momentum $\mathbf{p}$, q_k is the kth component of the free coordinate $\mathbf{q}$, t is the time.

the experience data, both in their naïve form, and in the rationalized and controlled form (the experiment),[48] are present both *at the beginning* of a theory as well as *in the development* and in its (final?) *verification*. It has been stated that it is not possible to conceive, set up, or carry out any experiment without having some ideas—being the reasons for performing it, or the achievements expected, or the observation forecast during its implementation. On the other hand, the theory itself suggests which experiments to develop and even triggers the whole experimental research lines.[49]

(8.9) But what is the source of knowledge? Where do the laws that are to be tested come from? Experiment, itself, helps to produce these laws, in the sense that it gives us hints. But also needed is imagination to create from these hints the great generalizations -to guess at the wonderful, simple, but very strange patterns beneath them all, and then to experiment to check again whether we have made the right guess. This imagining process is so difficult that there is a division of labor in physics: there are theoretical physicists who imagine, deduce, and guess at new laws, but do not experiment; and then there are experimental physicists who experiment, imagine, deduce, and guess.[50]

In another light, the problem of the relationship between physics and mathematics is thus proposed again in the relationship between theory and experiment. Let us remind the definition of science (physics) given by Galilei (quote 2.1 in Sect. 2.1) and that one given by Einstein (quote 1.7 in Sect. 1.2). Following Galilei, if mathematics is the pattern and the actual structure of the universe, then a *good theory* is a theory able to *explain* the universe, or at least a more or less conspicuous part of the phenomena composing it. Therefore, the exposition of the theory highlights the real structure of natural phenomena (cf. Sect. 2.1). It is known that Galileo has got a Platonic ideal: he is convinced that mathematics is effectively written within the world, that the fundamental structure of the world is mathematical. This is what he states in the famous passage 2.1 (Sect. 2.1). This *Galilei's postulate*, however suggestive, cannot be verified through experiments, since through them one has access only to the phenomena and not to the actual mathematical relationships. On the other hand, Einstein's statement 1.7 is more modern and certainly more weighted and nuanced.

(8.10) *[R]aramente il matematico riesce direttamente a trovare la soluzione del problema che gli viene posto dal fisico, dall'ingegnere, dal medico, ecc. Per lo più deve accontentarsi di capire qual è il tipo di problema che gli viene proposto, intuire quale tipo di modello matematico probabilmente si adatterà meglio, trasformare questa intuizione in proposizioni precise e poi dire al fisico o all'ingegnere "è probabile che il fenomeno che tu mi proponi possa essere schematizzato con questa equazione o con questa struttura matematica o con delle strutture matematiche simili a questa dal punto di vista qualitativo".*[51]

[T]he mathematician rarely manages in a straightforward way to find the solution to the problem posed by the physicist, engineer, or doctor, etc. He must more or less content himself with understanding what type of problem has been proposed to him, with guessing what type of mathematical model probably suits best, with transforming this intuition into precise

[48] Cf. e.g. quotations 1.5 and 4.5.

[49] Chapter 1 of Lindsay and Margenau (1957).

[50] R. Feynman, Sect. 1-1 in Feynman et al. (2005a). Following citation 1.4 in Sect. 1.1.

[51] Ennio De Giorgi in (2006a), English translation from De Giorgi (2006b).

propositions and then with saying to the physicist or engineer "it is likely that the phenomenon you proposed to me could be schematised with this equation or this mathematical structure or with mathematical structures similar to it from the qualitative point of view".

Therefore, the concept of a logical-mathematical *model* of a physical phenomenon assumes a decisive value. When such a model reaches a certain degree of breadth and complexity, it becomes a *theory*. Following this suggestion, the actual meaning of physical theories is then well summarized by Johan von Neumann.

(8.11) The sciences do not try to explain, they hardly even try to interpret, they mainly make models. By a model is meant a mathematical construct which, with the addition of certain verbal interpretations, describes observed phenomena. The justification of such a mathematical construct is solely and precisely that it is expected to work - that is correctly to describe phenomena from a reasonably wide area. Furthermore, it must satisfy certain esthetic criteria - that is, in relation to how much it describes, it must be rather simple.[52]

Beyond the last aesthetic-like consideration, with a rather philosophical flavour, we could call this statement *von Neumann's interpretation* of the sciences; from a certain point of view, it is the opposite of *Galilei's postulate*. Therefore, according to the practice of modern physical science, theories (with their articulated conceptual structures, continuously dialoguing with experiments) are *models* built with the specific language of the discipline, which depict the actual state of natural phenomena by means of schemes and *simplifications*.

8.3 Validity of Physical Laws and Theories

Therefore, a physical theory is a coherent scheme that is capable of providing, through logical-mathematical steps, concepts, systematized methods, both *descriptions* and *predictions* of a more or less large class of natural phenomena. If this happens, it is usually said that *the theory is true*. What exactly does this statement mean?

In common perception we often speak of *science*, meaning this term as what is included in the so-called "hard sciences", as opposed to the "soft sciences" or *humanities*. This sharp demarcation is based on the fact that the first ones are thought to be bearers of absolute truths, organized according to logically stringent and necessary theories, immediately provable through experiments and thus indisputable. This conception is essentially due to the old nineteenth-century positivism discussed in Sect. 5.2.1: it assumed *science* as the new religion bearer of indisputable truths of faith, as opposed to traditional religion. This vision of scientific knowledge is still widespread and a common belief: it is enough to pay attention to how the media report results of more or less exotic scientific studies or, worse, to how scientific knowledge is absolutized in discussions with apodictic sentences such as "*Science* states that…", most of the time by people who have got no real scientific knowledge. This is not only intellectually obsolete and superficial, it is profoundly wrong. For

[52] J. von Neumann in (1995a).

this reason, let us retrace the conceptual path by which a coherent and structured explanation of a class of natural phenomena, a theory, is said to be *true*.

As already stated, the theories discussed in the previous sections must be empirically *validated*, in order for them to be considered correct. Consequently, scholars prefer to speak of the "validity of a theory" rather than the "truth of a theory". Normally it is said that theories must be *verified*.[53] The measurements, performed within observations and experiments, provide the raw data that have to be systematized and schematized in the theoretical frameworks.

This schematization and validation process is the heart of the scientific knowledge construction, the heart of *science*. Not only it is a powerful method through which humanity has obtained important results, but it also remains a tool that allows us to reach truths, to really increase knowledge. It is important to remember that scientific knowledge is not made up of absolutely established elements, but of accredited knowledge relating to certain phenomenological and theoretical fields, and only within these limits it is valid.[54] Having dealt with these limits in Sect. 2.3, we have assessed that providing a *physical law* without specifying the limits of its validity is a meaningless operation.

> (8.12) It is a metaphysical doctrine that from the same antecedents follow the same consequents. No one can gainsay this. But it is not of much use in a world like this, in which the same antecedents never again concur, and nothing ever happens twice. Indeed, for aught we know, one of the antecedents might be the precise date and place of the event, in which case experience would go for nothing. [...] The physical axiom which has a somewhat similar aspect is "That from like antecedents follow like consequents." But here we have passed from sameness to likeness, from absolute accuracy to a more or less rough approximation.[55]

The experience of scientific investigation openly shows that what appears valid in the current conditions of experience and knowledge may no longer be valid, or be only partially valid, in modified conditions of experience and with other theories. Therefore, the gap between the reality of phenomena and the explanations provided by our interpretative schemes remains irreducible. This makes the approximation an intrinsic factor of all scientific construction.[56]

> (8.13) Each piece, or part, of the whole of nature is always merely an *approximation* to the complete truth, or the complete truth so far as we know it. In fact, everything we know is only some kind of approximation, because we know that*we do not know all the laws* as yet. Therefore, things must be learned only to be unlearned again or, more likely, to be corrected.[57]

The specific ways which can validate a theory are complex and varied. In particular, to confirm a given theory, the inference rules of the *modus ponens* are very often applied to its predictions (cf. Sect. 9.3); the conceptual procedures mostly used in scientific

53 A term not to be confused with the *verificationism* we talked about in Sect. 5.2.1.

54 Part II Chap. 5 of Duhem (2016).

55 J.C. Maxwell in (1882). Just before citation 6.10 in Sect. 6.3.1.

56 Part II Chaps. 4 and 5 of Duhem (2016).

57 R.P. Feynman, Sect. 1-1 in Feynman et al. (2005a).

practice are discussed in Sect. 9.1. Here we pose this question: if approximations are inevitable, how can we arrive at affirming that a given theory T_α is *true* and another T_β is *false*?

(8.14) We said that the laws of nature are approximate: that we first find the "wrong" ones, and then we find the "right" ones. [...] So [...] how can the results of an experiment be wrong? Only by being inaccurate. [...]

Now, what should we teach first? Should we teach the correct but unfamiliar law with its strange and difficult conceptual ideas, for example the theory of relativity, four-dimensional space-time, and so on? Or should we first teach the simple [...] law, which is only approximate, but does not involve such difficult ideas? The first is more exciting, more wonderful, and more fun, but the second is easier to get at first, and is a first step to a real understanding of the first idea. This point arises again and again in teaching physics. At different times we shall have to resolve it in different ways, but at each stage it is worth learning what is now known, how accurate it is, how it fits into everything else, and how it may be changed when we learn more.[58]

So what is the correct choice, the *true* choice? Should we say, for example, that classical mechanics is true and that quantum mechanics (or Einsteinian relativity) is false? If we assume that the notion of *scientific truth* must be valid in an absolute sense, without declaring its approximations, we risk reaching a dead end. As a matter of fact, while it is true that classical mechanics showed to be unsuitable to describe the motion of bodies near the speed of light, it is also true that even the most precise theory, Einstein's relativity, uses several concepts, tools, mathematical objects which it is not possible to experimentally confirm—the first of which are real numbers.[59] Therefore, we cannot say it is *true* either; unless saying that it is *at least* within the approximation for which real numbers can be approximated by rational numbers. In this sense, Feynman speaks of degree of precision: this concept must necessarily be associated with a theory.

(8.15) We no longer say "Newtonian mechanics is false and must be replaced by quantum mechanics which is correct." Instead we adopt the formula "Classical mechanics is a self-contained scientific theory. It is a strictly 'correct' description of nature wherever its concepts can be applied."[60]

It follows that the knowledge provided by a theory remains within the "structural" limits of that theory, i.e. within the conceptual, instrumental, linguistic, and modelling approximations on which it is based. If these limits are not declared, are not clarified, or are not recognized, there is the risk of confusing physical phenomena with their representation made of laws and theories, partly discussed in Sect. 2.3 and warned against by Wigner in the quote 2.15 (Sect. 2.3). Such an attitude leads to make absolute the theory, by interpreting it in a mystical key, as Laplace did in his famous citation 6.6 (Sect. 6.2.1). Approximations are thus the keystone of the validity and honesty of theories, as mentioned in the quotes 8.13 and 8.12. This fact can be

---

[58] R.P. Feynman, Sect. 1-1 in Feynman et al. (2005a).

[59] Remember that any quantitative measure, whatever precise, *always* provides a *finite* number of digits, i.e. in fact, a *rational number*.

[60] Werner Heisenberg, cited in Chap. 6 of Chibbaro et al. (2014).

faced according to three attitudes. Those who expect physics, or at least scientific knowledge, to be able, at least ideally, to provide a religiously *perfect* knowledge (like positivists) are bitterly *disappointed*. Some of them transform disappointment into more or less exasperated *scepticism*, ultimately believing that nothing can be known nor can one be sure about anything, *not even within science*. The third attitude is that of a pragmatic *realism*: recognizing that our knowledge and our "scientific certainties" are indeed consistent, albeit always *finite* and valid within certain areas.[61]

> (8.16) *Douter de tout ou tout croire, ce sont deux solutions également commodes, qui l'une et l'autre nous dispensent de réfléchir.* [62]
>
> Doubting everything or believing everything are two equally convenient solutions: both of them exempt us from thinking.

Acceptance and declaration of the *approximation degree* allows us to suitably discern the validity of a theory. As stated in quote 5.25 (Sect. 5.4), Poincaré reminds us that, in the advancement of knowledge, all theories always maintain *something* true and it is precisely in this *something* that *true knowledge* resides.

From this point of view, the criterion posed by falsificationism (Sect. 5.2.2), however subtle and suggestive, is strongly reduced. It is scientifically incorrect to say "this theory is true (false)". Instead, it is correct to state "this theory is true (false) *within these hypotheses* and *within these limits*". The limits and hypotheses can be of a numerical-mathematical, logical-mathematical, methodological, and of other kind. Although the declaration of the *numeric* limits of the theories, i.e. of their precision degree, is an essential component, often it is not the only element to specify the boundaries. In these cases, the mathematical language must often be integrated by other forms of language.

The declaration of validity of a theory, as a structured explanation of a class of phenomena, must always be accompanied by a clear statement of the considered validity limits, in the same way that the declaration of a measured value must *always* be accompanied by a clear statement of the found uncertainty (Sect. 8.1.2). From this perspective, the act of theorizing and the act of measuring are very similar.

It is worth noting how numerical limits can be formalized by adopting a relatively simple scheme.[63] Consider the theories T_1, T_2, T_3. Assume T_1 be grosser than T_2, and so T_2 grosser than T_3. By this we mean that the T_n theory is able to provide predictions with a lower precision degree than the T_{n+1} theory, which is thus more refined than the previous one. In the most important and broad theories of physics, one or more δ parameters can be identified that can act as a discriminant between T_{n+1} and T_n. Indeed, the relationship holding between the theories is

$$\lim_{\delta \to 0} T_{n+1}(\delta) = T_n. \tag{8.12}$$

[61] Cf. e.g. Zoglauer (2023).

[62] Henri Poincaré, Introduction of Poincaré (1902).

[63] Chapter 2 of Chibbaro et al. (2014).

We may, e.g. recognize

in the transition between the T_2 special relativity and the T_1 Newtonian mechanics we have $\delta = v/c$, where v is the speed of a body and c the speed of light;
in the transition between the T_3 general relativity and the T_2 special relativity we have $\delta = Gm/ac^2$, where G is the universal gravitational constant, m is the mass and a are the typical lengths of the involved phenomena;
in the transition between the T_2 statistical mechanics and the T_1 thermodynamics we have two parameters, $\delta_N = 1/N$ and $\delta_V = 1/V$, where N is the microscopic particle number, V is the volume they occupy, and we assume the condition $\delta_N/\delta_V \simeq$ constant;
in the transition between the T_2 wave optics and the T_1 geometric optics we have $\delta = \lambda/a$, where λ is the wavelength and a is the typical size of the considered objects;
in the transition between the T_2 quantum mechanics and the T_1 classical mechanics we have $\delta = h/\mathcal{S}$, where h is the Planck constant and $\mathcal{S}$ is some action involved in the system.

However, while this approach is mathematically valid, it is equally true that it does not exhaust the differences between one theory and another. While reductionism from one discipline to another is fallacious, as we saw in Sect. 6.2, it is often equally fallacious to change from one theory to another within the discipline of physics. In fact, the transition between T_1 and a T_2 at the limit $\delta \to 0$ is accompanied either by the introduction of new concepts and interpretations of the theory T_2 (cf. Definition 8.4), or the change in interpretation of the concepts of the theory T_1.[64]

(8.17) [I]t would be a misconception to believe that the difficulties of the atomic theory may be evaded by eventually replacing the concepts of classical physics by new conceptual forms [...] No more is it likely that the fundamental concepts of the classical theories will ever become superfluous for the description of physical experience.[65]

Moreover, a T_1 can be interpreted as the prerequisite of a T_2, without which it would not be intelligible.

(8.18) Newtonian mechanics is a kind of a priori for quantum theory. It is a priori in the sense that it is the language which enables us to say what we observe.[66]

Although physical theories are necessarily formulated in mathematical terms (cf. def. 1.7), there is however no mathematical algorithm, no calculation procedure to pass from one theory to another: there is always a complex process of interpretation, evaluation, and construction. Finally, given a validated theory, which is shown to be able to explain a large class of phenomena, under reasonable hypotheses and within large limits, this theory will hardly be abandoned: at the most, a sort of hierarchy of theories will be created.

---

[64] Chapter 6 of Chibbaro et al. (2014).

[65] Niels Bohr in (1934).

[66] Werner Heisenberg, cited in Chap. 6 of Chibbaro et al. (2014).

(8.19) [I]n the competition of schemata in the scientific enterprise, the triumph of one schema over another does not necessarily mean that the loser is abandoned and forgotten. In fact it may be utilized far more often than its more accurate and sophisticated successor. That is certainly true for Newtonian versus Einsteinian mechanics of the solar system. Victory in the competition between scientific theories may be more a matter of demotion of the old theory and promotion of the new one over its head than matter of the death of the vanquished.[67]

In our analysis it has been highlighted that the conceptual constructions of physics allow us to get closer to a description of the phenomena, working by successive approximations. With the approximation refinement, as pointed out by Poincaré in quote 5.25, *something* always remains of what has been built, and it is precisely in it that the core of validity (truth) of scientific explanations can be found.

References

G. Basti, *Filosofia della Natura e della Scienza*, vol. 1, 1a edn. (Pontificia Università Lateranense, Roma, Città del Vaticano, 2002)

G. Bellodi, *Gli strumenti di Alessandro Volta: il gabinetto di fisica dell'Università di Pavia* (Hoepli, Milano, Italia, 2002)

E. Bellone, *Caos e Armonia - Storia della Fisica* (UTET, Torino, Italia, 2004)

N. Bohr, *Atomic Theory and the Description of Nature* (Cambridge University Press, Cambridge, UK, 1934)

Bureau international des poids et mesures. *Le Système international d'unités (SI) - The International System of Units (SI)*. BIPM, Sèvres, France, 9 edition, 2019. https://www.bipm.org/utils/common/pdf/si-brochure/SI-Brochure-9.pdf. Disponibile online

S. Chibbaro, L. Rondoni, A. Vulpiani, *Reductionism, Emergence and Levels of Reality. The Importance of Being Borderline* (Springer, Cham, Schweiz, 2014)

E. De Giorgi, Lectio magistralis: "IL valore sapienziale della matematica", in *Ennio De Giorgi Selected Papers* (Springer, Berlin, Deutschland, 2006a)

E. De Giorgi, Lectio magistralis: "The sapiential value of mathematics", in *Ennio De Giorgi Selected Papers* (Springer, Berlin, Deutschland, 2006b)

P. Duhem, *La théorie physique. Son objet, sa structure*, 2 edn. (ENS Éditions, Lyon, France, 2016). http://books.openedition.org/enseditions/6077

F. Dyson, *Infinite in All Directions* (Harper & Row, New York, USA, 1989)

A. Einstein, B. Podolsky, N. Rosen, Can quantum-mechanical description of physical reality be considered complete? Phys. Rev. **47**, 777–780 (1935)

F. Enriques, *Problemi della Scienza* (Zanichelli, Bologna, Italia, 1906)

D. Faggiani, *La struttura logica della fisica* (Einaudi, Torino, Italia, 1957)

R.P. Feynman, R. Leighton, M. Sands, *The Feynman Lectures on Physics*, vol. 1, 2nd edn. (Addison-Wesley, Boston, USA, 2005)

P. Fornasini, *The Uncertainty in Physical Measurements. An Introduction to Data Analysis in the Physics Laboratory* (Springer, Cham, Schweiz, 2008)

M. Gell-Mann, *The Quark and the Jaguar. Adventures in the Simple and in the Complex* (W.H. Freeman & C., New York, USA, 1994)

M. Grabe, *Measurement Uncertainties in Science and Technology*, 2nd edn. (Springer, Cham, Schweiz, 2005)

A.N. Kolmogoroff, *Grundbegriffe der Wahrscheinlichkeitsrechnung* (Springer, Berlin, Deutschland, 1933)

[67] Murray Gell-Mann, Chap. 7 in Gell-Mann (1994).

A. Levy, *Basic Set Theory* (Springer, Berlin, Deutschland, 1979)

R.B. Lindsay, H. Margenau, *Foundations of Physics*, 2nd edn. (Dover, New York, USA, 1957)

J.K. Maxwell, I. Does the progress of Physical Science tend to give any advantage to the opinion of Necessity (or Determinism) over that of the Contingency of Events and the Freedom of the Will?, in *The Life of James Clerk Maxwell with a Selection from his Correspondence and Occasional Writings and A Sketch of his Contributions to Science, chapter XIV*, ed. by L. Campbell, W. Garnett (Last Essays at Cambridge. Macmillan & Co., London, UK, 1882)

R. Penrose, *The Emperor's New Mind. Concerning Computers, Minds, and The Laws of Physics* (Penguin Books, New York, USA, 1989)

H. Poincaré, *La Science et l'Hypothèse* (Flammarion, Paris, France, 1902)

E. Segrè, *From Falling Bodies to Radio Waves: Classical Physicists and Their Discoveries* (W.H. Freeman & C., New York, USA, 1984)

J. von Neumann, Method in the physical sciences, in *The Neumann Compendium*, World Scientific Series in 20th Century Mathematics, vol. 1, ed. by F. Bródy, T. Vamós (World Scientific, Singapore, 1995), p. 627

A.N. Whitehead, *An Introduction to Mathematics* (Dover Publication, Mineola, New York, USA, 2017). Unabridged republication of the work originally published in 1911

R.L. Wilder, *Introduction to the Foundations of Mathematics*, 2nd edn. (Dover Publications, Mineola, New York, USA, 2012)

T. Zoglauer, Truth relativism, science skepticism and the political consequences, in *Truth and Fake in the Post-Factual Digital Age: Distinctions in the Humanities and IT Sciences* (Springer Fachmedien Wiesbaden, Wiesbaden, Deutschland, 2023)

Part III
Practice of Knowledge

This part intends to analyse the effective working method of scientists, specifically physicists, in their research activity and, consequently, the main features of the community they form. These are the issues from which the discussion starts:

What are good scientific practices?
What are the most used logical-conceptual strategies in carrying out experiments?
How do physicists behave in front of a mistake?
Which role do mistakes play in scientific research?
How relevant is science fraud?
How does the scientific community react to such episodes?
How do scientists verify the correctness of other people's data, recognizing mistakes and frauds and discriminating the former ones from the latter?
Does the structure of the scientific community, as it was in the past or as it is currently, actually favour scientific good practices?
What are the main issues in the current configuration of scientific communities?
How do you combine unity and plurality of scientific results and interpretations?

Chapter 9
How "Science Should Be Done": Trying and Trying Again

Après avoir étudié les conditions dans lesquelles travaille le physicien, j'au cru qu'il faillait le montrer á l'oeuvre.
Henri Poincaré

Experimental physics is a discipline that develops through activities satisfying some *practices*. These practices have been consolidated through tests and reasoned selections performed by scientists over time. They have given rise to a *scientific community* which, in practicing physical science, shares procedural rules deduced from a *rationalized practice*; these norms are called *scientific practices*.

In their most general sense, the practical rules for accomplishing scientific research (beyond those coinciding with the foundations of the discipline discussed in the previous chapters) depend in detail on the kind of investigation to be carried out; for this reason, a specific analysis would be long and tedious. Therefore, we focus on those few practical rules with a highly general character, which can be constantly found within the multifaceted research world and the scientific community. These criteria, tacitly assumed by scientists, are *fruitful strategies* to increase knowledge in the field of physics. Specifically, we deal with two most important elements that trigger virtuous circles in research practice: on the one hand, the use of *good scientific practices* which exploit the most tested rules of logical inference; on the other hand, the emergence and identification of errors, as tools to advance in knowledge.

9.1 Good Scientific Practices

In our analysis so far, a large number of items concerning the *method* which is used for scientific research in the field of physics have been considered. The problem that we are going to face here is that of the *practical implementation* of these items. Physics is a discipline built by human beings called "scientists" or, more specifically, "physi-

J. Parravicini, *The Foundations of Experimental Physics*,
https://doi.org/10.1007/978-3-031-55452-0_9

cists", who *do research*, i.e. they perform the investigation of natural phenomena according to the methodological features of the discipline as illustrated previously (cf. in particular Chaps. 1, 3, 8). This activity requires the use of *practical* strategies. They have been empirically selected and tested in order to accomplish a rationally ordered research. These strategies become a part and parcel of the branch itself. These elements of performing research, which we call *good scientific practices*, are also conceptual, even though they have a specifically empirical character: consequently, they are practical methods added to the other conceptual methods already discussed.

9.1.1 Controlled Conditions

We know that the exploiting *experiment* is a minimally necessary condition for the explanation of a given natural phenomenon to be called "scientific". Experiment is the proof of the *scientific truth*.[1] We also saw that the fundamental feature of the experiment, which distinguishes it from a simple *observational investigation*,[2] is the fact that the former is carried out under *controlled conditions*.[3] It is precisely this control of the boundary conditions that is the key element to discern from the "chaotic variety of our sense-experience"[4] those elements that are useful to set up an explanatory structure of natural phenomena.

Specific procedures allowing to control the items of an experiment strongly depend on the circumstances and the actual investigated phenomena. For this reason, in our analysis it is neither useful nor possible to provide a close examination of it. In fact, a *full* control of *all* variables reasonably influencing a natural phenomenon is practically impossible. This is likely to have got deleterious effects on the reliability of the scientific work conclusions.[5] Then, the design of an experiment often starts from evaluating the undesirable variables which could actually influence the results (as they must be kept under a strict control) and those variables having a negligible influence.[6] This evaluation starts from the general observation of the interconnection among phenomena (quote 8.3 in Sect. 8.2.1). It can be more or less estimated, probabilistic, or reasoned, and it requires implementation through the knowledge and ability of the experimenter.

However, both in designing and performing the experiment, as well as in the development of the theories, gaining full control of all items still remains an impossible option. So, what is a solution to this problem?

---

[1] Cf. e.g. quote 1.4 in Sect. 1.1.

[2] Cf. Definition 4.1.

[3] Cf. Definition 4.2.

[4] A. Einstein, quote 1.1 in Sect. 1.1.

[5] Cf. e.g. Ioannidis (2005).

[6] Chapter 1 of Montgomery (2001).

9.1.2 Inference Rules

The most widely used method to determine causes and effects of phenomena, to validate, falsify, amend theories and explanations, is the use of the main logic *inference rules*.[7] In physical investigation they are often unconsciously used: these are logical schemes through which, given some premises, a specific conclusion is reached.[8] Such practically applied inference rules provide *concept schemes* which are widely used in scientific reasoning. The mainly used inference rules are

- the *modus tollens*[9];
- the *modus ponens*.[10]

Due to their importance in producing a concept framework to develop the scientific practice, the features of such schemes are worth discussing. As previously an extensive use of the most important physicists' quotations have been proposed, even in this case our study is guided by them, especially through their conceptual development in specific paradigmatic cases of their research activity. Therefore, we do not present the concept schemes of these rules in an abstract way, instead we cite cases of their effective application in achieving important results in physics.

9.2 The *Modus Tollens*

9.2.1 General Scheme

The *logical scheme* of the so-called *modus tollens* can be expressed as it follows.[11] Given two propositions A and B, the first one is a sufficient condition for the second one, i.e. in symbols:

$$A \rightarrow B.$$

In other words, if A holds, i.e. if it is true, then B holds too, i.e. it is true. The *modus tollens* states that, in these conditions, if B does not hold, then A does not as well. In symbols[12,13]

$$((A \rightarrow B) \wedge \neg B) \rightarrow \neg A. \tag{9.1}$$

[7] Cf. also Sect. B.2.3.

[8] Basti (2002), Copi et al. (2019), Strumia (2017).

[9] More precisely, *modus tollendo tollens*.

[10] More precisely, *modus ponendo ponens*.

[11] Chapter 9 of Copi et al. (2019), Church (1956).

[12] A list of the main logical symbols is reported in Appendix B.1.

[13] Or, with another notation

$$A \rightarrow B, \neg B \vdash \neg A.$$

A standard example of such inference kind is a syllogism like[14]:

All physicists are mammalian ($A \rightarrow B$).
All fishes are not mammalian ($\neg B$).
All fishes are not physicists ($\neg A$).

The *concept scheme* of the *modus tollens* is often used when there is a given physical phenomenon whose structure is to be investigated. If the physical phenomenon can be described with the proposition B, a model is constructed which hypothesizes that A is its cause, or, at least, a sufficient condition to occur (the condition $A \rightarrow B$ is assumed). In order to evaluate the correctness of the hypothesis, it is thus sufficient to apply the *modus tollens*, i.e. to "remove" A (i.e. assume the further condition $\neg A$) and to evaluate the occurrence of (9.1), i.e. make sure that B no longer occurs ($\neg B$). This logical scheme is generally applied when trying to find an explanation starting from established factual data.

9.2.2 Planck's Constant

The path that led Max Planck, "unwilling revolutionary" (E. Segrè) to formulate the quantum hypothesis and the law bearing his name.[15] Planck studies the spectral energy density u_ν of a radiation field. If we are in thermodynamic equilibrium conditions, it is given by[16]

$$u_\nu = \frac{8\pi\nu^2}{c^3}\langle\mathcal{E}\rangle. \tag{9.2}$$

To estimate this relationship, we need to know the value of the mean energy $\langle\mathcal{E}\rangle$ of the single oscillator at frequency ν. By the *equipartition theorem*, an energy of $(1/2)K_BT$ is associated to each degree of freedom of the system. According to Boltzmann, in the state of thermodynamic equilibrium a single oscillator has got p probability of having energy $\mathcal{E}$, $p = p(\mathcal{E})$, according to the relationship

$$p(\mathcal{E}) \propto e^{-\beta\mathcal{E}}, \tag{9.3}$$

with $\beta \equiv 1/K_BT$. Classical electromagnetism states that, for an oscillating magnetic field, $\mathcal{E}$ can assume *any* (real positive) value. Then, if this condition holds, as well as (9.3), the mean $\langle\mathcal{E}\rangle$ energy of a single oscillator *is necessarily given by an integral*, i.e.

$$\langle\mathcal{E}\rangle = \frac{\int_0^\infty \mathcal{E}e^{-\beta\mathcal{E}}d\mathcal{E}}{\int_0^\infty e^{-\beta\mathcal{E}}d\mathcal{E}} \tag{9.4}$$

---

[14] CAMESTRES syllogism of *second figure*, cf. Sect. B.2.2.

[15] Here we follow the exposition provided by Max Born in Chap. 8 of Born (1969).

[16] ibid.

from which, by doing calculations[17] we obtain *equipartition*

$$\langle \mathcal{E} \rangle = 1/\beta. \tag{9.5}$$

We note that nor $\mathcal{E}$, nor $\langle \mathcal{E} \rangle$ depend on ν; this implies

$$\frac{d\mathcal{E}}{d\nu} = 0, \tag{9.6a}$$

$$\frac{d\langle \mathcal{E} \rangle}{d\nu} = 0. \tag{9.6b}$$

In particular, by substituting Eq. (9.5) in (9.2) we have the *Rayleigh–Jeans law*

$$u_\nu = \frac{8\pi\nu^2}{c^3}\frac{1}{\beta} \tag{9.7}$$

according to which u_ν is an increasing monotonic function between 0 and $+\infty$. Therefore, the total radiation energy u, for volume unit, that is obtained by integrating (9.7) from 0 to ∞, gives the result

$$u = \int_0^\infty u_\nu d\nu = \int_0^\infty \frac{8\pi\nu^2}{c^3}\frac{d\nu}{\beta} = \infty, \tag{9.8}$$

which is sometimes called *ultraviolet catastrophe.*

According to the classical theory, therefore, the total spectral density, integrated over all frequencies, would be an *non-finite* quantity, which is evidently contradicted by the experimental data. Planck analyses the theory in detail and he realizes that the key point lies in the estimation of the mean energy $\langle \mathcal{E} \rangle$ obtained from (9.4). As a matter of fact, the result of the equipartition (9.5) does not depend on ν; this makes inevitable that the (9.8) diverges. Therefore, it is necessary to apply the *modus tollens* as follows:

> The (9.5) equipartition (proposition A) $\rightarrow$ Relationship (9.8) must hold, i.e. the *ultraviolet catastrophe* (proposition B).
> Experimental data show that there is no *ultraviolet catastrophe*, i.e. $\neg B$.
> Then we deduce that equipartition, at least in its (9.5) form, is not correct, i.e. $\rightarrow \neg A$.

From this logical reasoning, Planck modifies the A proposition, i.e. the relation (9.5), and he goes beyond the usual calculation of $\langle \mathcal{E} \rangle$. To construct a definition of $\langle \mathcal{E} \rangle$ that

---

17

$$\frac{\int_0^\infty \mathcal{E} e^{-\beta\mathcal{E}} d\mathcal{E}}{\int_0^\infty e^{-\beta\mathcal{E}} d\mathcal{E}} = -\frac{d}{d\beta}\ln\left(\int_0^\infty e^{-\beta\mathcal{E}} d\mathcal{E}\right) = -\frac{d}{d\beta}\ln\left(\frac{1}{\beta}\right) = \frac{1}{\beta}$$

is different from the classical equipartition ($\neg A$) but at the same time it takes into account the known experimental data,[18] Planck is forced to make *two assumptions*.

1. The first one is the *quanta* hypothesis, for which the energy $\mathcal{E}$ of the oscillator is able to assume *integer multiples* of a given fixed quantity

$$\mathcal{E} = n\mathcal{E}_0, \quad n = 0, 1, 2, \ldots \tag{9.9}$$

From this assumption, he changes the (9.4) in

$$\langle \mathcal{E} \rangle = \frac{\sum_{n=0}^{\infty} n\mathcal{E}_0 e^{-\beta n \mathcal{E}_0}}{\sum_{n=0}^{\infty} e^{-\beta n \mathcal{E}_0}} \tag{9.10}$$

then

$$\langle \mathcal{E} \rangle = \frac{\mathcal{E}_0}{e^{\beta \mathcal{E}_0} - 1}. \tag{9.11}$$

2. The second one is renunciation to the (9.6) property: he hypothesizes that the energy associated with each oscillation mode depends *in some way* on the frequency. Taking into account (9.9), the mathematically simplest dependency is

$$\mathcal{E}_0 \propto \nu. \tag{9.12}$$

By means of (9.9) and (9.12) assumptions, through the (9.11), derivative of the first one, we may go beyond the (9.5) prediction. Then, by inserting the first ones into (9.2), we obtain the well-known *radiation Planck's law* (of black body)

$$u_\nu = \frac{8\pi h \nu^3}{c^3} \frac{1}{e^{\beta h \nu} - 1}, \tag{9.13}$$

where *Planck's constant* h is introduced, playing the role of a *proportionality constant* of (9.12). Such relationship, which is in excellent agreement with experimental data,[19] expresses the first noteworthy result of *quantum theory*.

It is worth emphasizing again the evident use of inference. Deductions which, from the hypotheses, lead to the *ultraviolet catastrophe* are necessary and stringent. Since this phenomenon *does not* occur, we need to identify the cause of such an erroneous prediction (i.e. the proposition A). This cause is found in (9.5) and (9.6); therefore, a theory is constructed where the latter ones cease to hold. This theory allows to avoid the incorrect proposition (B) and, through an appropriate construction, to obtain a good agreement with the experimental data.

[18] Chapter 7 of Born (1969).

[19] ibid.

9.2.3 EPR Paradox

A significant, and at the same time very peculiar, example of the *modus tollens* use can be found in the famous article in which Albert Einstein, Boris Podolski, and Nathan Rosen introduce the so-called *EPR paradox*.[20] Since Einstein was deeply dissatisfied with quantum mechanics and its standard interpretations, in particular that by the *Copenhagen school*,[21] the logical pattern of this work is constructed with the express purpose of showing the *incompleteness* of quantum mechanics. The reasoning can be briefly summarized as follows.

Let I and II be two *quantum systems*, which may interact together in the time interval between $t = 0$ and $t = t_f$. *If Quantum Mechanics is complete* (proposition A of (9.1)), the combination of these systems, $I + II$, is *entirely described* by the wave function, which is obtained by solving the associated Schrödinger's equation,

$$\Psi = \Psi(x_1, x_2),$$

where x_1 and x_2 are the set of variables selected to describe, respectively, I and II systems. If we consider a given physical quantity Z, corresponding to $z_1, z_2 \ldots$ eigenvalues, and the associated eigenfunctions $u_1(x_1), u_2(x_1), \ldots$, we obtain

$$\Psi(x_1, x_2) = \sum_{n=1}^{\infty} \psi_n(x_2) u_n(x_1),$$

where ψ_n are the expansion coefficient on an orthogonal function basis. If we experimentally measure the Z quantity at a given $t = t_m > t_f$ time instant, due to the II Postulate of Quantum Mechanics (Eq. (4.5)), we obtain z_k value, corresponding to the eigenfunction $u_k(x_1)$. This determines that the second system, due to the III Postulate (Eq. (4.7)), is described for $t > t_m$ by $\psi_k(x_2)$. Conversely, if we decide to take into account a different physical quantity, let W, corresponding to the $w_1, w_2 \ldots$ eigenvalues and $v_1(x_2), v_2(x_2), \ldots$ eigenfunctions, following Quantum Mechanics we can write $\Psi(x_1, x_2)$ as a function of these last ones, that is,

$$\Psi(x_1, x_2) = \sum_{n=1}^{\infty} \phi_n(x_1) v_n(x_2),$$

where ϕ_n are the new expansion coefficients on an orthogonal function basis. Then, if we experimentally measure the W quantity at a given $t = t_m > t_f$ time, even due to the II Postulate, we obtain a w_k value, corresponding to the $v_k(x_2)$ eigenfunction. This implies that the I system, for $t > t_m$, is described by $\phi_k(x_1)$.

---

[20] Einstein et al. (1935).

[21] Cf. e.g. Chap. 29 of Penrose (2004).

In other words, Einstein and their collaborators say, if we assume that the I Postulate of Quantum Mechanics holds,[22] due to the Packet Reduction Postulate, one would obtain that the measurement on one system *necessarily* determines the state of the other system, *although the two systems do not interact* (proposition B of (9.1)). This would prove, they conclude, that the basic hypothesis is not correct: Quantum Mechanics *is not complete.*[23]

If we analyse the conceptual itinerary of this argument in the light of (9.1), Einstein and his collaborators use the scheme of the *modus tollens*, following the (9.1), i.e.

> Quantum Mechanics is complete (A) $\rightarrow$ Given two non-interacting systems, a measurement on the first one determines the state of the second one (B).
> The measurement of a physical system cannot determine the state of a second system *with which the first one does not interact* ($\neg B$).
> We must thus conclude that Quantum Mechanics is not complete ($\neg A$).

The work of the *EPR paradox* has got a capital importance in the development of physics. It highlights the phenomenon of *entanglement* as the pivot of all Quantum Mechanics. Important research lines arise, concerning the foundations and interpretation of the theory, involving issues related to its limits of validity, to quantum and classical information theory, to (quantum) cryptography, the problem of decoherence, the notion of relativistic space–time, quantum optics, atomic physics, and many other branches of physics. The EPR paradox gives origins to the theoretical works by David Bohm (1917–1992), John S. Bell (1928–1990) and the experimental ones by Alain Aspect (1947), Anton Zeilinger (1945), John F. Clauser (1942). The experimental evidence provided by the latter ones showed a phenomenology compatible with the idea that Quantum Mechanics is actually a *complete theory*, contrarily to Einstein's reasoning[24]; one would deduce that B is actually B and not $\neg B$; it follows that A is an unrefuted assumption. This is the interpretation which is generally considered definitive by the vast majority of physicists, although some do not consider it entirely satisfactory, since the effective correspondence of theoretical models with respect to real physical systems is not always guaranteed[25] and the relationship between Quantum Mechanics and other theories still remains a complex and conceptually very subtle issue.[26]

What is remarkable for our discussion is that such a peculiar application of the *modus tollens*, in a "negative" sense, by Einstein, Podolski, and Rosen greatly expanded scientific research and opened up entirely new research lines, which are still flourishing.[27] A big example of an inference rule application which has proved to be such a useful and extremely fruitful tool in spite of its simplicity.

---

[22] I.e. that the state of the system is *completely described* by the Ψ wave function, cf. Sect. 4.3.2.

[23] Or, in other words, it is not true that Ψ is able to provide a *complete* description of the system.

[24] Aspect (2015), Picasso (2016).

[25] Carati and Galgani (1999, 2001, 2014), Carati et al. (2018), Lerose et al. (2014).

[26] Cf. e.g. L. Landau in the quote 6.13 in Sect. 6.4.

[27] Just as an example, we mention that the aforementioned A. Aspect, J.F. Clauser and A. Zeilinger have won the Nobel Prize in Physics in 2022.

9.3 The *Modus Ponens*

9.3.1 General Scheme

The *modus tollens* inference rule is accompanied by that of the *modus ponens*.[28] The *logical scheme* is as follows. If we consider again two propositions, A and B of which, as in the previous case, the first one is a sufficient condition for the second one:

$$A \rightarrow B.$$

The *modus ponens* rule states that, in these conditions, if A is true, then B is necessarily true, too.[29]

$$((A \rightarrow B) \wedge A) \rightarrow B \tag{9.14}$$

A standard example is a syllogism like[30]:

All feline cats are mammalian ($A \rightarrow B$).
All cats are feline (A).
All cats are mammalian (B).

The *conceptual scheme* of the *modus ponens* is often used to confirm, or refute, a theory, evaluating its implications, i.e. the predictions of not yet observed phenomena. Assuming that the investigated physical phenomena can be described through a proposition, or a coherent set of propositions, A, we employ the specific characteristic of physical science to be *explanatory and deductive* (Sect. 4.1). We deduce one or more consequences, B, for which A is a sufficient condition. In other words, as before, we have got the condition where $A \rightarrow B$. To evaluate the correctness of the A hypothesis, the actual presence of the various B implications deriving from A is evaluated. This logical scheme is often applied when trying to find confirmations for certain theories.

It is also worth noting that, given an A, the occurrence of a single deduction B is not sufficient to assure the correctness of A, i.e. the occurrence of (9.14). If we reasoned in this way, we would apply an incorrect use of the inference rules for which we would exchange a necessary condition with a sufficient one.[31] On the other hand, the *practical* confirmation of a theory always occurs by testing the correctness, first logical-mathematical and then experimental, of a certain number of implications like B. There is no general rule to determine what are a reasonable number of implications that must be satisfied before a theory A is considered *true*.

[28] Chapter 9 of Copi et al. (2019), Church (1956).

[29] Or, with another notation

$$A \rightarrow B, A \vdash B.$$

[30] BARBARA syllogism of *second figure*, cf. Sect. B.2.2.

[31] "*Fallacy to state the consequence*", cf. Sect. B.2.3.

This is a matter that strictly depends on the considered physical phenomena and problems as well as on the refinement degree and certainty of the starting theories. Again, alongside stating the developed law or theory, it is always necessary to specify the limits within which they should hold and the limits of the checks to which they have been subjected (Sect. 8.3).

9.3.2 The Electromagnetic Waves

In order to highlight the practical application of *modus ponens* in physics, here we consider again the example of a great discovery. An important case is the path that led from *Maxwell's equations* to the prediction and subsequent discovery of electromagnetic waves. The Scottish scientist introduced 20 relations which, later rewritten in the peculiar form which is known today, would then be called *Maxwell's equations*.[32] Using a modern vector notation,[33] the relations written by Maxwell can be expressed in the following form[34]

$$\mathbf{J}_T = \mathbf{J} + \frac{\partial \mathbf{D}}{\partial t}, \tag{9.15a}$$

$$\mu \mathbf{H} = \mathbf{B} = \nabla \times \mathbf{A}, \tag{9.15b}$$

$$\nabla \times \mathbf{H} = 4\pi \mathbf{J}_T = 4\pi \left(\mathbf{J} + \frac{\partial \mathbf{D}}{\partial t} \right), \tag{9.15c}$$

$$\mathbf{E} = \mu(\mathbf{v} \times \mathbf{H}) - \frac{\partial \mathbf{A}}{\partial t} - \nabla \psi = (\mathbf{v} \times \mathbf{B}) - \frac{\partial \mathbf{A}}{\partial t} - \nabla \psi, \tag{9.15d}$$

$$\mathbf{E} = k\mathbf{D}, \tag{9.15e}$$

$$\mathbf{E} = \varrho' \mathbf{J}, \tag{9.15f}$$

$$\nabla \cdot \mathbf{D} = \rho, \tag{9.15g}$$

$$\nabla \cdot \mathbf{J} + \frac{\partial \rho}{\partial t} = 0. \tag{9.15h}$$

Equations (9.15), published in 1865, are commented by Maxwell himself in his original paper.[35] Among other considerations, Maxwell remarks that the quantity k,

---

[32] Maxwell (1865).

[33] Sengupta and Sarkar (2003).

[34] Where $\mathbf{J}$ is the current density; t is the time; $\mathbf{D}$ is the dielectric displacement vector; $\mathbf{J}_T$ is the total current density (sum of $\mathbf{J}$ and displacement current); $\mathbf{H}$ is the magnetizing field; μ is the magnetic permeability of the considered medium; $\mathbf{B} = \mu \mathbf{H}$ is the magnetic induction field; $\mathbf{A}$ is the magnetic potential vector; $\mathbf{v}$ is the velocity of the considered conductive medium; ψ is the electrostatic potential; k is the inverse of the dielectric function, modernly denoted by ε; ϱ' is the resistivity of the material.

[35] Maxwell and Jenkin (1865).

which he calls *electric elasticity*,[36] can be calculated through the simple relationship

$$k = 4\pi u^2, \tag{9.16}$$

where u is a quantity, with the velocity dimensions, which can be experimentally measured by exploiting electrostatic effects.

Maxwell's subsequent arguments are an excellent example of the *modus ponens* use. Still using modern notation, we see that Maxwell inserts the (9.15e) in the (9.15c)

$$\nabla \times \mathbf{H} = 4\pi \mathbf{J} + 4\pi \frac{1}{k}\frac{\partial \mathbf{E}}{\partial t}$$

and adding the (9.15b) he obtains

$$\nabla \times (\nabla \times \mathbf{A}) = 4\pi \mu \mathbf{J} + 4\pi \frac{\mu}{k}\frac{\partial \mathbf{E}}{\partial t} = \nabla(\nabla \cdot \mathbf{A}) - \nabla^2 \mathbf{A}.$$

So, the Britannic physicist assumes the physical hypothesis of dealing an *ideal dielectric*, which is a material with a null conductance. Therefore, we have $\mathbf{J} = 0$,

$$k\left[\nabla(\nabla \cdot \mathbf{A}) - \nabla^2 \mathbf{A}\right] = 4\pi \mu \frac{\partial \mathbf{E}}{\partial t},$$

and, by also substituting $\mathbf{E}$ with the expression (9.15d), we obtain

$$k\left[\nabla(\nabla \cdot \mathbf{A}) - \nabla^2 \mathbf{A}\right] + 4\pi \mu \left(\frac{\partial^2}{\partial t^2}\mathbf{A} + \frac{\partial}{\partial t}\nabla \psi\right) = 0.$$

Then, through some differentiations and substitutions,[37] Maxwell places himself in the conditions where $\nabla \cdot \mathbf{A} = 0$ and $\psi = 0$,[38] so he gets

$$k\left(-\nabla^2 \mathbf{A}\right) + 4\pi \mu \frac{\partial^2}{\partial t^2}\mathbf{A} = 0, \tag{9.17}$$

that is, by passing from the potential to the field[39] and applying again the (9.15b)

$$\nabla^2 \mathbf{B} - 4\pi \frac{\mu}{k}\frac{\partial^2}{\partial t^2}\mathbf{B} = 0$$

---

[36] Part IV of Maxwell (1865).

[37] Part VI of Maxwell (1865).

[38] As a matter of fact, it employs a procedure very similar to modern *gauge transformations*.

[39] The intermediate step is to go from Eq. (9.17) to

$$k\nabla^2(\mu \mathbf{H}) = 4\pi \mu \frac{\partial^2}{\partial t^2}(\mu \mathbf{H}).$$

which allows us to define the constant[40]

$$V = \pm\sqrt{\frac{k}{4\pi\mu}}. \tag{9.18}$$

So, we obtain the *wave equation*

$$\nabla^2\mathbf{B} - \frac{1}{V^2}\frac{\partial^2}{\partial t^2}\mathbf{B} = 0. \tag{9.19}$$

Therefore, Maxwell deduces for the **B** magnetic field a relation that is identical to wave equation, and he notes that

> (9.1) [T]he wave is propagated in either direction with a velocity V.
>
> This wave consistes entirely of magnetic disturbances, the direction of magnetization being in the plane of the wave [...]
>
> Hence magnetic disturbances propagated through the electromagnetic field agree with light in this, that the disturbance at any point is transverse to the direction of propagation, and such waves may have all the properties of polarized light.[41]

The parallelism with light continues. Maxwell notes that, where $\mu = 1$, as in air, one can match the value of u given by (9.16) with that of V, i.e. $V = u$. He points out that there are experimental measurements of u, obtained through electrostatic procedures; independently of them, there are measurements for the speed of light, obtained through optical experimental procedures. Maxwell stresses that the values of u, V and the speed of light are nearly the same within the measurement uncertainties of his age. Therefore, he concludes

> (9.2) The agreement of the results seems to show that light and magnetism are affections of the same substance, and that light is an electromagnetic disturbance propagated through the field according to electromagnetic laws.[42]

Subsequently, equations (9.15) are rewritten in the more modern forms known today, to highlight the symmetries therein. This mainly occurs thanks to the works by Oliver Heaviside (1850–1925) and Heinrich Hertz (1857–1894); the latter one, in particular, arrives to four *vectorial equations* in the *rectangular form* known today[43]:

$$\begin{aligned} \nabla\cdot\mathbf{E} &= \frac{\rho}{\varepsilon_0}, & \nabla\times\mathbf{E} &= -\frac{\partial\mathbf{B}}{\partial t}, \\ \nabla\cdot\mathbf{B} &= 0, & \nabla\times\mathbf{B} &= \mu_0\mathbf{J} + \mu_0\varepsilon_0\frac{\partial\mathbf{E}}{\partial t}. \end{aligned} \tag{9.20}$$

[40] The $\pm$ sign indication is expressly indicated and commented by Maxwell himself in his original work.

[41] James K. Maxwell in Part VI of Maxwell (1865).

[42] ibid.

[43] As usual, we indicate with μ_0 the magnetic permeability and with ε_0 the dielectric permittivity of the vacuum.

This form makes it much easier to mathematically treat the relations (9.20) in order to obtain an equation like (9.19), or similar ones, with or without sources. Thanks to the interaction between mathematical manipulations of the theoretical relationships and experimental skills, it is Hertz who finally obtains the experimental demonstration of the *electromagnetic waves* existence.[44]

Here too, we think back to the conceptual itinerary leading to the discovery of electromagnetic waves and the fact that light coincides with them.

Equations (9.15) are true, in the sense that they are a good logical-mathematical description of electromagnetic phenomena (A).
With appropriate passages, from (9.15) we obtain the wave equation (9.19); this is interpreted according to quote 9.1, for which electromagnetic phenomena and light are different manifestations of the same phenomenon ($A \rightarrow B$).
There are theoretical considerations (Maxwell, Heaviside) confirming the A theory. Furthermore, there are several experimental data, including the one providing the values of the variables in (9.16), which confirm the theory (A).
The light features (commented by Maxwell) and Hertz's experiments confirm the correctness of all predictions (B).

Although this itemization is logically correct, it is the research practice in physics to allow us to affirm that the truthfulness of the B propositions (i.e. the verification of the theory's predictions) and to give *confirmations* of the theory (A). This is only *probabilistically* true,[45] in the sense referred to by Maxwell himself in the quote 8.12, under the specific limits discussed in Sect. 1.2 and Chap. 8. In this sense, however, the *modus ponens* is an efficient strategy to *confirm* interpretations, schemes, and theories.

9.4 The Value of the Error

Since scientific research is carried out by human beings, *error* is present. The error discussed here is evidently a different thing from the *uncertainty* mentioned in Sect. 8.1.2. Here we mean the actual *mistakes* done by physicists in their research work, in particular, the mistakes in interpreting and explaining the studied phenomena. Among those who have not got a technical-scientific training, i.e. in public opinion, there is a widespread belief that "science" is virtually immune from errors. If there may be some errors, they can only be found *during* the scientific investigation process; finally, when the explanation of a phenomenon is provided in its definitive form, there may be limits and/or incompleteness, but not errors. "Scientific explanation" would somehow be synonymous with "certain, exact, definitive explanation".

[44] Elliott (1988), Sengupta and Sarkar (2003).

[45] If one believed this confirmation to be true in an absolute sense, one would fall into the already mentioned *fallacy of affirming the consequence*, cf. Sect. B.2.3.

Nevertheless, those involved in research know very well that this is not the case. On the one hand, the specific limits of any *scientific* explanation cannot be overlooked; on the other hand, they must be explicitly stated alongside the deemed valid explanation. Notwithstanding, the experience of error is still present within the daily practice of scientific research and it is more than a risk to beware of. The error is a component of the research process which, after being evaluated, recognized, discussed, i.e. when adequately addressed, it becomes a fruitful occasion for increasing knowledge. The error is not a fraud, and, most of the time, it is not the result of incompetence.

The fact that scientists are unlikely to explicitly admit their mistakes is probably one of the major problems encountered in the present context of scientific research. In those that are in the early stages of their career, it would be seen as a sign of inexperience, jeopardizing their career. In those who are already established, it would be an admission of limitation and inaccuracy.[46] On the contrary, from its beginnings the whole scientific enterprise is full of errors, corrections, and second thoughts.[47] These elements are present in the work of the *great scientists* who built the physics discipline. If one follows their path, it emerges how often great advances in knowledge have arisen *precisely* from their errors. Among these there are wrong interpretations, incomplete or inexact theories, partially or totally incorrect explanations, theories or explanations not adequately supported by data, experimental artefacts mistaken for genuine phenomena, etc. More than abstractly analysing all the cases that a researcher may run into, following the general method of this present text, examples of great physicists at work are proposed, highlighting that their *mistakes* have often proved to be extremely fruitful. We analyse here three revealing episodes not with the aim of a historical account, but to learn a *practical methodological approach*. In this way, the mistakes of the most important physicists become archetypes of *good scientific practices*. They teach us that "science should be done" by *making mistakes and trying again*.

9.4.1 Galilei's Mistakes: Observation, Simplicity, Beauty

The first "scientific error" case proposed belongs to the first physicist in the modern sense of the term, Galileo Galilei. It is certainly to be expected such a profound innovator to make several mistakes in his scientific interpretations—in an age when the tools are still very limited, both experimental and conceptual errors. However, it is remarkable to evaluate what these errors generate in the subsequent physics development. Therefore, here we briefly analyse, without claiming to be complete, three of the main errors by Galilei.

1. **Trigeminal Saturn**—During the summer of 1610, after the studies on Jupiter which the previous year led him to discover the Medici Satellites, Galilei begins to observe the planet Saturn. Probably he expects to see something similar to

46 We deal more deeply with this problem in the following chapter.

47 Chapter 12 of Ossicini (2012).

what he has seen observing Jupiter , i.e. satellites. Today it is known that Saturn has indeed satellites, one of which, Titan, is even larger than the Mercury planet. Galilei's instrument, however, does not allow him to see these satellites. On the contrary, the telescope aimed at Saturn allows the Pisan physicist to see a celestial body with an absolutely peculiar shape. He describes what he sees stating that the planet Saturn is *triple*, since it is composed of *three celestial bodies* which, contrary to Jupiter's satellites, always maintain the same mutual distance; between them, the central one appears three times larger than the laterals. Galilei calls the planet "Trigeminal Saturn", writing in a letter to Kepler: "Altissimum planetam tergeminum observavi". Observing the planet again after a few months, however, Galilei finds that the lateral bodies are apparently no longer present. Today we know that Saturn is surrounded by huge rings lying on its equatorial plane: in a completely casual way, Galilei has observed Saturn in the time interval just before rings, being edge-on, are invisible from the earth.[48]

In the following decades several astronomers point out again that Saturn has a peculiar shape. However, we need to wait almost half a century after Galilei's first observations for the publication of *Systema Saturnium* (1659) by Christiaan Huygens (1629–1695), where the Dutch scientist states that the observed shape is due to the presence of a ring surrounding the planet.[49] A few years later (1675), the Italian astronomer Gian Domenico Cassini (1625–1712) discovers that the ring is not unique, but is composed of several rings interspersed with areas of separation, among which the largest one will be named after him (Cassini division).[50]

On the one hand, we can say that Galilei completely got wrong his interpretation of the planet morphology: Saturn is not trigeminal at all. The error made by the Pisan is due to several reasons: certainly both the limitation of his observation instruments and bad luck (the visible cutting rings) contribute to it. On the other hand, what is worth noting about this error is that, while providing an erroneous interpretation, Galilei makes aware scholars that *the shape of Saturn is particular*; it is different from the shape of the other planets, there is *something else*. This is the real phenomenon that Galilei discovers and *precisely this* is what proves fruitful, stimulating for decades subsequent scientists to try to understand what was particular about Saturn. We can thus say that an erroneous observation by Galilei was able to open up a whole new research line.

2. **Circles and not ellipses**—Noteworthy is the fact that Galilei rejects a large part of the work by Johannes Kepler (1571–1630). Despite Kepler expresses scientific esteem and appreciation in various letters sent to Galilei, the latter always maintains a rather cold tone, replying very rarely and ignoring collaboration requests. Specifically, Galilei carefully avoids answering to Kepler's request for a telescope specimen (an instrument that the German astronomer is later able to obtain through other ways).[51] This position of the Pisan is probably due to the fact that he looks

[48] Chapter 1 of Segrè (1984), Ossicini (2012).

[49] Chapter 1 of Segrè (1984).

[50] Hoskin (1999).

[51] Chapter 5 of Hoskin (1999), Tornaghi and (a cura di) (1997).

at the German colleague as a formidable rival. Galilei approves his colleague's adherence to Copernicanism, but does not recognize his scientific achievements. Among Kepler's results, in particular, Galilei systematically ignores the *three laws*, obstinately rejecting the idea of *elliptical orbits* and remaining until the end of his days strictly anchored to the theory according to which the motion of the planets draws *circles*.[52]
The rejection of the orbit ellipticity and Kepler's laws is an evident error. It can be related to two factors. First of all, the astronomical theories that are being elaborated at the time of the two astronomers are not yet firmly founded, neither from an experimental point of view (see next section) nor from a theoretical one (for which it is necessary to wait for Newton). Therefore, both elliptical trajectories and Keplerian mathematical laws appear to be abstruse.[53] The second reason should probably be recognized in the *human limits*: Galilei's rejection highlights fear or envy for a colleague who is recognized as valid and thus perceived (perhaps rightly) as a formidable rival. However, this attitude deprives the Pisan scholar of useful tools, such as the predictions provided by the laws of the German, useful to validate the heliocentric model he embraces.

3. **Heliocentrism**—Certainly it looks paradoxical to consider *heliocentrism* among Galilei's scientific errors, but a careful evaluation of his writings shows the meaning of this statement. What is meant by *heliocentrism*? Any general physics course says that, in modern language, heliocentrism means assuming that a *Sun reference frame* is the privileged one to describe the motion of the solar system planets. This assumption is mainly based

    (a) on the validity of Newton's *Universal Gravitation law*;
    (b) on the evidence of the Earth motion with respect to the stars.

    At Galilei's time the first proposition is not known, the second has not got an observational confirmation yet. Leaving aside the gravitation law, it should be noted that Galilei has got absolutely no proof of the Earth motion with respect to the Sun: this is because at that time the measurement precision is not sufficient. Therefore, on which experimental data does Galilei found his conviction of the Copernican model validity? In his *Dialogo dei Massimi Sistemi* Galilei states that the proof of the Earth motion is given by the tides.

    (9.3) *[S]ará ormai tempo che venghiamo all'accidente massimo, dal quale presero origine i nostri ragionamenti; parlo del flusso e reflusso del mare, la cagione del quale pare che assai probabilmente si possa referire a i movimenti della Terra.*[54]

    It is by now be time that we come to the greatest occurrence, from which our reasoning originated; I am speaking about the ebb and flow of the sea, whose cause it seems most probably can be referred to the movements of the Earth.

---

[52] Hoskin (1999), Tornaghi and (a cura di) (1997).

[53] Chapter 5 of Hoskin (1999).

[54] G. Galilei, Dialogo dei Massimi sistemi, Giornata Terza, in Galilei (1897). Also cf. texts in Tornaghi and (a cura di) (1997).

Galilei is so convinced that the tides are proof of the Earth motion, he plans to title his book *On the flow and ebb* (of the tides). However, his assumption is completely incorrect: today we know that the tide phenomenon is due to the gravitational attraction of the Moon.

So, what experimental evidences would Galilei have needed to prove the heliocentrism validity? A first way would have been to compare the *predictions* between the most advanced geocentric model, the Ptolemaic one, with the Copernican model. Since astronomy is the discipline that was mathematized before any other (in Galilei's time, as well as today), it is ideally possible to apply a deductive method, try to develop predictions and check them.[55] From this point of view, the Ptolemaic model at that time universally accepted, which has been perfected by the *epicycles* assumption which accounted for the *retrograde motion* of the planets, is an advanced phenomenological theory. As a matter of fact, it makes possible to predict the position of the planets in excellent agreement with the experimental precision then available; this is why the model has been considered true for millennia. However, the agreement of this model with the observational data, in the form available at the beginning of the seventeenth century, is so good that it is not possible to distinguish between a geocentric model and a heliocentric model. Therefore, another experimental datum is needed.

From a general point of view, the confirmation of the heliocentrism was given by Newton's gravitation theory, thanks to its amazing successes in predicting the positions of the celestial bodies with excellent precision.[56] The mathematical form of the Universal Gravitation law provides a description in terms of forces, what is now called *dynamics*, as opposed to the mere *kinematics*. This form allows us to identify in the Sun reference frame a *privileged one* confirming the correctness of a heliocentric model.

On the other hand, the validity of the Newtonian theory does not give a *direct* experimental confirmation. A specific observational proof is provided by the evidence of an apparent motion of the fixed stars through the measurement of an annual *stellar parallax*,[57] which requires many efforts and arrives far beyond the Galilei's death, which occurs in 1642. A first phenomenon is attributed to the Earth motion with respect to the "fixed stars" in 1729, when the English astronomer James Bradley (1693–1762) accomplishes the measurement of a *stellar aberration*, noting that the stars observed during the year do not have a fixed position, instead they describe small ellipses.[58] Bradley is able to discriminate shifts on the sky of the order of 20 arc seconds.[59] Such observational phenomena, however, must be associated with estimations of stellar distances, in order to represent a definitive proof of cosmological models. It is in this direction that

[55] Cf. Sect. 4.1.

[56] Cf. Sect. 4.2, but also, e.g. quote 5.10 in Sect. 5.2.3.

[57] This consideration had already been made by Galilei himself index[2]Galilei, Galileo (Chap. 7 of Hoskin (1999)).

[58] Chapter 7 of Hoskin (1999).

[59] Kovalevsky and Seidelmann (2004).

several astronomers follow, in spite of the Newtonian paradigm imposed during the eighteenth century. Sketches of measurement of the fixed stars apparent motion, in order to obtain parallax measurements, are made at the beginning of the nineteenth century. The first scientists claiming to measure relatively large stellar motions are

> in 1804 Giuseppe Piazzi (1746–1826), Director of the Palermo Astronomical Observatory, who breaks the news of a great motion of the star *61 Cygni*, estimating it no less than 5 arc seconds per year[60];
> in 1806 Giuseppe Calandrelli (1749–1827), Director of the Vatican Observatory in Rome, who reports the motion of the star *Vega*[61];
> in 1812 Friedrich W. Bessel (1784–1846), Director of the Königsberg Observatory, who publishes the *61 Cygni* motion estimation, independently on the aforementioned Piazzi.[62]

The first two are affected by the peripheral role assumed by Italian scientists at the end of the eighteenth century; the last one, on the other hand, has got the opportunity to develop his own observations and research. Starting from the preliminary observations published in 1812, Bessel is able to finally provide, in 1838, a measure of the parallax angle of *61 Cygni*, of about 0.3 arc seconds.[63] Although in Bessel's time no scientist questions the heliocentric model in its Newtonian formulation, it is a fact that the direct and definitive experimental confirmation of the helicentrism is achieved almost two centuries after Galilei's death.

It is appropriate to analyse this fact in relation to our discussion about errors. From a strictly empirical-rational point of view, we have seen Galilei had not *the proofs* of heliocentrism. What he considered a *proof* (the motion of the tides) is not a proof; what would have been the *main test*, the stellar parallax, is not yet available. Indirect demonstration by Newtonian dynamics has not discovered yet; the other indirect proof, the comparison between the predictions by the heliocentric and geocentric models, is unable to provide a distinction.[64] So what remains of this *error*?

---

[60] Chapter 7 of Hoskin (1999).

[61] Settele (1992).

[62] Chapter 7 of Hoskin (1999).

[63] ibid.

[64] There were further data of the *Venus phases*, discovered by Galilei thanks to the use of the telescope (Bellone 2004; Segrè 1984). This discovery provided a strong argument against the Ptolemaic model, but it was also compatible with Tycho Brahe's hybrid model which was neither heliocentric nor geocentric (Segrè 1984). Precisely for this reason, at the beginning of the seventeenth century, this model was supported by the majority of scholars of the Roman College, one of the main astronomical research institutions of the time (Boyer 1968; Hoskin 1999; Tornaghi and (a cura di) 1997). It was the Roman College to reform the calendar and produced the *Gregorian Calendar* still in use.

Although he has not got *all* the elements, Galilei has got a powerful *insight*, which he out-and-out follows. He recognizes that a Copernican model[65] would have been *simpler* and *more elegant* to explain the motion of the celstial bodies, although, at a purely observational level, there is no evidence at that time. Today we may say that the *empirical error* committed by Galilei makes him grasping an even deeper feature of nature than adherence to experimental data: the beauty and simplicity of the mathematical descriptions.

> (9.4) We now see that we have to change the principle of simplicity into a principle of mathematical beauty. The research worker, in his efforts to express the fundamental laws of Nature in mathematical form, should strive mainly for mathematical beauty. He should still take simplicity into consideration in a subordinate way to beauty. [...] It often happens that the requirements of simplicity and of beauty are the same, but where they clash the latter must take precedence.[66]

Thus, Galilei's "mistake" turns out to be the occasion to introduce a new guiding criterion in scientific investigation, *beauty*. A criterion that is not easy to be defined with a logical and unambiguous reasoning,[67] but which is present in a large number of researches and it has proved to be fruitful and contributing, balanced with the necessary experimental criteria, to the development of knowledge.

9.4.2 Stern and Gerlach: When Experiment Is More **Robust** *than Theory*

The *Stern and Gerlach experiment* is considered one of the key experiments of Quantum Mechanics.[68] It is probably the most important in its field: it is reported and discussed in all Quantum Mechanics books, it is generally introduced as the experiment that *demonstrated the existence of the spin*, i.e. of the *intrinsic angular momentum* of electrons.[69] However, the facts are much more complex than such an interpretation, which is actually quite simplistic.

Usually, the experiment, carried out by Otto Stern (1888–1969) and Walther Gerlach (1889–1979) in 1922, is described in correlation with the *spin of electron*.[70] The considered experiment actually gives the possibility to demonstrate the existence of an *intrinsic angular momentum* (*spin*) of electrons. Indeed, a particle with charge q

[65] Although the term *heliocentric* and the term *Copernican* are often interchangeably used, it is important to note that the model developed by Copernicus is only one of the several heliocentric models conceived. As a matter of fact, Nicolaus Copernicus (1473–1543) proposes a model closely linked to the previous ones: according to this the planets still move with *circular motions* as they are set on *spheres*. Only with Kepler the concept of *orbit* and *elliptical* is introduced.

[66] Paul M. Dirac in (1938–39).

[67] Dirac himself recognizes that it is almost impossible to give a convincing definition of *mathematical beauty* but that it is nevertheless evident to those who do mathematics (Dirac 1938–39).

[68] Chapter 5 of Feynman et al. (2005), Chap. 1 of Sakurai (1994).

[69] ibid.

[70] Chapter 5 of Feynman et al. (2005), Ossicini (2012), Sakurai (1994).

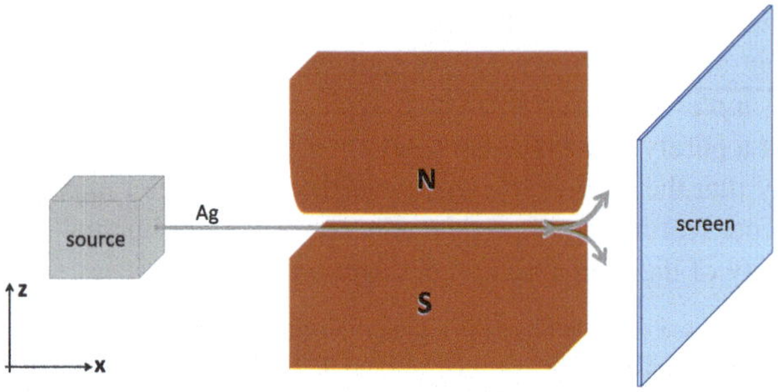

Fig. 9.1 Simple scheme of the *Stern and Gerlach experiment*. A *source* emits a beam of atoms in the x-direction (*Ag* in the original experiment), passing between the expansions of a suitably shaped magnet (*N-S*) so to produce a gradient of magnetic field; this gradient is able to produce a spatial separation of the atoms along the z-direction, according to their respective *magnetic moment*, which can be recognized from their distribution on the *screen*

and mass m, with an angular momentum $\mathbf{L}$, is associated with a magnetic moment $\boldsymbol{\mu}$ of

$$\boldsymbol{\mu} = \frac{q}{2m}\mathbf{L}, \tag{9.21}$$

making the measurement of μ a method to estimate $\mathbf{L}$. Being the force acting on a dipole μ in a magnetic field $\mathbf{B}_{\text{ext}}$ is given by $\mathbf{F} = \nabla\,(\mu \cdot \mathbf{B}_{\text{ext}})$. In particular, if just B_z varies, that is, the component along z-direction, we have[71]

$$F_z = \mu \cdot \nabla B_z \simeq \mu_z \frac{\partial}{\partial z} B_z.$$

In the Stern and Gerlach experiment a highly anisotropic magnetic field is used, in such a way to maximize the force felt by the magnetic dipoles (Fig. 9.1). Basic Quantum Mechanics states that the angular moment is quantized, its strength is given by $L = \hbar\sqrt{l(l+1)}$ and the z-direction component is

$$L_z = m_l \hbar$$

with the usual quantum numbers $l = 0, 1, \ldots$ e $m_l = -l, -l+1, \ldots, l$. In an experiment like this, this causes the atomic beam to produce $2l+1$ distinct traces on the screen at the exit of the magnet. Since for the Ag atoms used by Stern and Gerlach $\mathbf{L} = 0$, there is a null magnetic moment and, the factor ∇B_z should not act on the atoms. This determines that the Ag atoms should draw a single continuous trace on the screen. Conversely, Stern and Gerlach see *two* discrete traces. This phenomenon is explained through the assumption that the total angular momentum of an atom does not coincide with its orbital momentum, but there is another term, an *intrinsic angular momentum* $\mathbf{S}$ of the electrons, which is the *spin*. Therefore, we have the total

[71] Cf. e.g. Chap. 1 of Sakurai (1994).

momentum which is

$$\mathbf{J} = \mathbf{L} + \mathbf{S}, \tag{9.22}$$

so that the (9.21) should be suitably modified, with all consequences for Quantum Mechanics. Quantization rules state that, being the total angular momentum depending on the quantum number j, we get $2j + 1$ traces on the experimental screen. The presence of two traces when $L = 0$ demonstrates that *spin* exists, that the associated s quantum number is $1/2$, i.e. *the electrons have half-integer spin*, to which an angular momentum, $\pm\hbar/2$, corresponds to z and produces $2s + 1$ traces.

This is the modern interpretation of the experimental data. However, this is not the line of thought followed by Stern and Gerlach when they accomplish the experiment.[72] There are two elements of the modern interpretation which are completely absent at the time of the experiment.

1. In 1921 Otto Stern suggests his next experimentation and states the aim of proposing a technique to study the *space quantization*: "Ein Weg zur experimentellen Prüfung der Richtungsquantelung im Magnetfeld".[73] The experiment wants to study the implications of a specific concept, that of *space quantization*, which is a tassel of the *old* quantum theory by Arnold Sommerfeld (1868–1951). On the contrary, the *spin* concept is introduced only in 1925 by the theorists George E. Uhlenbeck (1900–1988) and Samuel A. Goudsmit (1902–1978)[74]. In no way, do Stern and Gerlach conduct their experiment with the aim of visualizing or studying *spin*.
2. The theory by Arnold Sommerfeld (1868–1951), from which Stern starts, predicts $J = L = \hbar$, i.e. the angular momentum, coinciding with the orbital one, is $l = 1$.

Today we know that the Stern and Gerlach experiment is wrong, both in its theoretical assumptions and in the interpretation immediately given by the scientific community. When *spin* is introduced a few years later, it is related to the *anomalous Zeeman effect*.[75] However, a little later the British physicist Ronald G. Fraser (1899–1985) demonstrates that the orbital angular momentum of Ag atoms is $L = 0$[76]: it follows that Stern and Gerlach should have inspected no separation between the traces on the screen. To explain the result by Stern and Gerlach it is necessary to introduce the *total angular momentum*, in the form of (9.22), into the atomic Quantum Mechanics. The experiment thus becomes a demonstration of the *spin* existence. Even more: this observation is today considered a cornerstone of Quantum Mechanics: it is able of highlighting, in a relatively simple way, the peculiar features of the theory.[77] The identification of the *spin* has provided solid experimental bases to explain many kinds of atomic and molecular bonds and symmetries, to study the *entanglement*

[72] Ossicini (2012), Weinert (1995).

[73] "A way to experimentally demonstrate space quantization in the magnetic field", Stern (1921).

[74] Ossicini (2012), Uhlenbeck and Goudsmit (1925).

[75] Inguscio and Fallani (2013), Ossicini (2012).

[76] Fraser (1927).

[77] Chapter 1 of Sakurai (1994).

concept, the teleportation as well as the quantum information theory, and finally to investigate the branch of *spintronics*, the development of atomic clocks, etc.[78] How many problems and errors does this experiment solve: its effects are truly noteworthy!

The fruitful results of this experiment can be associated with an interesting empirical concept. Though founded on an *erroneous* theory, the experiment is what is technically called to be *robust*[79]: it is well implemented, able to provide clear and reproducible data, even able to survive the theories from which it started.[80]

What has been presented shows that there are scientific experiments which, in some way, have got a much deeper meaning than the theories to which they are related. The experiments have not got the sole aim of *confirming* or, in Popper's words, *falsifying* the predictions of a theoretical structure.[81] They can be able to suggest research directions and developments of theories, as if they have got a life of their own.[82] Although based on an *incorrect* theory, Stern and Gerlach experiment has been able to produce valid, important and fruitful discoveries. Carrying out a good experiment is similar to unearthing a buried treasure: if the experiment is really *great* (which can only be known a posteriori), the opening of the casket can show not only great riches, but unexpected and previously unthinkable valuable finds.

9.4.3 The Cosmological Constant: Problem, Mistake, Insight, Prediction

Gravitational theories have always been considered among the most elegant and coherent in physics. Thus, in the era immediately after their reception, both *Newton's theory*, with the grandiose astronomical results that it made possible, and Einstein's *General Relativity*, which started the modern cosmology. The development of related concepts, however, is far from orderly and linear.[83] On the contrary, they are characterized by adjustments, errors, and afterthoughts: concerning these, it is instructive, as well as fascinating, to evaluate, very briefly, the essential items.

Newton has already realized that the form of his *Universal Gravitation law* raises issues about the very structure of the universe. The main issue is the following: if, among all the celestial bodies, there is an exclusively attractive force, such as the Newtonian gravitational one, why don't they collapse on themselves? Newton's answer to this question is that the universe should be *infinite* in all directions: it is the presence of an infinite number of celestial bodies, approximately isotropically distributed in the entire infinite space, able to produce a total gravitational force that maintain the general stability of the universe and avoid a general gravitational

[78] Cf. e.g. Ossicini (2012).

[79] Stegenga (2009).

[80] Ossicini (2012), Stegenga (2009).

[81] Cf. Sect. 5.2.2.

[82] Weinert (1995).

[83] Carroll (2001).

collapse.[84] This vision is consistent with the Newtonian conception, according to which space is a sort of *infinite box* that exists in itself independently on the bodies.[85]

Similar to all concepts involving a *factual infinite*, many do not consider satisfactory Newton's vision: as a matter of fact, it poses great conceptual and theoretical problems to formally define and measure quantities such as force or *potential* gravitational, when they would need to be calculated taking into account an infinite number of factors.[86]

Theoretically, this problem is directly addressed at the end of the nineteenth century by the mathematician Carl G. Neumann (1832–1925) and the astronomer Hugo von Seeliger (1849–1924). They believe that Newton is right in asking why the universe does not gravitationally collapse.[87] They take into account the Poisson equation for the gravitational potential

$$\nabla^2\phi = 4\pi G\rho, \tag{9.23}$$

where $\rho = \rho(\mathbf{r})$ is the density of mass. Neumann and Seeliger proposes to add a term to the Equation in order to depict some *repulsion* between bodies. By introducing such element, that would be called *cosmological term*, the equation becomes[88]

$$\nabla^2\phi - \Lambda\phi = 4\pi G\rho, \tag{9.24}$$

where Λ is called *cosmological constant*. If $\Lambda = 0$, the (9.24) is reduced to the (9.23); assuming a suitable $\Lambda > 0$, the term $-\Lambda\phi$ is exactly aimed at balancing the attractive gravitational force through a repulsive cosmological one, in order to avoid collapse.

In 1915 Einstein publishes the *Theory of General Relativity* where he presents the *field equation* (4.3) in a formulation that *does not* include a cosmological term

$$\frac{1}{2}\mathcal{G}_{jk} = -4\pi G\frac{T_{jk}}{c^4}.$$

As above mentioned, the resulting Einstein's idea is, in some way, the opposite of Newton's[89]: the structure of space (and time) is by no means absolute, instead is determined by the bodies, i.e. by the mass–energy distribution represented by the tensor T_{jk}. The solutions of the (4.3) are studied in the following years.[90] Einstein realizes that his field equation predicts a *dynamic universe*, i.e. a universe in which its own structure, space–time, can expand or contract. This prediction does not agree

[84] Ray (1990).

[85] Cf. Sect. 4.2.2 and in particular quote 4.6 in Sect. 4.2.2.

[86] Ray (1990).

[87] ibid.

[88] Harvey and Schucking (2000), Ray (1990).

[89] It recovers Aristotle's space concept according to which space is not absolute but it is determined by bodies (Strumia 2017).

[90] According to Dyson's observation in quote 8.7 in Sect. 8.2.1.

with the astronomical measurements available at that time.[91] For this, and for other considerations, more conceptual than technical,[92] in 1917 Einstein decides to introduce a *cosmological term* into his equation.[93] By adding this term in the (4.3) in order to highlight the similarities with the (9.24)[94] we obtain

$$\frac{1}{2}\mathcal{G}_{jk} - \frac{1}{2}\Lambda g_{jk} = -4\pi G \frac{T_{jk}}{c^4}, \tag{9.25}$$

which nevertheless is usually written[95] in its compact form[96]

$$\mathcal{G}_{jk} = \Lambda g_{jk} - \kappa T_{jk}. \tag{9.26}$$

Thanks to the introduction of the new term, the solutions of Einstein's equations are able to predict a *static universe*. The great physicist is satisfied of this solution, which is called *static Einstein's universe*.

However, during the 1920s, mainly thanks to the works by Alexander A. Friedmann (1888–1925), George Lemaître (1894–1966) and Edwin Hubble (1889–1953), the scientific community realizes that, contrary to what has been previously believed, the universe it is actually expanding.[97] The crucial work is published in 1927 by Lemaître.[98] Combining the predictions from general relativity with the most recent astronomical observations, he arrives at the following conclusions:

> (9.5) *Nous avons obtenu une solution qui vérifie les conditions suivantes:*
> *1. La masse de l'univers est constante et est liée á la constante cosmologique [...]*
> *2. Le rayon de l'univers croit sans cesse depuis une valeur asymptotique R_0 pour $t = -\infty$. [...]*
> *4. Le rayon de l'univers est du même ordre de grandeur que le rayon R_E déduit de la densité par la formule d'Einstein. On a*
>
> $$R = R_E \sqrt[3]{\frac{R_0}{R_E}} = \frac{1}{5} R_E$$
>
> *[...] Il resterait á se rendre compte de la cause de l'expansion de l'univers. Nous avons vu que la pression de radiation travaille lors de l'expansion. [...] Il semble que lá doit être cherchée l'origine de la vitesse d'expansion R'/R qu'Einstein supposait nulle et qui dans notre interprétation est observée comme vitesse radiale des nébuleuses extragalactiques.*[99]

[91] Lemaître (1949), Ray (1990).

[92] Ray (1990).

[93] Carroll (2001), Harvey and Schucking (2000), Lemaître (1949).

[94] Chapter 7 of Ohanian and Ruffini (2013), Carroll (2001).

[95] Carroll (2001), Harvey and Schucking (2000).

[96] With $\kappa = 8\pi G/c^4$.

[97] Lemaître (1949), Ray (1990).

[98] Hoskin (1999), Lemaître (1927).

[99] George Lemaître in (1927).

> We obtained a solution that satisfies the following conditions:
> 1. The mass of the universe is constant and is related to the cosmological constant [...]
> 2. The radius of the universe continuously grows from an asymptotic value R_0 for $t = -\infty$. [...]
> 4. The radius of the universe is of the same order of magnitude as the radius R_E deduced from the density by Einstein's formula. We have
>
> $$R = R_E \sqrt[3]{\frac{R_0}{R_E}} = \frac{1}{5} R_E$$
>
> [...] The cause of the expansion of the universe remains to be understood. We have seen that radiation pressure works during expansion. [...] It seems that the origin of the speed of expansion must be sought here. R'/R which Einstein assumed to be null and which in our interpretation is observed as the radial speed of extragalactic nebulae.

As a result of this work and the accumulating experimental evidences, Einstein recognizes that the expansion of the universe is indeed present. Between 1931 and 1932 he proposes a model of an expanding universe where the *cosmological constant* is zero. Consequently, the German physicist, seeking the greatest simplicity, suggests to eliminate Λ from his field equations.[100] Some believe that Einstein claims that the cosmological constant is his "biggest mistake"[101]: although such a categorical statement is very doubtful, since it is not present in Einstein's writings. Nevertheless, it is certain that these have never been satisfied with the cosmological term, probably considering it a sort of *oversight*.[102]

Field equations in the form of (9.26), however, have got much larger generality than in the form (4.3)[103]; for this reason, the term Λ will never be completely eliminated from *General Relativity*, playing an important role in the development of cosmological models.[104] Introduced for the purpose of obtaining a static universe model, Λ opens Einstein's relativity to modern cosmology.[105]

> (9.6) The history of science provides many instances of discoveries which have been made for reasons which are no longer considered satisfactory. It may be that the discovery of the cosmological constant is such a case.[106]

In the years following Einstein's death, several models are developed, in which the assumption $\Lambda \neq 0$ plays a more or less important role.[107] At the end of the twentieth century, observational studies accomplished on supernovae at great distances

[100] Einstein (1931), Ray (1990).

[101] Sentence quoted by his friend George Gamow, who, however, was known for his humour, his puns, and his use of hyperbole (O'Raifeartaigh and Mitton 2018).

[102] O'Raifeartaigh and Mitton (2018).

[103] Ray (1990).

[104] Cf. e.g. Chap. 9 of Ohanian and Ruffini (2013), Carroll (2001), Lemaître (1949).

[105] ibid.

[106] G. Lemaître in (1949).

[107] Cf. e.g. Ray (1990), Carroll (2001), Harvey and Schucking (2000), Ohanian and Ruffini (2013).

demonstrate an acceleration of the universe expansion.[108] Following the most trusted cosmological model, the ΛCDM, this behaviour is interpreted as a confirmation of a non-zero and strictly positive value of the *cosmological constant*.[109]

Therefore, it is here appropriate to think back to the conceptual itinerary followed by the *cosmological constant*.

Initially it is suggested as a *possible solution to a problem* to which Newton had not given a satisfactory answer.
It is introduced in the formulation of *General Relativity* to generalize its equations and make the theory more consistent with the experimental data known at that time.
It is used for the development, more or less theoretical, of cosmological models.
It is abandoned and expunged from General Relativity for simplicity, since it is *no longer necessary* to explain the new (correct) observational data.
It is kept, in the *field equation*, by some theoretical astrophysicists to develop possible models of the universe with greater generality.
Finally, it is used to explain the most recent cosmological observational data.

We summarize all these items in some questions and answers, paradigmatic of scientific research proceedings.

Did introducing Λ solve any problems? It solved some problems, it gave rise to new ones.
Was introducing Λ a *mistake*? It was for some reasons, it wasn't for others.
Was introducing Λ an *insight*? Certainly yes, it was a profound insight, although not adequately supported by data at that time.
Was the introduction of Λ somehow a *prediction*? With today's knowledge we can say it was: it opened a new research field and anticipated phenomena that would be experimentally confirmed many decades later.

This highlights that the research development is a highly *nonlinear* process,[110] where each element is able to provide a completely unexpected contribution and for which even the most brilliant mind cannot deduce all the implications.

References

A. Aspect, Closing the door on Einstein and Bohr's quantum debate. Physics **8**, 123 (2015)
G. Basti, *Filosofia della Natura e della Scienza*, vol. 1, 1a edn. (Pontificia Università Lateranense, Roma, Città del Vaticano, 2002)

[108] For this result, the US astrophysicists Saul Perlmutter (1959), Brian Schmidt (1967), and Adam Riess (1969) were awarded with the *Nobel Prize in Physics* in 2011.

[109] Chapter 9 of Ohanian and Ruffini (2013), Hoskin (1999).

[110] In this *nonlinearity* we can find an echo of the rather iconoclastic but acute considerations of Feyerabend in quote 5.20 in Sect. 5.3.2.

E. Bellone, *Caos e Armonia - Storia della Fisica* (UTET, Torino, Italia, 2004)
M. Born, *Atomic Physics*, 8th edn. (Blackie & Son, London & Glasgow, UK, 1969)
C.B. Boyer, *A History of Mathematics*, 2nd edn. (Wiley, New York, USA, 1968)
A. Carati, L. Galgani, Non locality of classical electrodynamics of point particles and violation of Bell's inequalities. Nuovo Cimento B **114**, 489 (1999)
A. Carati, L. Galgani, Theory of dynamical systems and the relations between classical and quantum mechanics. Found. Phys. **31**, 69pp (2001)
A. Carati, L. Galgani, Classical microscopic theory of dispersion, emission and absorption of light in dielectrics. Eur. Phys. J. D **68**, 307pp (2014)
A. Carati, L. Galgani, A. Maiocchi, F. Gangemi, R. Gangemi, Classical infrared spectra of ionic crystals and their relevance for statistical mechanics. Phys. A **506**, 1–10 (2018)
S.M. Carroll, The cosmological constant. Living Rev. Relativ. **4**, 1 (2001)
A. Church, *Introduction to Mathematical Logic* (Princeton University Press, Princeton, New Jersey, USA, 1956)
I.M. Copi, C. Cohen, V. Rodych, *Introduction to Logic*, 15th edn. (Taylor & Francis, New York, USA, 2019)
P.A.M. Dirac, The relation between mathematics and physics. Proc. R. Soc. (Edinb.) **59**, 122–129 (1938–39)
A. Einstein, Zum kosmologischen Problem der allgemeinen Relativitätstheorie. *Sitzungsber. Preuss. Akad. Wiss.*, pp. 235–237 (1931)
A. Einstein, B. Podolsky, N. Rosen, Can quantum-mechanical description of physical reality be considered complete? Phys. Rev. **47**, 777–780 (1935)
R.S. Elliott, The history of electromagnetics as Hertz would have known it. IEEE Antennas Propag. Soc. Newsl. **30**, 5–18 (1988)
R.P. Feynman, R. Leighton, M. Sands, *The Feynman Lectures on Physics*, vol. 3, 2nd edn. (Addison-Wesley, Boston, USA, 2005)
R.G.J. Fraser, The effective cross section of the oriented hydrogen atom. Proc. R. Soc. A **114**, 212–221 (1927)
G. Galilei, Dialogo sopra i due massimi sistemi del mondo tolemaico e copernicano, in *Le opere di Galileo Galilei, sotto gli auspici di Sua Maestà il Re d'Italia* (Tipografia G. Barbera, Firenze, Italia, 1897)
A. Harvey, E. Schucking, Einstein's mistake and the cosmological constant. Am. J. Phys. **68**, 723–727 (2000)
M. Hoskin, *The Cambridge Concise History of Astronomy* (Cambridge Unviversity Press, Cambridge, UK, 1999)
M. Inguscio, L. Fallani, *Atomic Physics - Precise Measurements & Ultracold Matter* (Oxford University Press, Oxford, UK, 2013)
J.P.A. Ioannidis, Why most published research findings are false. PLoS Med. **2**(8), e124 (2005)
J. Kovalevsky, P.K. Seidelmann, *Fundamentals of Astrometry* (Cambridge University Press, Cambridge, UK, 2004)
G.E. Lemaître, The cosmological constant, in *Albert Einstein Philosopher-Scientist* (MJF Books, New York, USA, 1949)
G.E. Lemaître, Un univers homogène de masse constante et de rayon croissant rendant compte de la vitesse radiale des nébuleuses extra-galactiques. Annales de la Société Scientifique de Bruxelles **47**, 49 (1927)
A. Lerose, A. Sanzeni, A. Carati, L. Galgani, Classical microscopic theory of polaritons in ionic crystals. Eur. Phys. J. D **68**, 35pp (2014)
J.C. Maxwell, F. Jenkin, On the elementary relations between electrical measurements. Lond. Edinb. Dublin Philos. Mag. J. Sci. **29**(198), 436–460 (1865)
J.K. Maxwell, A dynamical theory of the electromagnetic field. Philos. Trans. R. Soc. Lond. **155**, 459–512 (1865)
D.C. Montgomery, *Design and Analysis of Experiments*, 5th edn. (Wiley, New York, USA, 2001)

H. Ohanian, R. Ruffini, *Gravitation and Spacetime*, 3rd edn. (Cambridge University Press, Cambridge, UK, 2013)
C. O'Raifeartaigh, S. Mitton, Interrogating the legend of Einstein's "biggest blunder". Phys. Perspect. **20** (2018)
S. Ossicini, *L'universo è fatto di storie non solo di atomi - Breve storia delle truffe scientifiche* (Neri Pozza, Vicenza, Italia, 2012)
R. Penrose, *The Road to Reality. A Complete Guide to the Laws of the Universe* (Jonathan Cape, London, UK, 2004)
L. Picasso, *Lectures in Quantum Mechanics* (Springer, Cham, Schweiz, 2016)
C. Ray, The cosmological constant: Einstein's greatest mistake? Slud. Hist. Phil. Sci. **21**, 589–604 (1990)
J.J. Sakurai, *Modern Quantum Mechanics*, 2nd edn. (Addison-Weseley, Boston, USA, 1994)
E. Segrè, *From Falling Bodies to Radio Waves: Classical Physicists and Their Discoveries* (W.H. Freeman & C., New York, USA, 1984)
D.L. Sengupta, T.K. Sarkar, Maxwell, Hertz, the Maxwellians, and the early history of electromagnetic waves. IEEE Antennas Propag. Mag. **45**, 13–19 (2003)
G. Settele, Richiesta di imprimatur per il manuale Elementi di Astronomia, in *Copernico Galilei e la Chiesa, fine della controversia (1820), gli atti del Sant'Uffizio*, ed. by W. Brandmüller, E.J. Greipl (Leo S. Olschki Editore, Firenze, Italia, 1992), pp. 167–177. https://disf.org/settele-richiesta-approvazione-eliocentrismo
J. Stegenga, Robustness, discordance, and relevance. Philos. Sci. **76**, 650–661 (2009)
O. Stern, Ein Weg zur experimentellen Prüfung der Richtungsquantelung im Magnetfeld. Z. Phys. **7**, 249 (1921)
A. Strumia, *Dalla filosofia della scienza alla filosofia nella scienza*, 1a edn. (ESC - Edizioni Santa Croce, Roma, Italia, 2017)
F. Tornaghi, G. Mangiarotti (a cura di), *Galileo Galilei mito e raltá - Itinerario antologico* (CE.Se.D. Edizioni, Milano, Italia, 1997)
G.E. Uhlenbeck, S.A. Goudsmit, Ersetzung der Hypothese von unmechanischen Zwang durch eine Forderung bezüglich des unneren Verhaltens jedes einzelnen Elektron. Naturwiss **13**, 953 (1925)
F. Weinert, Wrong theory-Right experiment: the significance of the Stern-Gerlach experiments. Stud. Hist. Phil. Mod. Phys. **26**(1), 75–86 (1995)

Chapter 10
How "Science Should Not Be Done": Frauds and Distortions

> *A possible explanation of the physicist's use of mathematics to formulate his laws of nature is that he is a somewhat irresponsible person.*
>
> *Eugene Wigner*

This chapter may result a little unpleasant for some, similarly to the previous one Chap. 5 even though for quite different reasons. This chapter is concerned with the *scientific community* and when its members misbehave. As a matter of fact, the aim of this text is to precisely lighting up those elements which are generally not visible, the "engine room" of physics. It is advisable to become aware of how in this machine there may be breakdowns and jams, which may be due to the willful misconduct of individuals, the inexperience of the controllers, and the poor management of the community organization. Therefore, frauds and misrepresentations are a full part of *what is taken for granted.*

Definition 10.1 (*Scientific community*) *Scientific community* is defined as that group of individuals who, by sharing knowledge, methods, and languages, carry out research and evaluate the results in the field of mathematical, physical, and natural sciences.

Despite the fact that our analysis is focused on that part of the community specifically dealing with physics (that we could call *physicists community*), the main features and customs are shared with researchers of all other disciplines. Among the characteristics of the aforementioned scientific community, and within the actual daily activity of those belonging to it, this chapter focus on the pathological or problematic aspects. It would seem trivial to highlight that the behaviour of individual physicists can be correct or incorrect. This seemingly obvious statement reminds us that, contrary to popular opinion, the practice of physics is not a pristine ritual celebrated by angelic creatures in white robes, but a daily activity performed by passionate human beings. For this reason, both human strengths and weaknesses can be found in physicists.

J. Parravicini, *The Foundations of Experimental Physics*,
https://doi.org/10.1007/978-3-031-55452-0_10

Although the *physicists' community* has spontaneously structured itself in such a way as to identify, stigmatize, and punish scientifically incorrect behaviours, especially *scientific frauds*, this practice has got numerous flaws. Sometimes the forms of the scientific community seem to favour, rather than hinder, the emergence of misbehaviour.

Following the general method of this discussion, we try to privilege the analysis of concrete examples and facts, before conducting a more general and abstract discourse.

10.1 Frauds in Physics

There is a famous sentence by the American physicist Steven Weinberg (1933–2021) which is quoted as an example of an overly optimistic statement.

> (10.1) To the best of my knowledge there never has been an important case of outright falsification of data in physics.[1]

In this imprudent statement we may read a position tending to present physics as an immaculate discipline according to a scientism view, an erroneous position easily made by scientists. It would be nice if Weinberg were right. Unfortunately, we must notice that facts deny it: there have been several frauds and forgeries, in all the natural sciences, including physics. Even worse: some of these are not secondary at all, but sometimes they involve leading figures and research institutions. If, on the one hand, this could be unpleasant, on the other it shouldn't come as a surprise: as already stated, being a manifestation of the human spirit, in physicists it is possible to find both the merits and the *defects* of human beings, including dishonesty.

The issue of fraud in sciences is not recent at all: many authors cite a 1830 text[2] written by the English mathematician Charles Babbage (1791–1871) as the first one where this problem is rationally analysed.[3] Providing a precise definition of *scientific fraud* depends on the adopted criteria. The following definition can be given in a strict sense[4]:

Definition 10.2 (Scientific fraud *in a strict sense*) *Scientific fraud* is a violation of *scientific integrity* by practices such as data fabrication, falsification, and plagiarism, in drafting projects, carrying out research, publishing results, and/or reviewing.

The practices of *fabrication, forgery, and plagiarism* are often referred to by the acronym FFP.[5] To the aforementioned *frauds in a strict sense*, practices euphemistically called *questionable* (not to say *unfair*) can be added.[6]

[1] Steven Weinberg in (1992).

[2] Babbage (1830).

[3] Cf. e.g. Reich (2009), Chevassus-au-Louis (2016).

[4] Cf. Chap. 1 of Chevassus-au-Louis (2016), whose English translation is published in Chevassus-au-Louis (2019).

[5] ibid.

[6] ibid.

Definition 10.3 (*Unfair scientific practices*) We considered *scientifically unfair* ("*questionable*") those practices such as: presence of conflict of interest; one-sided choice of data presented in the articles; adding "ghost" authors to publications to which they have not contributed; refusal to make raw data available; non-recognition of some authors, especially members of the technical staff; psychological abuse of students or subordinates; violations of regulations, ethical and otherwise, in carrying out the experiments.

Therefore, it is possible to provide a definition of *scientific fraud in a broad sense* including both the practices of Definition 10.2 as those of the Definition 10.3.[7]

Definition 10.4 (Scientific frauds *in a broad sense*) *Scientific frauds in a broad sense* includes both frauds *in a strict sense*, in particular FFP practices, as well as the aforementioned unfair practices.

If the determination provided by Definition 10.2 is perhaps too restrictive, on the other hand considering as *frauds* all the unfairnesses listed by the Definition 10.3 (and analogous ones) is perhaps equally excessive. Similarly, if FFP-type cases are particularly serious, and thus limited,[8] any researcher of some experience has at least heard from his working contacts about some "questionable" practices among those listed in Definition 10.3. Therefore, our analysis here focuses on *frauds* in a strict sense (those enumerated in the FFP) with special attention to some paradigmatic cases where the principle of *submission to judgement of the experiment* enunciated by Feynman[9] is completely disregarded and even overturned. Nevertheless, it is important to know that it is possible to encounter both fraud and unfair practices as well as to know the main mechanisms facilitating them.

In recent decades, the research world has enormously grown in quantity, in the number of researchers, institutions, publications, fields of investigation, data, models.[10] For this reason, it is our aim to raise the awareness of what can be encountered within scientific research and to help developing the aptitude for a critical examination of all results claiming to be scientific.

10.1.1 Scientists of the Past in Front of a Historical Episode

The case of the German physicist Emil Rupp (1898–1979) is probably the most important blatant episode of scientific fraud in physics. Rupp's name is unknown among physicists today, but during the "years that shook physics" (Gamow), he was one of the most important and esteemed experimentalists. The quotation by Louis de Broglie in his *Nobel Lecture* of 1929 is significant, as in it the name of Rupp is

[7] ibid.

[8] The following sections describe better to what extent they are limited.

[9] Definition 1.1 and related discussion.

[10] Ossicini (2012).

combined with that of the major experimental physicists of that time, such as the American Lester Germer (1896–1971), Clinton J. Davisson (1881–1958), and the British George P. Thomson (1892–1975).

> (10.2) Now, experiment which is the final judge of theories, has shown that the phenomenon of electron diffraction by crystals actually exists and that it obeys exactly and quantitatively the laws of wave mechanics. To Davisson and Germer [...] falls the honour of being the first to observe the phenomenon [...] Professor G. P. Thomson of Aberdeen [...] found the same phenomena. Then Rupp in Germany, Kikuchi in Japan, Ponte in France and others reproduced them, varying the experimental conditions. Today, the existence of the phenomenon is beyond doubt and the slight difficulties of interpretation posed by the first experiments of Davisson and Germer appear to have been satisfactorily solved.
>
> Rupp has even managed to bring about electron diffraction in a particularly striking form.[11]

Among the aforementioned, Davisson and Thomson are awarded the Nobel Prize for Physics in 1937 for their experiments on the *wave-particle duality* mentioned by De Broglie, while Rupp's name is unknown.

Emil Rupp begins his career in 1923 at the Heidelberg University, where he works under the guidance of Philipp Lenard (1862–1947), Nobel Prize for Physics 1905. The latter gets the Prize for his studies on *cathode rays*, namely, for the investigation of electron beams, i.e. of negatively charged particles. About 20 years later, Lenard's group extends its line of research to the study of what the physicist Eugene Goldstein (1850–1930) calls *Kanalstrahlen* (*canal rays*), i.e. beams of positively charged ions. Rupp specializes in the study of the visible radiation emitted by these beams, using interferometric techniques. In 1926 he publishes his first work on the subject,[12] where the crucial experimental parameter for the experiment success is the *coherence length*. This is the largest difference of optical path for which the light produced at different points or moments of the *channel rays* beam still remains able to produce interference fringes.

At the time, the issue about the *wave nature* of particles is still open. In 1924 De Broglie advances his hypothesis on the wave nature of electrons, while several scientists such as Bohr are still convinced that quantization is an effect due to *statistics*, rather than an intrinsic feature of microscopic matter.[13] For this reason, experiments able to clearly show the wave nature of electrons are a hot topic. Einstein hears of the work published by Rupp and is convinced that the colleague is able to accomplish some experiments he has in mind. At the beginning of 1926 he writes to Rupp proposing a collaboration; the latter replies positively. Einstein is already one of the most famous and established physicists, he is already winner of the Nobel Prize; Rupp, 20 years younger, has just begun his scientific career.

Although this could be an interesting topic, this is not the place for analysing the extensive correspondence between Einstein and Rupp during 1926[14] and we limit

---

[11] Louis de Broglie in (1965).

[12] French (1999).

[13] French (1999).

[14] This correspondence is reconstructed in several publications, including Ossicini (2012), van Dongen (2007).

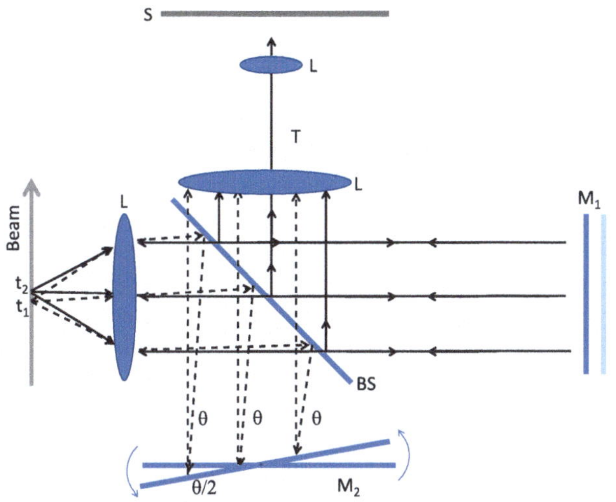

Fig. 10.1 Scheme of the *rotated mirror experiment* (*Spiegeldrehversuch*) by Einstein and Rupp (van Dongen 2007). The beam of *channel rays* (*Beam*) lies in the focal plane of a lens; transversally to the motion of its particles, it emits light, which is collected by the lens and sent to a Michelson interferometer containing, in addition to suitable lenses (*L*), a beam-splitter (*BS*), mobile mirrors (M_1 and M_2), a telescope (*T*) and a screen (*S*) where the interference of fringes is visualized. Einstein notes that light emitted at different times, t_1 and t_2, can produce interference only if their optical paths are suitably modified. For this, he suggests to implement the rotation of the mirror M_2 by an angle $\theta/2$. Note that this rotation is (correctly) drawn in a *clockwise* sense

to give an essential report. For a couple of years Rupp is Einstein's *experimental collaborator*: the latter is "the mind", the former is "his arm". The guiding idea is to demonstrate the wave nature of electrons through interferometric measurements. Einstein writes proposing the experimental schemes and the measurements to be performed, Rupp implements everything in his laboratory, records the data, illustrates them by replying in writing. However, the collaboration between these two poses a problem: formally Rupp's boss and head of his laboratory in Heidelberg is still Lenard, who is a bitter enemy of Einstein, being a fervent supporter of Nazism. Because of this, Einstein has got no opportunity to visit Rupp's laboratory and witness his measurements.[15] Einstein conceives and discusses different configurations, until he comes to the conclusion that the correct scheme is that of the so-called *Spiegeldrehversuch*, i.e. the *experiment of the rotated mirror* (Fig. 10.1).

There are two key elements to inspect the interference fringes, which are correlated with the wave nature of the electrons: having a sufficient *coherence length* of the emitted radiation and rotating the mirror by the *right angle* $\theta/2$ in the *right direction*.

For this reason Einstein never attends the experiments: he and Rupp communicate exclusively by mail. In their letters, the former always tells the latter what to do, who replies mostly within a few days (often the following day), writing to have per-

[15] Ossicini (2012), van Dongen (2007).

formed the experiments as requested and reporting the data of what he has inspected. Knowing what would have happened, it is curious that Einstein does not find suspicious the speed with which Rupp carries out complex and very delicate experiments, especially performed without any help. When Einstein complains that some data are inconclusive, Rupp promptly responds with new data that closely confirms the theory predictions.[16] However, the idea that such predictions could constitute a *conditioning* for the young researcher does not cross Einstein. Rupp confirms that the suggested experimental configuration is the correct one: he writes that if the mirror is not at the right angle, in fact, no interference fringes are formed, while if it is, the fringes appear, demonstrating the wave nature of the electrons. In the autumn of that same 1926, Einstein and Rupp publish each a work with almost the same title in the proceedings of the *Preußische Akademie der Wissenschaften* (*Prussian Academy of Sciences*). Einstein's work is theoretical—it proposes and discusses the experimental scheme[17]; Rupp's writing is experimental—it examines the obtained data.[18] Each article presents a scheme of the experimental configuration (similar to the one in Fig. 10.1): the first as an ideal project, the second as an actually employed apparatus. The two works are interpreted as confirmations of the intrinsic wave nature of electrons and they place Rupp among the greatest physicists of the time. The citation by De Broglie in his *Nobel Lecture* 3 years later is the most evident sign.[19] In 1930 Heisenberg summarizes the cornerstones of the Quantum Theory up to then developed in the text *Die physikalischen Prinzipien der Quantentheorie* (*The Physical Principles of Quantum Theory*), where he highlights the *Einstein and Rupp experiment* as one of the most important and decisive.[20]

Strong of his great results, Rupp leaves his master Lenard and obtains a position at the prestigious University of Göttingen, in the group led by physicist Robert W. Pohl[21] (1884–1976). Meanwhile, some scientific articles begin to be published where doubts are raised about the reliability of Rupp's experiments. It is worth mentioning the first of these, written by Robert Atkinson (1898–1982), which states that the large lengths of coherence claimed by Rupp are not realistic.[22] Interestingly, the British physicist publishes in a German journal writing in German. From then on, a heated debate is triggered in scientific journals, where several scientists claim that they have *failed* to obtain the same results as Rupp for various reasons.[23] Some physicists question whether Rupp, with an apparatus like the one he describes,[24] has actually been able to produce the interference fringes. They accomplish control experiments, placing in the conditions declared by Rupp and trying to reproduce the results. They write about it

[16] van Dongen (2007).

[17] Einstein (1926).

[18] Rupp (1926).

[19] Cf. quote 10.2 in Sect. 10.1.1.

[20] Chapter 1 of Heisenberg (1930).

[21] One of the fathers of modern experimental solid-state physics (Ossicini 2012).

[22] Atkinson (1926).

[23] French (1999), Ossicini (2012).

[24] In the aforementioned work (Rupp 1926).

in scientific journals, declaring that, under those conditions, there is no way to obtain Rupp's results. Lenard himself states that he does not consider his old pupil's way of working reliable at all.[25] The debate that ignites, carried out in scientific journals, in communications and in conference proceedings, in letters between physicists, is part of the wider controversy between scientists of a more conservative nature and those more innovative. At first, Rupp is supported by physicists like Einstein and Max von Laue (1879–1960), mainly theorists of the Berlin school; he is contested by Lenard, Wilhelm Wien (1864–1928), his assistant Eduard Rüchardt (1888–1962) and his pupil Harald Straub, as well as Walther Gerlach, mainly experimentalists from the Munich school.

Rupp responds by minimizing objections and providing large quantity of data, graphs, tables, and above all photographs.[26] Meanwhile, in 1928 he moves from the University of Göttingen to Berlin, to the research centre of the AEG (*Allgemeine Elektricitäts Gesellschaft*) company. This is one of the best equipped and best funded private research institutes of the time, headed by Carl Ramsauer (1879–1955). Rupp is left with great independence. He addresses two new topics, of great interest at the time: the polarization of electrons and the production of positrons.[27] To carry out such kind of experiments Rupp needs equipment capable of generating very high voltages, tens or hundreds of kV. Between 1928 and 1934 he publishes dozens of papers, almost all as sole author, each accompanied by figures and graphs to support the data. These works also arise doubts and debates, both among experimentalists and theorists, especially in Germany and the United Kingdom.[28]

Nevertheless, the objections are becoming less and less elusive. In the positron production experiments, Rupp writes that he has used voltages of 500 kV: the most powerful equipment of that time is capable of producing about a tenth of the reported voltages.[29] At the end of 1934, two AEG researchers, skilled in high-voltage equipment, Arno Brasch and Fritz Lange, harbour very strong suspicions; they break into Rupp's laboratory in his absence and find that the equipment has got no signs of use. Director Ramsauer, made aware of what happened, orders Rupp to carry out the experiments with witnesses, but the latter refuses. In early 1935, just before being fired, Rupp is forced to write a retraction of his latest work in *Zeitschrift für Physik*, one of the most prestigious scientific journals of the time. The same year, Gerlach and Rüchardt publish a reply where they note how, in the 1926 article in which he proposes the *Spiegeldrehversuch*, Einstein has made a mistake by drawing the mirror rotated *clockwise*.[30] With such an incorrect rotation sense, replicated by Rupp in his article,[31] there is no way to obtain the claimed interference fringes. It is the final blow. Ramsauer sends a personal letter to about 20 directors of the most important

[25] van Dongen (2007).

[26] Ossicini (2012), van Dongen (2007).

[27] French (1999).

[28] In addition to the already mentioned physicists, P. Dirac and Nevill Mott (1905–1996).

[29] van Dongen (2007).

[30] Einstein (1926).

[31] Rupp (1926).

European physics research institutes, explaining the situation, defending the work of his institute and recommending that they no longer consider Rupp's publications.[32] Finally, the annual congress of the *Deutsche Physikalische Gesellschaft* (*DPG, German Physics Society*) of 1935 issues a statement recommending that Emil Rupp's results and publications no longer be cited.[33] All this causes a *damnatio memoriæ* to fall on the whole story, so that in a short time nobody talks about Rupp and his experiments anymore.[34] For some time, German-speaking physicists have used the term "*geruppt*",[35] which can be translated as "rupped", as a synonym for "turbid", "tangled ", "not very reliable", but in a few years a total oblivion on the part of all[36] takes over. Such a profound forgetfulness on the part of the scientific community is remarkable, taking into account the fact that Rupp's story takes place precisely in the golden age of physics; involves crucial experiments for the development of quantum mechanics; and crosses many of the greatest physicists of the history, with whom he collaborates or discusses.

It is worth summarizing the essential items of the events.

- The scientific and human story of Emil Rupp is part of the *heroic era of physics* and involves many of the most important and famous physicists of all time. For this reason it is remarkable and indicative that it has been modestly removed from the *scientific community*; for the same reason, it is important the story to be known and analysed.
- The *experiment of the rotated mirror* is actually correct, as is the theory elaborated by Einstein: some years later, using different conditions and a *correct* mirror rotation direction (counterclockwise, as shown in Fig. 10.1), it gives the expected results.[37]
- The relationship between Einstein and Rupp, between a famous and established professor and a researcher eager to emerge, plays its role.
- A certain *theoretical bias*, on the part of Einstein and other physicists, also plays a role. Since both Einstein and, reasonably, Rupp are both convinced of the goodness of the theory, the experiments *necessarily had* to provide a confirmation.
- In the *Spiegeldrehversuch*, Einstein's impossibility of viewing the laboratory means that he cannot account for the situation.
- Political, social, and historical aspects have got an impact on the whole story.
- First, there are aspects of *research policy*, in particular the clash between two schools of thought: the more conservative and experimental one, closely linked to academia in southwestern Germany (Munich and Heidelberg), against the most modern, and the theoretical one, open to industrial collaborations in the Northeast (Göttingen and Berlin).

---

[32] French (1999).

[33] van Dongen (2007).

[34] Even texts that explicitly deal with scientific fraud often do not mention it (cf. e.g. Ascheron (2023), Chevassus-au-Louis (2016)).

[35] French (1999).

[36] Ossicini (2012).

[37] Ossicini (2012).

Finally, there are political aspects, with the clash between pro-Nazis and anti-Nazis in the years of Hitler's seizure of power.

The fact that Lenard embraces a conservative position in research and the Nazism in social life means that his and the Munich-based group's critique of Rupp's data are interpreted as dictated more by political prejudices than by scientific assessment. Although in the case of Lenard this may have been true,[38] it is not for other great scientists, such as Walther Gerlach.[39]

Certainly, this story of scientific fraud shows how non-scientific elements of various kinds influence the development of scientific research: prejudice, preconception, politics, envy, social climbing, dishonesty, and partisanship. On the other hand, there are more strictly scientific elements to play a role: the characteristics of the *scientific community*, the difficulties of understanding between theory and experiment, and the difficulties of carrying out experiments and reproducing them. However, it is even more interesting to compare it with contemporary cases of scientific fraud, as this might highlight more clearly several peculiar key traits of scientists and of the scientific community. This topic is discussed in the following sections.

10.1.2 Contemporary Scientists Facing a Recent Affair

The largest case of fraud in post-war physics involves again a German physicist, Jan Hendrik Schön (1970). The case is interesting because it allows to highlight the main distortions within modern scientific research in physics, especially regarding the behaviour of the *scientific community* as a whole. Even in this case a detailed story would take much time,[40] so here we limit to summarize the main facts.[41]

The theatre of this scientific fraud is the *Bell Laboratories*. Established in 1925 and located in Murray Hill, New Jersey (USA), they are not academic laboratories, but a private research centre, probably the most important and recognized in the field of matter physics. Their ownership has always been in the hands of telecommunications companies. During the twentieth century, 11 Nobel Prizes have been awarded for research carried out there.[42] Among them, we find[43]:

The aforementioned Clinton J. Davisson for the demonstration of the wave nature of the electron (1937).

[38] French (1999).

[39] van Dongen (2007).

[40] There is an extensive investigative book on the subject, written a few years after the events on the basis of direct testimonies and interviews (Reich 2009).

[41] In addition to the aforementioned (Reich 2009) cf. e.g. Ossicini (2012), Gelfert (2023) and Chap. 4 of Ascheron (2023).

[42] Reich (2009).

[43] Ossicini (2012), Reich (2009).

John Bardeen (1908–1991), Walter Brattain (1902–1987), and William B. Schockley (1910–1989) for the invention of the *transistor* (1956), which makes possible the development of modern electronics.
Philip W. Anderson for the development of theories on antiferromagnetism, superconductivity, and symmetry breaking (1977).
Steven Chu for the development of *laser cooling* (1997) which allows to trap and manipulate atoms[44].
Horst Störmer (1949), Robert B. Laughlin (1950), and Daniel Tsui (1939) for the theory of quantum fluids and the *Hall effect* (1998).

This overview gives an idea of how the Bell Laboratories represent a highest level, very stimulating, and competitive environment.

The protagonist of the frauds, Schön, arrives from the University of Konstanz in Germany, where he obtains his physics Ph.D. in 1997. Even before accomplishing his doctorate, he begins working at Bell Laboratories in the Bertram Batlogg's group, a highly esteemed Austrian physicist and a superconductivity authority.

As a matter of fact, Batlogg hires some researches and sets up a new group to study organic materials for electronics. During the first months of Schön's work in the group, Störmer, Laughlin, and Tsui are awarded of the Nobel Prize; the year before, *Bell Laboratories* earns a Nobel Prize, with the recognition of Chu. Consequently, at Murray Hill the mood is euphoric during the last months of 1997: it is full of satisfaction and expectations. At his first job experience in the place appearing the most important physics centre in the world, Schön is certainly strongly impressed by such a climate.[45]

The topic on which the group is working on has got great application importance: the development of organic materials, in particular polymeric ones, with efficient conductive properties would make it possible to re-found electronics with huge gains in terms of lightness and cheapness. Being able to make efficient electronic devices, such as *transistors*, with organic materials (and in particular polymeric ones) would be a decisive step towards a profitable application, especially in industrial research.

At the end of 1999 Schön claims to have found the fractional quantum Hall effect in a molecular organic material; this discovery takes place exactly in the same research centre where the fractional quantum Hall effect has been explained by physicists who have then been awarded the Nobel Prize. The German physicist shows to be an excellent researcher, a collaborative and cordial colleague, exhibiting affability, discussing with colleagues, hearing suggestions on the most interesting topics, and on the most promising predictions. After a few weeks, or even a few days, he is able to present experimental data confirming the predictions. His collaborators are amazed. In two and a half years he makes a series of discoveries which are considered exceptional, most of them published in the most prestigious journals. Among these[46] there are the aforementioned fractional quantum Hall effect in an organic semicon-

[44] Inguscio and Fallani (2013).
[45] Ossicini (2012).
[46] Ossicini (2012).

ductor, the superconductivity in organic compounds such as anthracene, tetracene, and pentacene, as well as the superconductivity in organic polymer.

In 2000, Nobel-Prize-winner Klaus von Klitzing, who first has discovered the *quantum Hall effect* in silicon, proposes Schön a job position in Stuttgart for the development of field-effect transistors (*FET*) based on organic crystals, but the young researcher refuses, preferring to stay with the Bell.[47] Schön's discoveries seem have got indirect confirmation from the announcement of the Nobel Prize in Chemistry to Alan J. Heeger (1936), Alan G. MacDiarmin (1927–2007), and Hideki Shirikawa (1936), who in October 2000 are awarded for their discovery of conductive polymers.[48]

In the same year Batlogg plans the transfer to the highly lucrative *ETH Zurich*, where he moves by the end of the year. Then, Schön is supervised by John A. Rogers, new head of the *Department of Materials Science*, and Federico Capasso, who heads the entire *Physical Research Laboratory* of Bell Laboratories. Capasso is considered among the most important physicists of matter, recognized as the main architect of the *quantum cascade laser*. Schön's interest then turns to the development of organic light-emitting semiconductors. The young researcher quickly demonstrates the laser emission from an organic semiconductor and the light emission from an organic *FET*. Actually, no one wonders if at Murray Hill Schön has got equipment for photonic and not just electrical measurements in his laboratory. Both for scientific and managerial interests, the department's heads strongly support the German physicist. Executives of Bell Laboratories have got a higher hierarchy rank and cherish more commercial interests than scientific ones.[49]

During year 2000, the owner company of Bell Laboratories announces a drastic spending review. Many employees fear being fired. In such a mood, Bell executives find in the successes of Schön an opportunity to promote their business and demonstrate that they are not an expense but a profitable investment. Meanwhile, Schön seems unstoppable, always dedicating himself to new research topics.[50] He has got an impressive productivity: between 1998 and 2001 Schön publishes 94 articles. In 80% of the articles he is the first author and only in 2001 he publishes an average of an article every 8 days.[51] Some of his works appears in the highest *impact factor* journals: between 2000 and 2001 he publishes 2 articles on *Advanced Materials*, 6 articles on *Nature*, and 9 articles on *Science*.

Nevertheless, someone begins to notice inconsistencies: in certain works an unrealistically high number of measured samples is indicated, the values of the involved electric fields seem too large, the trends of some experimental data are not in agreement with already largely consolidated theories. When critiques arrive from article reviewers, they are slyly exploited. Schön is very good at writing and fits into the competition between the most famous journals: if the reviewers of one are too metic-

[47] Chapter 5 of Reich (2009).

[48] ibid.

[49] ibid.

[50] ibid.

[51] Ossicini (2012).

ulous, the German physicist threatens to submit the work to a competing journal. He almost always manages to get publication: not following the journal's internal policy, the *editors* often accept to publish his works although Schön has not answered the reviewers.[52]

In physicists' environment Schön is extremely seductive precisely because he confirms the hypotheses of his colleagues, telling them what they would like to hear.[53] In addition to the inconsistencies, which are starting to become more and more a topic of discussion among researchers, there is also another important fact: all the scientists, laboratories, and groups that try to replicate Schön's experiments run into failure. This represents a weak point of the whole construction. Because contrary to the time of Rupp, the failures are not published, so the debate is confined to an informal level. The majority of scientists silence doubts because they trust Schön's co-authors, being esteemed and established, and highly regard the Bell Laboratories. Those who point out that even the number of Schön publications is unrealistically high are silenced as envious: if Schön is particularly strong and productive, why should anyone care? And why should doubts be raised, given that, as Capasso says, *Science* is based on trust[54]?

In 2001 Schön is hired by the Bells Laboratories, the place where the transistor has been invented, i.e. modern electronics: from where else can a new revolution in electronics start? There he dedicates to a topic with enormous potential application: nanotechnologies applied to transistors. That year he publishes two papers, one on *Nature*[55] and one on *Science*,[56] where he claims of doing a *single-molecule* organic field-effect transistor (*self-assembled monolayer field-effect transistor*). The discovery seems to pave the way for an extreme miniaturization of electronic components.

At the end of 2001, while Schön receives prestigious awards, he is publicly accused of fraud for the first time. A physicist from the Bell Laboratories notes that an experimental curve published by Schön has a *too perfectly Gaussian* trend; through the test of χ^2 he deduces that it is virtually impossible that the measurement of a natural phenomenon produces such a trend.[57]

On the internal front, the German researcher is defended by Bell's managers, but at the beginning of 2002 many scientists no longer hide their scepticism, although caution is always maintained. Among them, there is Lydia Sohn from Princeton University. She is not a famous physicist: she carries out her work with great attention and she has got good publications, but, having never obtained exceptional results, she

---

52 Chapter 6 of Reich (2009).

53 Cf. Sect. 1.1 and particularly quote 1.4 in Sect. 1.1, Sects. 8.1, 8.2.2 and also quote 10.2 in Sect. 10.1.1.

54 Chapter 10 of Reich (2009).

55 Schön et al. (2001b).

56 Schön et al. (2001a).

57 Chapter 10 of Reich (2009).

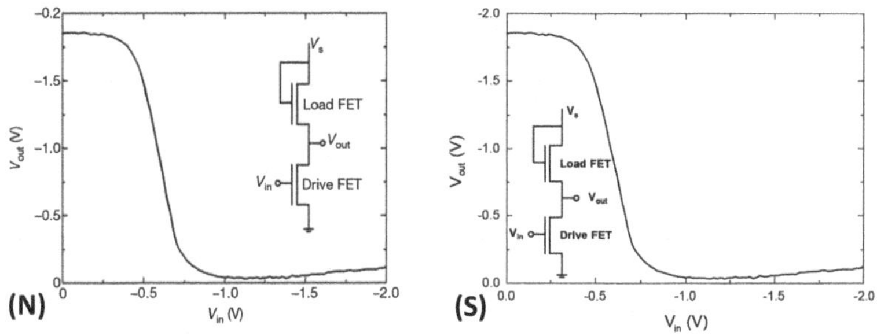

Fig. 10.2 Plots of alleged experimental data of organic diode (SAMFET) operation, taken from the *withdrawn* publications by Schön et al. (N) Plot adapted from *Nature* (Schön et al. 2001b): claimed measurement *at room temperature* on a sample composed of hundreds of electrically active *4,4'-biphenyldithiol* molecules; (S) plot adapted from *Science* (Schön et al. 2001a): declared measurement *at 4 K* on a layer composed of *alkanedithiol mixed with 4,4'-biphenyldithiol or 5,5'-terthiophenedithiol* in which there would be a couple of electrically active molecules. (Courtesy of Springer-Nature (N) and AAAS (S) publishers ©)

has never achieved the publication on journals such as *Nature* or *Science*.[58] In Schön's two articles on the single-molecule transistor she notices two plots, respectively, published in different journals. Each of them refers to a different measurement, of a different sample, at a different temperature (the first at room, the second at 4 K). Sohn realizes that the two experimental plots are *identical* (Fig. 10.2), confirming that the data have been replicated. Along with some colleagues, including Paul McEuen of Cornell, she begins to analyse other publications, discovering numerous instances of replicated data.

When Sohn tries to publicly express her doubts, a colleague sarcastically asks her how many articles on *Nature* and *Science* she had published that year.[59] However, the results of the analyses reach Bell executives. After considerable opposition, in mid-2002 they are forced to appoint an investigative commission, made up of five highly esteemed scientists, including Störmer himself. Schön claims that he has not committed fraud; however, he is unable to show either the original measurement samples, the raw data, nor a laboratory register or notebook. In September, the Commission publishes a nearly 130-page report, which is distributed by the *American Physical Society*.[60] The investigation confirms that the overwhelming majority of the German physicist's experimental results are the outcome of fraud. All publications in *Nature* and *Science* are withdrawn, along with a large number of other works. Schön is fired and forced to return the awards. The University of Konstanz revokes his doctorate, employing an old law, no longer applied since before the War, according to which the degree can be taken away from anyone who proves unworthy.[61]

---

[58] Chapter 11 of Reich (2009).

[59] Chapter 11 of Reich (2009).

[60] Beasley et al. (2002).

[61] Ossicini (2012).

Here again, the main items of the affair are outlined.

Jan Hendrik Schön's career develops within the most prestigious research environments. It involves researchers among the most esteemed in the world. His career culminates in coveted awards and publications in the most highly ranked scientific journals. For this reason, the story strongly shocks a large part of the *scientific community*, posing embarrassing issues about how it is possible that a similar fraud can take place.
Several are the elements that have got a role in this case: careerism, the exaggerated competitiveness of certain scientific circles, the rush to publish (the notorious motto "publish or perish"[62]), and the pressure to produce *quickly* important results.
A critical item is also the sectoralization of the research work: although almost all the works by Schön are multi-author, none of the co-authors has got the skills of, or thinks about, evaluating and checking the effective work of the colleague. This is particularly evident for those who occupy a supervisory position in the hierarchy—i.e. head and coordination of the working group.
Schön is the only one indicted for the whole affair; his co-authors emerge almost unscathed by the scandal. The Commission's report only underlines that Batlogg has not adequately controlled his collaborator in spite of his supervisory role.[63]
Batlogg's behaviour shows the tendency of many scientists, heads of important groups, to move away from the actual research activity to become *research managers*. The role played by scientific managers as well as their responsibilities remain an open question for the scientific community.[64]
The *scientific expectation* also plays a marked role. Since so many scientists think, hope, hypothesize, and expect to observe certain phenomena, Schön's statements have got the charm of *dreams come true*: this is a feature to which researchers are easily suggestible.
Other factors plays a role: commercial; economic; media exploitation; and, last but not least, the desire to make use of exceptional scientific discoveries to justify and keep one's job. As in the case of Rupp, social and historical aspects have got an impact on the whole story.
The capability of the *science system* to "self-correct" is insufficient. The fraud is discovered by the attention and insistence of individual scientists, not by the *scientific community*. *Science*, understood in its sense of a *system*, gives a bad test of itself.
Moreover, in the story there are elements related to *research policy*. Laughlin almost immediately notices inconsistencies and expresses perplexities,[65] certainly thanks to the authority and freedom that the Nobel Prize confers on him. However, apart from this important exception, the researchers who spot the inconsistencies, discover the forgeries, and have got the courage to expose the fraud are gener-

[62] Neill (2008), Nature (2010).

[63] Beasley et al. (2002).

[64] Norman (2009).

[65] Chapter 7 of Reich (2009).

ally not first-rank scientists. They are attentive and conscientious physicists, but relatively unknown. Several of these are at the post-doc or *tenure track* level or at the beginning of their academic career, like Sohn, or they are known to have got peculiar positions and are considered as a "black sheep", like McEuen.[66] In other words, those understanding the situation are those who are not entirely part of the *system*, who do not bow to the principle of *auctoritas* and believe their eyes, actual knowledge, genuine data, rather than the reputation of prestigious journals and institutions.

Conversely, the largest part of the most awarded and famous scientists and researchers have shown that they fideisticly believe in the *science system*, unwittingly contributing to perpetrating the fraud. The presence of esteemed co-authors, the importance of affiliation, and the prestige of the publishing journals have led to overlook the various inconsistencies and numerous scientific problems for a long time.

A noteworthy element is the role played by scientific journals, which have often followed commercial and promotional criteria instead of strictly scientific ones.

The demonstration of a single-molecule transistor is provided[67] a few years after the affair. In spite of this, if some results had been confirmed, the risk is that no one would heve realized the frauds perpetrated and that the inconsistencies, however present, could perhaps be classified as venial errors. As mentioned in the previous chapter, errors are not only inherent, but often necessary components for the advancement of physics.

On the other hand, in modern research there is a "grey area" between error and fraud which the Americans call "*beautification*" of data.[68] The term is an obvious play on words between *beauty* and *beatification*. It refers to the practice of selecting scientific data and embellishing them, i.e. forcing them in order to highlight that they support one's (important) scientific claims. This practice has been widely discussed for decades and it is recognized to be more widespread in the biomedical field.[69] It is favoured by many critical elements present in the modern scientific practice,[70] some of which are analysed in Sect. 10.2.

10.1.3 Past and Present Circumstances: A Comparison

The comparison between the *ancient* case of Rupp and the *modern* case of Schön is particularly interesting. It is useful for highlighting what has changed and what has not in scientific practice and community.

---

[66] Chapter 10 of Reich (2009).

[67] Kubatkin et al. (2003).

[68] Nature Cell Biology (2006), Chap. 3 of Chevassus-au-Louis (2016).

[69] ibid.

[70] Ossicini (2012).

At both the time of Rupp and of Schön (which is our time), there is not a pre-established method for detecting fraud, nor a recognized procedure, such as a *Code of Criminal Procedure*, to investigate, process, and impose penalties on the responsible for dishonest scientific conduct. For the most serious cases, such as those of Rupp and Schön, the penalties are the expulsion from the world of research; for less serious cases, it is the colleagues and scientific community's discredit and censure. A judgement on one's scientific behaviour is given in different ways: for the case of Schön, an *ad hoc* commission is set up, a sort of tribunal; for the case of Rupp, the evaluation of his manager and a few researchers is enough; for other cases, a more informal and less hierarchical way is preferred. Similarly, the investigation is carried out with diverse approaches: in most cases, the investigation and judgement are accomplished by peers, in a similar way to the *peer-review* process.

The affairs of Rupp and Schön share many similarities. Both physicists work at the dawn of their careers, dealing with some of the hottest scientific issues. They work in some of the most prestigious institutions of their age: in particular, they both spend some years in private industrial research laboratories, during years of economic difficulty (the former at AEG and the latter at Lucient Technologies). Moreover, they both are strongly influenced by suggestions from their brightest colleagues and show a willingness to meet their expectations. In both cases, their colleagues notice inconsistencies and unconvincing aspects, which trigger a heated scientific debate. Both physicists carry out an extensive work of falsifying the experimental data to meet the expectations of their scientific community, but they fail to avoid a large number of contradictions. Finally, both end their careers with the dismissal and the sudden removal from the research circle.

There are also many differences between the two cases. Besides the peculiar historical-political conditions of Rupp's Germany and the change of physics gravity centre from Europe to the USA, the major difference lies in both the scientific community and the way physicists work. While more than 80% of Rupp's publications are single-authored, almost all of Schön's have got 3 or more co-authors.[71] Nowadays, the two publications about the *Einstein and Rupp experiment*, each, respectively, signed by its own author,[72] would certainly be a single co-authoured paper: a work presenting experimental data based on solid theoretical bases. In the past, it is preferred to keep contributions of different nature separated; today there is a tendency to provide an explanation as much complete as possible especially in the most ambitious works. Rupp's vicissitude takes place in about 10 years, Schön's in about half the time. Nevertheless, the number of publications is decidedly higher for the latter one: certainly it is the well-known modern tendency towards publication proliferation.[73] To carry out his fraud, Rupp needs to employ great skill in forging numbers and photographic plates.[74] Conversely, much more limited technical knowledge is suffi-

[71] Ossicini (2012).

[72] Einstein (1926) and Rupp (1926).

[73] Chevassus-au-Louis (2016), Nature (2010), Neill (2008), Ossicini (2012).

[74] This skill turns to be useful to Rupp as after the War he finds a job in East Germany in a photographic material company (Ossicini 2012).

cient in the modern age: Schön just uses a common software running on a standard computer.

Another great difference is that in Rupp's time the journal prestige is a much less marked factor than in the contemporary age: journals such as *Nature* and *Science* already exist, but there is not such a tight race to publish on them. The role of journals changes a lot between the two ages: before the War they give space to discussion papers, openly doubtful about already published materials, as well as to reports of control experiments, even when unsuccessful. On the other hand, in Schön's time most of the debates take place at an informal level: papers for the sole discussion of other people's results are almost non-existent in physics. Today it is unthinkable to submit a paper where it is explicitly declared not to have been able to reproduce someone else's experiments. At most there is the possibility of sending some warning letters to the journal *editors*, who can decide whether to publish them or not (only in special cases). This is an important item. After the discovery of Schön's frauds, the *editors* of the most popular journals where he has published are called into question and declare that the purpose of their journals is *science* advancement; they remark that advancement happens not only through correct results, but also through errors.[75] This is certainly true, as stated in Sect. 9.4. Nevertheless, the problem lies in the asymmetry between agreeing to publish a little (or not at all) controlled data, potentially wrong or even falsified, and *not* publishing, or publishing moderately doubts, disputes, discussions, and refutations. In short, the big journals provide great visibility to those who, in the rush to publish, make ambitious and demanding scientific claims, and do not provide as much visibility to those who practice the *organized scepticism*[76] which should be at the basis of a good scientific research.

This is also demonstrated by how the scientific community has received the news of their misbehaviour. In Rupp's case, a letter sent to a few directors of the main physics institutes is enough to erase his work and his memory; on the other hand, in Schön's case, he still receives dozens of citations every year despite his publications have been retracted,[77] a demonstration that many researchers do not know or superficially take no account of the withdrawal. It is true that today the scientific community is enormously wider than in the 1930s: there are thousands and thousands of publications every year and being updated on everything is virtually impossible.

Therefore, has got such a large community effective tools to deal with fraud? Whoever gives credit to a scientific fraud is not aware of it, but whoever gives credit to the myth of *Science* capable of self-correcting, mechanically and necessarily, is aware; a similar position is still supported, more or less openly, by a vast majority. If any scientist is asked whether they trust another researcher, they answer that they must evaluate the other's work and method; they certainly does not grant blind trust.[78] On the contrary, if you ask the same scientist whether he trusts *Science*, he probably

---

[75] Reich (2009).

[76] Cf. Table 10.2.

[77] Many citations assume the results presented as still valid (Ossicini 2012).

[78] Chapter 10 and Epilogue of Reich (2009).

answers positively. However *Science* does not exist: there are data, theories, scientific explanations, hypotheses, forecasts, the scientific method, the scientific community, and *scientists*.[79] It is their *community* that creates the conditions for the production of genuine scientific knowledge. Human beings *do science*: they can behave honestly or not. The *system of science* is not a perfect mechanism: it can only create more or less favourable conditions for the development of knowledge.

10.2 Scientific Community: Open Issues

At the beginning of this chapter, Definition 10.1 of *scientific community* is provided. That is because it is precisely the human environment in which the daily activity of a physicist takes place. Ideally, this community has got the task of validating scientific knowledge, of instructing researchers, of guaranteeing methodological correctness; in a word, it is up to all the physicists who work in synergy with each other to achieve and continue the development of knowledge in the field. The form of this community is not planned by an external regulatory body on the basis of a project (as dictated by a state law), but it is spontaneously established through a general consensus around some recognized *right* methodological factors. The adjective *right* reminds what has been written in Sect. 1.1: the criterion by which an element is considered *right* is again absolutely practical and phenomenological. In this sense, as in Chap. 1, what is *right* leads to a knowledge increase. Therefore, the following question arises:

Is the structure of the physicists' scientific community and the presently adopted practices the most suitable to support the advancement of natural phenomena knowledge?

While suggesting possible answers to this question, the *human element* of scientific activity (highlighted in the introduction to this text) markedly stands out as a strength and a weakness. In the so-called *physics* the human element must always be taken into account. It follows that in the organization and practices of the scientific community there are several open issues, some of which are here briefly considered.

The previous discussion on frauds has highlighted some big issues:

- the community *consensus* issue;
- the presence of partially rational or completely irrational factors;
- the interaction between scientific research and historical-political-social factors.

The discussion in Chap. 5 has already dealt with these issues from a strictly theoretical and general point of view. They are here presented in a practical perspective.

Rational elements (Popper, Lakatos, Sect. 5.2), historical and/or partially rational elements (Kuhn, Sect. 5.3.1), and potential irrational elements (Feyerabend, Sect. 5.3.2) all influence the evaluations of the scientific community. Scientific results

[79] Parravicini (2022).

must be adequately disseminated, i.e. made public, in order to allow other scientists' judgement. Dissemination and evaluation are thus at the basis of obtaining the necessary *validation* of the results by researchers. The mechanism of this validation must be able to reach a *consensus* within the scientific community.

10.2.1 Publications and the **Publish or Perish** *Practice*

Dissemination of scientific results is done in various ways. The main ones are

informal communications;
discussions;
seminars;
conferences and workshops;
printed publications.[80]

Among these, publication is by far the most important. Since the dawn of physics, the printing of a text, more or less long, has been the privileged form to convey a scientific content. Publishing is also aimed at signalling the authorship of the obtained discoveries and results. Leaving out the historical methods, the first goals of a scientific publication are[81]

1. recording and documenting knowledge;
2. transmitting knowledge and information;
3. providing a first check;
4. setting intellectual priority rights;
5. ensuring a minimum methodological and qualitative standard.

The editorial process of scientific journals uses *peer review* and pledges that the aforementioned objectives are met. There are several opinions and studies on peer review. Ideally, this should be the process through which experts critically analyse scientific results before their publication[82]; in practice, a large number of problems and distortions have been highlighted: superficiality of the reviews, *bias* on the part of reviewers, strong gap between the opinions of different reviewers on the same work, etc.[83] Beyond the intentions, effective and shared strategies have never been devised to remedy these problems.[84]

To the fundamental objectives of the publications listed above, others (more or less additional) can be added[85]:

[80] Of course, this also includes printing on electronic media.
[81] Cf. e.g. Chap. 4 of Ascheron (2023).
[82] Alberts et al. (2008).
[83] Bornmann (2011), Santos (2023).
[84] ibid.
[85] Cf. e.g. Chap. 4 of Ascheron (2023).

6. opening a discussion space;
7. stimulating further progress;
8. teaching;
9. supporting scientific careers.

All these aims are correct, although over time their priority order has reversed. Often, the transmission of knowledge is only a marginal objective of publications: their main purpose is to justify and enhance scientific careers. Therefore, the goal of many scientists' work is not focused on increasing scientific knowledge, but just on publishing. In other words, publishing risks becoming not a *means* used by researchers, but their *end*. The Schön affair, an extreme case, well illustrates this dynamics. In this scenario, Anglo-Saxons coin the motto "*publish or perish*".[86] In a scientific community with an increasing number of precarious workers, where researches are increasingly financed by temporary projects, publication becomes a *currency* on which not only career progresses depend, but also one's job, salary, and funding. As a matter of fact, some states (including China) have adopted the policy of granting the researcher a cash bonus for each published paper.[87] In spite of these extreme examples, in Western countries—mainly in Europe and in North America—the majority of the scientists do not get funding (i.e. *money*, a salary, and a job) if they do not publish.[88]

In this context, the pressure to publish, especially in high-level journals, is constantly growing.[89] Regarding this topic, some iconic episodes highlight the main shortcomings of the peer-review process and the biases due to the race to publish. An interesting case is the development of the *SCIgen* software. This program is developed in 2005 by some researchers of the *Massachusetts Institute of Technology* for playing, in the field of theoretical computer science. It is able to produce a grammatically correct text, full of specific discipline jargon, completed with figures and references, all having the appearance of a scientific article; however, the result is completely meaningless, as it is built by juxtaposing concepts in a casual way. It is noteworthy that some papers generated by this program have been able to pass peer review and be accepted as conference papers or publications in mainstream scientific journals. Even worse it is the fact that the program, created as a game, has been used by some researchers to rapidly increase the number of their publications.[90] *SCIgen* is not the only program of such kind[91]: although these episodes have taken place in the field of theoretical computer science, the rest of scientific research is not immune from similar practices, indeed it is reasonable to think that the risk of such (mis)conduct can increase in the future.[92]

In the field of physics, as in many other branches, there are different kinds of publication. Beyond the monographs (books), the main kinds of scientific publication

---

[86] Fanelli (2010), Nature (2010), Neill (2008).

[87] Nature (2010), Sample (2013).

[88] Ossicini (2012).

[89] Ossicini (2012), Van Noorden (2021).

[90] Bohannon (2015), Van Noorden (2021).

[91] Among others, there are even programs able to write (pseudo)research projects.

[92] ibid.

Table 10.1 Main kinds of scientific publication on physics journals, with their *typical* length (neglecting the space for authors' name)

Kind of publication	Number of pages
*Communication**	from 1 to 2-3
*Letter** †	from 2 to 5-6
Research paper‡	from 4 to 15–18
Review paper	from $\simeq$ 10 to $\gtrsim$ 50

*Both the *communication* and the *letter* can be called *notes* or *short notes*
†As well *research letter, original letter* or similar
‡As well *research article, original article, original work, original paper*, etc.

in physics[93] are reported in Table 10.1. Where does this pattern arrive from? Today the real reason is unknown to many. Until the 1960s and 1970s, the ideal and usual scheme is approximately the following. A first highly preliminary result is very briefly communicated through a *communication*, which often coincides with the publication of a communication at a conference. A *communication* also has got the purpose of getting comments and corrections from colleagues. Once a sufficiently stable and important objective is reached, it is good practice to publish a *letter*, where the fundamental items of the specific discovered phenomenon are highlighted, thus claiming the discovery authorship. It is common practice, and often recommended, for a *letter* to be followed by a *research paper*, i.e. the actual article, where an in-depth and much more detailed investigation of the discovered phenomenon is reported. After a few years of work about a certain topic, a *review* is drafted: a small essay where the *state of the art* of research on the subject is summarized in a reasoned manner. Of course, this is just an ideal model with a lot of exceptions, but that is the pattern to follow. A general expository clarity is obtained, so it is relatively easy to follow the development of a given research topic carried out by a scientist or a group.

Only a shadow of this scheme remains today, together with the different kinds of articles in Table 10.1. Except the *review*, each kind of publication needs to remark that what is reported is absolutely *new* and *never seen before*. In physics it is today unusual for a *research paper* to openly declare that it reports a deeper investigation of what it has been previously reported in a *letter*. Nowadays there is no more time to carry out a step-by-step, in-depth, and detailed research: it is mandatory to continuously publish, under penalty of scientific *death—publish or perish*!

In the pre-War era an even more careful physics is developed: as shown in the story of Rupp, there was a widespread practice of *repeating* the experiments of others, of checking them and even of publishing articles where a *failure* in carrying out an experiment can be openly declared, whose reasons were discussed in detail.[94] Today, no one would ever write an article reporting a *failure* in accomplishing experiments, nor would any journal accept it. Journals show a series of "exceptional" successes[95]!

---

93 Cf. e.g. Chap. 4 of Ascheron (2023).

94 Cf. Sect. 10.1.1.

95 Ossicini (2012).

Who could possibly make a career by doing "checking experiments"? Reporting failures? Who would ever write a research project asking for funding to *repeat* someone else's experiment? Which funding body would grant it?

The speed at which researchers need to produce results, and strictly *new* results (the only ones that can be published), means that they cannot spend too much time analysing other people's data. Much more than in the past, today physicists are forced to *believe* in their colleagues. Capasso says that the science system is based on trust[96]: this statement is not wrong at all, since trust has always been one of the basic elements of science. If each scientist had to repeat all the experiments of his predecessors, no progress would ever have been made. So, the arising question is: what are the *reasons* that create trust? The *Schön scandal* highlights how the reasons for believing in certain results can be unsuitable or inconsistent (reputation of the institution, reputation of the researcher, community expectations, etc.). Unfortunately, believing without having suitable reasons is not a reasonable trust, but an unreasonable fideism.

Each fraud scandal has encouraged the adoption of countermeasures. For example, today many scientific journals, and, in particular, the most ranked ones, require the raw experimental data to be made available in special online repositories. The same journals can appoint supervisors to randomly visit some laboratories where the published research has been conducted. Other journals have activated or expanded specific spaces for comments and discussions.[97] However, the various implemented countermeasures have often obtained modest results[98]: the persistence of the *publish or perish* model keeps the problem alive.

In Sect. 9.4 it is shown how errors are part of scientific research and how they have often be fruitful. Mistakes are today hardly admitted, since they are often perceived as *incapacity and incompetence*; if you are not competent, you don't publish, nor get funding, nor get a career. Perhaps today someone would consider incompetent Galilei, Stern, Gerlach, Sommerfeld, or Einstein.[99]

In a contemporary episode such as the Schön affair, all these elements remarkably emerge. Scientists who discover the fraud are those who are not completely submitted to this pattern and carry out their research less aggressively, more slowly, more carefully—although this practice often produces a less brilliant career, it is certainly based on more solid results. From this perspective, the sarcastic comment that Lydia Sohn receives from a colleague (who demands her how many articles she has published in *Nature* and *Science* that year[100]) is highly indicative of the *publish or perish* practice to the extreme of assessing a scientist's capability through quantifying the articles published in titled journals.

(10.3) We all know what distorting incentives have done to finance and banking. The incentives my colleagues [scientists] face are not huge bonuses, but the professional rewards that accompany publication in prestigious journals [...]

[96] Chapter 10 of Reich (2009).

[97] Ossicini (2012).

[98] Nature (2015).

[99] ibid.

[100] See Sect. 10.1.3 and Chap. 11 of Reich (2009).

> These luxury journals are supposed to be the epitome of quality, publishing only the best research. Because funding and appointment panels often use place of publication as a proxy for quality of science, appearing in these titles often leads to grants and professorships. But the big journals' reputations are only partly warranted. While they publish many outstanding papers, they do not publish only outstanding papers. Neither are they the only publishers of outstanding research.
>
> These journals aggressively curate their brands, in ways more conducive to selling subscriptions than to stimulating the most important research.[101]

When Peter Higgs wins the Physics Nobel Prize, he publicly declares that today he would never have made a career, and almost certainly not even entered the academy, because his "productivity", in terms of articles and publications, would have been considered totally insufficient.[102] Certainly, a Nobel Prize winner is, by definition, an exception, but the extent of his critique of the *publish or perish* system has been understood much more by the laymen rather than by scientific community. In other words, the fact that a physicist is *publishing a lot* does not guarantee that he is actually *hard working*, nor that he is *well working*.

In addition to encouraging the temptation to defraud, the pressure to publish *too much* produces a further damage to scientific knowledge. Any result today is published even if it is partial or poorly understood. There is a large number of publications based on still largely incomplete data and theories. A worse practice has taken root: it is the tendency to divide a single result into several publications that would find its natural place in a single wide-ranging work. All this makes it virtually impossible to stay updated on all the publications in one's investigation field, increasing the difficulty of orienting within the scientific literature and of easily grasping the different value of the different results.[103] John P.A. Ioannidis, one of the most important researchers and one of the founders of the *metascience* discipline, in 2005 demonstrates by a simple statistical reasoning that, at least in the biomedical field, most of this huge number of articles are false, since, despite the high-flown contained claims, their data do not significantly deviate from standard random statistical fluctuations.[104] Although physical sciences deal with very different experiments, employ different methodologies, and generally collect much more solid data, this work is certainly an alarm for the entire scientific community.

Against the distortions related to *publish or perish*, several initiatives have been undertaken. They aim at proposing a new paradigm according to which:

1. *quality* of publications matters much more than *quantity*;
2. under no circumstances this last one can be quantified using the *impact factor* (or similar indicators) of a journal.

The evaluation of the *value* of a considered work should be undertaken with judgement, accepting all the risks that this entails.

---

[101] Randy Schekman in Schekman (2013).

[102] Aitkenhead (2013), Ferreira (2022).

[103] Ossicini (2012).

[104] The paper exposing this result has had a large impact, as evidenced by the over 6000 citations collected so far (Ioannidis 2005).

(10.4) Funders and universities, too, have a role to play. They must tell the committees that decide on grants and positions not to judge papers by where they are published. It is the quality of the science, not the journal's brand, that matters.[105]

Among the initiatives following this line, there is, for example, the *San Francisco Declaration on Research Assessment*, known as *DORA*, where thousands of researchers and institutions of various sciences claim that it is not acceptable to evaluate the skills of scientists on the basis of the *impact factor*[106] of their publications.[107] Another interesting proposal is the so-called *"slow science" manifesto*.[108] These as well as other initiatives are certainly worthy in their intent to arouse a debate in the scientific world and to propose attempts to reform usual practices; nevertheless, it is still unclear whether the scientific community will accept suggestions to overcome the distortions correlated with the rush to publish.

10.2.2 CUDOS *and* PLACE

Throughout most of the twentieth century, scientific research in Western countries has been ordered according to some principles called *Mertonian norms*, named after the sociologist Robert Merton (1910–2003), probably the greatest researcher on scientific community.[109] These principles are identified by the acronym CUDOS, built on the Greek word κῦδος, meaning "glory", "honour". The attitude that arises from these norms is called *academic research* and has got the following features:[110]

Communalist Scientific inquiry develops within a large scientific community and its results are shared among all.

Universal Scientific investigation and its results do not belong to a group, but they have got the greatest possible sharing, which goes beyond institutions, politics, states, etc.

Disinterested Research is not driven by specific economic or political interests, at least in principle. If it is true that there will always be funding bodies, and that there is an applied research with its legitimate aims in line with the funding agencies, however, the scientist's activity is ultimately be driven by love for knowledge and not by other interests.

Original Research is the result of the creativity, ability, genius of a scientist, or a group; these are free to choose, within reasonable limits, the paths to follow in their work. The results of their research are recognized as theirs and no one else's.

---

105 R. Schekman in (2013).

106 It is noteworthy the fact that this indicator has been conceived as a support to librarians, to help them in classifying scientific journals; later it has been increasingly employed to evaluate researchers. The *impact factor*, however, has not been created with this purpose (Alberts 2013).

107 Alberts (2013).

108 Owens (2013), Rosen (2011).

109 Merton (1942).

110 Kellog (2006), Ziman (2000).

Sceptic The *scientific community*, made up by all scientists, follows an *organized scepticism*, according to which data and results must be shared for the purpose of allowing cross-checks; reliability is gained exclusively by the data and results quality and not by anything else, such as eminent positions, awards, or memberships in prestigious institutions. All scientists must constantly be scrutinized: only scientific results matter, not where they come from.

By carrying out a research according to the aforementioned principles, a scientist exactly pursues the κῦδος, glory, fame, recognition, approved by all, this is the only ideal reward for his work. Obviously, this model is an ideal scheme that depicts scientific research in the middle of the twentieth century. Like all models, it does not claim to be applied literally, but it represents a valid pattern that identifies the key elements of scientific practice. This scheme is elaborated in relation to a specific historical age, since it is explicitly developed to describe scientific practice *in democratic societies*, in open contrast to the research system in the totalitarian regimes of that time.[111] In short, in the ideal model that the scientific practice follows, strong correlations can be identified with the order of society as a whole, even in its most delicate aspect of social freedom preservation.[112] This system goes into crisis in the last quarter of the twentieth century, so that some speak about the "collapse of Mertonian norms".[113] An *academic science* that pursues the CUDOS is being replaced by an *industrial* or *post-academic science*, driven by different principles. These are identified by the acronym PLACE[114]: the features of the new modern industrial-style scientific research, replacing the old criteria, are the following[115]:

Proprietary Scientific research is funded by industrial or state bodies which, pursuing their specific objectives, intend to make exclusive use of the obtained results; then they have got no interest in sharing but, on the contrary, they hold the rights of economic or political exploitation.

Local Besides the great possibilities of communication and exchange of our age, scientific investigation is carried out in a well-defined sphere, so the results remain within the stakeholder aimed at exploiting it. Sharing the achieved knowledge is usually not considered, and is often forbidden or discouraged.

Authoritarian Scientific research is not moved by the free creativity and ability of scientists, instead it is hierarchically *a priori* established, by funding bodies. The researcher is not free to pursue what he deems the most interesting paths, instead he must follow established policies.

Commissioned Research is not driven by the desire for knowledge, instead it is commissioned with specific goals to solve specific issues.

---

[111] This intention is recognizable right from the title of Merton's work, "Science and technology in a democratic order", cf. Merton (1942). See also Kellog (2006), Ossicini (2012), Parravicini (2022).

[112] Cf. e.g. Introduction by Popper (1962), Parravicini (2022), etc.

[113] Karachalios (2008), Kellog (2006), Ziman (2000).

[114] ibid.

[115] Kellog (2006).

Table 10.2 Comparison between *academic science*, following the CUDOS paradigm, and the post-academic *industrial science*, following the PLACE paradigm (from Kellog (2006))

Academic science	Industrial science
COMMUNALIST	**P**ROPRIETARY
UNIVERSAL	**L**OCAL
DISINTERESTED	**A**UTHORITARIAN
ORIGINAL	**C**OMMISSIONED
SCEPTICAL	**E**XPERT

Expert The researcher ceases to be a free scholar and becomes an expert, to be hired when needed only for the time necessary to solve a given issue.

On the one hand, in Mertonian *academic science* the researcher pursues the κῦδος, on the other hand, in post-academic *industrial science* the researcher may at most achieve a *place*, a job, generally precarious, and little else.[116] Again, this is a model. Just as no scientific environment perfectly reproduces the norms of the CUDOS model, in the same way no scientific environment perfectly reproduces the norms of the PLACE model. Therefore, the two models are mixed in practice, with the prevalence of one or the other (cf. Table 10.2).[117] For sure, the Mertonian norms are close to a more idealist (one could say more romantic) approach to scientific research. In this sense, it follows the definition of *science* provided by the *American Physical Society*:

> (10.5) Science extends and enriches our lives, expands our imagination and liberates us from the bonds of ignorance and superstition. The American Physical Society affirms the precepts of modern science that are responsible for its success.
>
> Science is the systematic enterprise of gathering knowledge about the universe and organizing and condensing that knowledge into testable laws and theories.
>
> The success and credibility of science are anchored in the willingness of scientists to:
>
> 1. Expose their ideas and results to independent testing and replication by others. This requires the open exchange of data, procedures and materials.
> 2. Abandon or modify previously accepted conclusions when confronted with more complete or reliable experimental or observational evidence.
>
> Adherence to these principles provides a mechanism for self-correction that is the foundation of the credibility of science.[118]

This declaration is, clearly, strongly inspired by a Mertonian scientific practice. They are noble statements and lofty principles, with which all physicists state to agree. However, if we analyse the effective running of the modern research world, we find a very different practice: this is pointed out by the reactions of the scientific

[116] Ossicini (2012).

[117] Ossicini (2012).

[118] APS (1999).

community during the events of the Schön scandal, which happens in a mainly *place* paradigm context. This is the character trait of our age, where this research model is prevalent.[119]

It is also important to remark that the *industrial science* expression does not necessarily mean that it is decided and financed by private industries. The funding system based on research projects (the *grants*) obeys, to a greater or lesser extent, the pattern of the PLACE even when the funding entity is a state body. The EU research support mechanism, a highly complicated and bureaucratized structure, largely responds to the scheme of industrial science. On the other hand, the two models cannot be contrasted by stating that *academic science* is good and *industrial science* is bad: there are advantages and disadvantages in both models, the discussion of which is proposed in more specific analyses.[120] However, it is a fact that Mertonian norms depict a system where research improves because scientists enjoy great freedom, and the scientific community, made up of highly selected scholars, is largely independent on political-economic interference, thus having strong correction tools.

Conversely, in the PLACE model, dominant in twenty-first century research in Western countries, most of scientists have got little or no freedom in their activity.[121] In our age, daily scientific work is carried out by a large mass of precarious and underpaid researchers within more or less large groups, led by more or less well-known and important scholars.[122] As mentioned previously, the more the latter ones have got a prominent position, the more they risk to neglect their specific scientist activity to play the role of *research manager*. It is precisely the latter ones, however, who take the overwhelming part of the *glory*, those who obtain the κῦδος for the achieved results; the work of their subordinates is not aimed at the κῦδος, but hoping to obtain a modest *place*, even though they know that only a small part of them succeeds. It is reasonable to argue that in such conditions the risks of forgery, fraud, *beautification*, and forcing of data are high.[123]

In order to get a place you need to publish and get funding. However, the Nobel Prize for Medicine Randy Schekman (1948) states how funding is often not assigned on the basis of ideas and innovation, but on the basis of the proponent previous successes.[124] Although this system is aimed at favouring the talent and merit of the individual scientist, the reality is not so simple. Some studies employing complex system models, applied to human structures, highlight the following fact: in society, the skills of individuals present typically a Gaussian distribution, with a few very

---

119 Kellog (2006), Ossicini (2012), Parravicini (2022), Ziman (2000).

120 Cf. e.g. Kellog (2006), Ziman (2000).

121 Kellog (2006), Ossicini (2012).

122 Zhang et al. (2022).

123 It is known that the branches most subject to incorrect practices (cf. Definitions 10.2, 10.3, 10.4) are the biomedical ones, where, although open fraud seems to remain relatively limited, the data *beautification* is certainly very widespread (Nature Cell Biology 2006; Chevassus-au-Louis 2016). At the opposite, unfair practices are absent in mathematics. Physics, chemistry, material science, engineering, etc. present few cases (Chevassus-au-Louis 2016; Marshall 2000).

124 Quote 10.4 in Sect. 10.2.1.

scarce individuals, the same number of very good ones, and a large majority on average.[125] On the other hand, *success*, which can be measured through appropriate *figures of merit*, follows a power law, with a completely different trend. The discrepancy between the two trends demonstrates that *fate* plays a much more important role than it is generally believed: this feature should be taken into account when establishing research funding strategies through public, national or international funds.[126]

Finally, some recent studies also demonstrate how many "great scholars" cease to produce noteworthy results when they have got no way of exploiting the work of a mass of researchers (poorly paid and even less recognized).[127] In other words, the skill of some esteemed scholars would be based more on the ability to exploit the work of others, obtaining funds and employing the work of subordinates, rather than in their actual *scientific abilities*.[128]

Although what have been said certainly cannot be generalized, it nonetheless raises serious issues regarding the ordering, customs, and above all the limits of the scientific community as well as the fair recognition of the contribution and work. These limits thus require a deep reflection by all those belonging to the scientific community.

10.2.3 Politics and Parties

The influence of politics is the last problem here analysed within the scientific research practice. "Politics" is meant as any exercise of power within a human community. As far as the scientific community is concerned, two forms of political interference can be found:

1. an *external* influence, by the *political power in a strict sense*: governments, parties, professional politicians, etc.;
2. an *internal* influence, by those in the community who hold a some power to direct.

The *first form* is more brash and, at least in words, all researchers agree that it is able to cause great damage to scientific progress. The *second* form is more subtle and difficult to judge objectively as it is *internal* to the scientists' community and thus it might result slightly treacherous.

1. Being scientific research a human practice, it is subject to the regulatory activity of the law and government: claiming that *science* should have got *absolute* freedom means claiming a different *status* from all other human activities, *legibus soluta*, a condition which is unacceptable in a legal state. On the other hand, there is external interference when the political power states to *embrace*, or to *condemn*, a scientific theory or a group of scientists. Political power keeps an oversize

---

125 Pluchino et al. (2018).

126 ibid.

127 Zhang et al. (2022).

128 ibid.

strength compared to that of the scientific community and it is driven by criteria and pursues goals that are completely different.

It is unlikely to have got political interference on a particularly abstract, exotic, and specialized subject. Conversely, inference arises when power needs scientific knowledge to exploit it. Then, politics manipulates scientific results to opportunistically justify choices; the result is that the ordinary discussion among scientists is compromised. Namely, political power is able to stifle the voice of scientists who legitimately do not agree with the interpretation embraced. It is known that in totalitarian regimes this power is exercised up to persecution, violence, and physical elimination.[129] Today, political power does not use equally bloody methods, at least in the West, but it equally has got effective tools to interfere and force reluctant scientists to comply: inciting public opinion, cutting funding, disciplinary inspections, and so on. Today as yesterday, political power can easily find scientists ready to support it and to denigrate colleagues raising doubts.

Moreover, sometimes there are enthusiasts who turn themselves into scientists, supporting rambling theories, which can gain some notoriety in our hyper-connected age. For this reason, it is not surprising and often correct to overlook "too heterodox" theories formulated by some pseudoscientists who are in bad faith and cast specious doubts about well-established knowledge. However, in the field of scientific knowledge there is a "reasonable doubt" which should always be legitimate, especially when dealing with objectively very complex and patterned issues: a sound *organized scepticism* is an essential cornerstone of knowledge.[130] As highlighted a few years ago in a paper on the authoritative *American scientist* magazine, it is possible to legitimately employ intellectual tools to reject bad faith doubts, but it is equally possible (and is actually done) to use the same tools to discredit those who raise quite reasonable doubts.[131] The problem is that there is not an agreement on what a "reasonable doubt" is: as a matter of fact, scholars who have got a given political opinion can affirm that some doubts are unreasonable, even if they are actually absolutely reasonable if impartially judged.[132] Fiery positions of many researchers come to mind about the most debated issues involving the whole of society, such as energy sources, climate change, pollution, epidemics, health policies, food sources, biodiversity, and new drug development, in which the technical-scientific items play an essential role. All these topics, and others, are extremely complex or, to put it formally, they deal with systems where

---

[129] A brief mention of a few most famous persecution cases against scientists is here provided. In Nazi Germany, scientists like Philipp Lenard and Johannes Stark (1874–1957) fomented the persecution of physicists who supported Einstein's relativity: the most illustrious was Einstein himself, forced into exile (Medawar and D. Pyke 2000). In fascist Italy, Enrico Fermi and Emilio Segrè (Segrè 1980) run away. In Soviet Union, Stalin supported the *Lysenkoism* theory, where concepts such as *natural selection* and *Mendelian genetics* were rejected for ideological reasons; the scientists with opposite views were eliminated (Fisher 1948; Mosterìn 2008). Among the victims, there is Nikolaj I. Vavilov, the most important Russian geneticist of the time (Birstein 2001).

[130] Cf. Table 10.2.

[131] Allison et al. (2006).

[132] ibid.

a strong nonlinearity is coupled with a huge number of variables. Therefore, these should be evaluated with great caution, rationality, and attention, especially in front of public opinion. On the contrary, the interference of politics fits into such a complex and patterned scenario, demanding very simple answers to very complex problems; today as yesterday, most of the scientists are unfortunately influenced by this tendency to simplification.[133] The result is a very deteriorated image of the scientific enterprise.[134]

When *media* question a scientist on a technical-scientific subject with *political* importance, such as the aforementioned topics, the scholar usually states, or suggests, that the esposed opinion is based on scientific *absolutely certain* results[135]; actually, the consulted scientist never specifies the experimental uncertainty and the validity limits of the explanatory theories, without which *there is no science*[136] as already seen. Marked contradictions often have emerged, so that some of those *absolutely certain* scientific results are contradicted after some time. Or else, the same *certain and scientific* results are not considered trustworthy in different countries or political contexts; unfortunately, it is easy to find new compliant scientists ready to approve the theses embraced by temporary leaders or a political party they support.[137] Some important scholars regret that "trust in science" is declining in society.[138] If this trend were confirmed, we would not be surprised as in accordance with the scenario just depicted.[139] Maybe, the whole scientific community should examine their conscience.

Politics can influence research not only imperatively, embracing and supporting a theory, but also ideologically. In this case, scientists support a theory due to their own political preference instead of obedience to data from reality. These risks are highlighted by some affairs happened among scientific publications. One of the most important is the *Sokal affair*. In 1996 Alan Sokal, Professor of Statistical Physics at the *University College of London*, writes a paper to be submitted to a scientific journal of *cultural studies*. The article presents a language which mixes the jargon of quantum physics and postmodern philosophy. The result is a virtually meaningless text, which any physicist would find hilarious, but structured in such a way to highlight keywords close to the ideological setting of the journal editors.[140] The purpose of as well as of that research line. The result is remarkable: the nonsensical article is accepted with compliments from the reviewers! After revealing the hoax, a heated debate is triggered on the effective reliability of

---

[133] Abbasi (2020), Büntgen (2024), Parravicini (2022), Zoglauer (2023).

[134] ibid.

[135] Cf. e.g. Abbasi (2020).

[136] Cf. especially Sects. 2.3, 8.1.2, 8.3, and quotations 2.16 in Sect. 2.3, 8.12 in Sect. 8.3, 8.13 in Sect. 8.3.

[137] Allison et al. (2006).

[138] Cf. e.g. Fox (2023), Funk (2017).

[139] Mind that the expression "trust in science" is very questionable: what exactly is meant by *science*? We address this issue in the next chapter.

[140] Sokal (2008).

some kinds of studies, especially in contrast to natural sciences. The episode raises the issue about *conformism* in academic journals: scholars in a given field are necessarily encouraged to accept, recognize, appreciate works with a setting more congenial to them.[141] Since the hoax is carried out by a physicist, the result highlights possible distortions of the whole academic journals system, scientific accuracy checks, peer review, etc.[142] The event also questions natural sciences: are we absolutely sure that such a dynamic could not be reproduced in a discipline less ideologically (and politically) connoted as physics?
The problem is that scholars, academics, and scientists are not angelic and impartial human beings, immune from passions and detached from their environment. This statement seems trivial: if the phenomena underlying scientific research are objective, the scientists who carry out the research are subjective. For this reason they have the same strengths and weaknesses as all human beings and they are affected by the environment and the historical circumstances they live in. Therefore, their research cannot be considered a pure and aseptic activity. Undoubtedly, research asymptotically aims at an objective knowledge, which is recognized as possible; however, it must constantly admit the interferences to which it may be subjected, first of all by the researchers themselves, as stated by Penrose (quote 5.24). A scientist, like any true scholar, should have got the intellectual honesty to recognize that he can never be completely impartial, not even in the most detached activity, and that there is always the risk that *non-scientific criteria* have influence *on him*.[143] On this issue scientists should constantly warn themselves and agree to be warned.[144]

2. In this section, it is analysed the *second form* of political interference in scientific research. Even among scientists there are power groups, trends, and *parties*. Parties have got the tendency to embrace an ideology and so to uncritically defend a given position. Unfortunately, this can also be found in the scientific community, where a *conformism* mood[145] is often established.

> (10.6) The excellence of science is *assumed*, it is not *argued for*. Here scientists and philosophers of science act like the defenders of the One and Only [doctrine] [...]: doctrine is true, everything else is Pagan nonsense.[146]

Among many examples of *parties* in scientific research, an interesting example is found in astrophysics, in the so-called MOND *theory*. This is proposed by the Israeli physicist Mordehai Milgrom (1946) in the early 1980s; MOND means *modified*

[141] Also in the so-called *social sciences*, a similar but more extensive hoax is carried out between 2017 and 2018, in what is known as the *Grievance studies affair*.

[142] Baddeley (2015).

[143] There is a remarkable observation by the Italian historian Gaetano Salvemini, which is also valid when a scientist has to do with politics: "We cannot be impartial. We can only be intellectually honest: that means to realize our passions, to guard against them, and to warn our readers against the dangers of our bias. Impartiality is a dream, probity is a duty." in Salvemini (1952).

[144] Baddeley (2015).

[145] Baddeley (2015).

[146] P. Feyerabend, Part 2 Chap. 1 of Feyerabend (1978).

Newtonian dynamics. The theory starts from the observational data of inconsistencies between the galaxy dynamics and the corresponding measured masses.[147] The vast majority of astrophysicists believe that this discrepancy is due to a certain amount of non-visible mass, the so-called *dark matter*, according to a pattern which is consistent with the *standard model* and the cosmologic ΛCDM theory. Milgrom proposes a totally different approach: he hypothesizes that at the spatial scales under study, Newton's equation of dynamics (8.9) assumes a modified form such as

$$\mathbf{F} = m\mu(\xi)\mathbf{a}, \quad \text{with} \quad \mu(\xi) = \begin{cases} 1 \text{ if } |\xi| \gg 1 \\ \xi \text{ if } |\xi| \ll 1 \end{cases}$$

where **F** is the force, **a** is the corresponding acceleration, m is the considered mass, $\mu = \mu(\xi)$ is a function of the variable $\xi = |a|/a_0$, with a_0 a parameter to be determined.[148] On the one hand, this theory has got excellent agreement with a large number of observational data.[149] On the other hand, it attracts a lot of criticism, due to several weaknesses it shows: the main one is that currently the MOND does not present a consistent framework with General Relativity and the standard model,[150] contrary to the ΛCDM model.

However, what is most interesting for the present discussion is that astrophysicists side with *two parties*: a strong minority actively supports the MOND and a vast majority which considers it a *heresy*.[151] The MOND theory is surely incomplete, but it has also got reasonable strengths due to the excellent agreement with many experimental data. On the other hand, the astrophysicists supporting the *dark matter* model certainly have got excellent scientific reasons. Nevertheless, the issues involved and having an impact on the discussion between these two views are clearly not scientific. First, as astrophysicists are used to face pseudoscientific theories, they have welcomed the theory with scepticism. Moreover, *social conformism* elements have also played a role in the prevailing *paradigm*[152] as well as the tendency to remain within solid conceptual structures well established, such as Newton's equation. In addition, the competition for careers, academic positions, funding[153] also affected the debate between the two parties, which remains highly polarized and often taking heated tones.[154] As a matter of fact, scientists wanting to work on this theory at the beginning of their career do it *at their own risk*.

---

147 Sanders (2015).

148 Helbig (2021), Sanders (2015).

149 Sanders (2015).

150 ibid.

151 Helbig (2021).

152 The use of the word *paradigm*, suggested by Sanders (2015), is not casual, as the dynamics involved easily follows Kuhn's framework (cf. Sect. 5.3.1 and in particular the quoted 5.12 in Sect. 5.3.1).

153 Sanders (2015).

154 Helbig (2021).

Any researcher proposing some idea, which is not revolutionary such as the aforementioned theory, but which is markedly *new*, i.e. not classifiable in a specific *trend* or a recognized and consolidated *research line*, knows that it is a challenge to get the idea recognized. First and foremost, this happens for a difficulty in *understanding* and a tendency to scientific conformism. Supposing a researcher has developed an idea, an interpretation or an entire research project whose results question, even only partially, a more or less consolidated research line. Those experts who have worked on the previous research line have got an established career, professorships, work positions, funded grants that are so endangered. How can the innovative idea not find strong opposition? In such conditions one is often not willing to have got the humility to be *judged*, as stated by Feynman.[155] Although ideally extraneous to the scientific method, these dynamics do play a role and must be taken into account.
A strong presence of *parties* is to be found also in the field of fundamental theoretical physics and in very abstract contexts where sorts of *ideological trends* arise.

> (10.7) If Einstein were alive today, he would be horrified at this state of affairs. He would upbraid the profession for allowing this mess to develop and fly into a blind rage over the transformation of his beautiful creations into ideologies and the resulting proliferation of logical inconsistencies.[156]

The *titled scientific journals* are also specific *centres of power* able to affect scientific research on the basis of their own criteria. In the same year in which R. Schekman wins the Nobel Prize for Medicine, his position causes a sensation: he launches a harsh attack against journals, declaring that they have got a harmful distorting effect on scientific research, because they follow their own commercial criteria and yet exert an enormous influence on it.

> (10.8) These [luxury] journals aggressively curate their brands, in ways more conducive to selling subscriptions than to stimulating the most important research. Like fashion designers who create limited-edition handbags or suits, they know scarcity stokes demand, so they artificially restrict the number of papers they accept. The exclusive brands are then marketed with a gimmick called "impact factor" [...] Better papers, the theory goes, are cited more often, so better journals boast higher scores. Yet it [the "impact factor"] is a deeply flawed measure, pursuing which has become an end in itself - and is as damaging to science as the bonus culture is to banking.[157]

[155] Cf. quote 1.4 in Sect. 1.1 and related discussion.

[156] Robert Laughlin, Chap. 10 of Laughlin (2005).

[157] R. Schekman in (2013).

Schekman concludes his stinging blow by calling for research to free itself from this kind of unscientific conditioning.

> (10.9) [S]cience must break the tyranny of the luxury journals. The result will be better research that better serves science and society.[158]

As many other human structures, the scientific community has defects such as partisanship, conformism, and enslavement to power. These factors stand opposite to the virginal idea that society has about "*Science*" and they are often in striking contradiction with the principles which should inspire research. Although these elements prove to be an obstacle to the progress of knowledge, they are often underestimated by scientists in their daily work. However, gaining awareness of their presence is the only way for the scientific community to correct itself. It is certainly a long and ambitious work, but also beneficial.

References

K. Abbasi, Covid-19: politicisation, "corruption," and suppression of science. BMJ **371**, m4425 (2020)

D. Aitkenhead, Peter Higgs: I wouldn't be productive enough for today's academic system. The Guardian (2013)

B. Alberts, Impact factor distortions. Science **340**, 787 (2013)

B. Alberts, B. Hanson, K.L. Kelner, Reviewing peer review. Science **321**, 15 (2008)

B.D. Allison, G. Pavela, I. Oransky, Reasonable versus unreasonable doubt. Am. Sci. **106**, 84 (2006)

APS. What is science? https://www.aps.org/policy/statements/99_6.cfm, 1999. Statement of American Physical Society, adopted by the Council on Nov 14, 1999

C. Ascheron, *Scientific Publishing and Presentation: A Practical Guide with Advice on Doctoral Studies and Career Planning* (Springer, Berlin, Heidelberg, Deutschland, 2023)

C. Babbage, *Reflections on the Decline of Science in England and on Some of its Causes* (B. Fellowes, London, UK, 1830)

M. Baddeley, Herding, social influences and behavioural bias in scientific research. EMBO Rep. **16**, 902 (2015)

M.R. Beasley, S. Datta, H. Kogelnik, H. Kroemer, D. Monroe, Report of the Investigation Committee on the Possibility of Scientific Misconduct in the Work of Hendrik Schön and Coauthors (2002)

V.J. Birstein, *The Perversion of Knowledge - The True Story of the Soviet Science* (Westview Press - Perseus Books Group, Cambridge, MA, USA, 2001)

J. Bohannon, Hoax-detecting software spots fake papers. Science (2015). https://doi.org/10.1126/science.aab0381

L. Bornmann, Scientific peer review. Annu. Rev. Inf. Sci. Technol. **45**, 197–245 (2011)

U. Büntgen, The importance of distinguishing climate science from climate activism. npj Clim. Action **3**, 36 (2024)

N. Chevassus-au-Louis, *Malscience - De la fraude dans les labos* (Éditions du Seuil, Paris, France, 2016)

N. Chevassus-au-Louis, *Fraud in the Lab - The High Stakes of Scientific Research* (Harvard University Press, Cambridge, Massachusetts, USA, 2019)

R. d'E. Atkinson, Über interferenz von kanalstrahlenlicht. Naturwissenschaften **14**, 599–600 (1926)

[158] R. Schekman, ibid.

L. De Broglie, The wave nature of the electron, in *Nobel Lectures, Physics 1922–1941* (Elsevier Publishing Company, Amsterdam, The Nederlands, 1965)
A. Einstein, Über die Interferenzeigenschaften des durch Kanalstrahlen emittierten lichtes. Sitzungsber. Preuss. Akad. Wiss. **25**, 334–340 (1926)
D. Fanelli, Do pressures to publish increase scientists' bias? An empirical support from us states data. PLoS One **21**, e10271 (2010)
J.B. Ferreira, Exhausted and not doing enough? the productivity paradox of contemporary academia. She Ji: J. Design Econ. Innov. **8**, 181–191 (2022)
P. Feyerabend, *Science in a Free Society* (New Left Books, London, UK, 1978)
R.A. Fisher, What sort of man is Lysenko? Occasional Pamphlet of the Society for Freedom in Science **9**, 6–9 (1948)
K. Fox, Nobel prize winner Giorgio Parisi: 'There's a lack of trust in science - we need to show how it's done'. The Guardian (2023)
A.P. French, The strange case of Emil Rupp. Phys. Perspect. **1**, 3–21 (1999)
C. Funk, Mixed messages about public trust in science. Issues Sci. Technol. **34**, 86–88 (2017)
A. Gelfert, Of fakes and frauds: can scientific "hoaxes" be a legitimate tool of inquiry? in *Truth and Fake in the Post-Factual Digital Age: Distinctions in the Humanities and IT Sciences* (Springer Fachmedien Wiesbaden, Wiesbaden, Deutschland, 2023)
W. Heisenberg, *Die physikalischen Prinzipien der Quantentheorie* (Verlag S. Hirzel, Leipzig, Deutschland, 1930)
P. Helbig, Sonne und Mond, or, the good, the bad, and the ugly: comments on the debate between MOND and ΛCDM. The Observatory **140**(1279), 225–247 (2021). http://philsci-archive.pitt.edu/18810/
M. Inguscio, L. Fallani, *Atomic Physics - Precise Measurements & Ultracold Matter* (Oxford University Press, Oxford, UK, 2013)
J.P.A. Ioannidis, Why most published research findings are false. PLoS Med. **2**(8), e124 (2005)
K. Karachalios, Managing science: is the Cudos still in place? Biotechnol. J. **3**, 306–310 (2008)
D. Kellog, Toward a post-academic science policy: scientific communication and the collapse of the Mertonian norms. Int. J. Commun. Law Policy (2006). https://ssrn.com/abstract=900042
S. Kubatkin, A. Danilov, M. Hjort, J. Cornil, J. Brédas, N. Stuhr-Hansen, P. Hedegård, T. Bjørnholm, Single-electron transistor of a single organic molecule with access to several redox states. Nature **425**, 698–701 (2003)
R. Laughlin, *A Different Universe: Reinventing Physics from the Bottom Down* (Basic Books, New York, USA, 2005)
E. Marshall, How prevalent is fraud? that's a million dollar question. Science **290**, 1662–1663 (2000)
J. Medawar, D. Pyke, *Hitler's Gift - The True Story of the Scientists Expelled by the Nazi Regime* (Arcade Publishing, New York, USA, 2000)
R. Merton, Science and technology in a democratic order. J. Legal Polit. Sociol. **1**, 115–126 (1942)
J. Mosterìn, Social factors in the development of genetics and the Lysenko affair, in *Epistemology and the Social* (Brill, Leiden, The Netherlands, 2008)
Nature Cell Biology, Beautification and fraud. Nat. Cell. Biol. **8**, 101–102 (2006)
Nature. Publish or perish. Nature **467**, 252 (2010)
Nature. Publish or perish. Nature **521**, 159 (2015)
U.S. Neill, Publish or perish, but at what cost? J. Clin. Invest. **118**, 2368 (2008)
M.R. Norman, Cargo-cult science redux. Nat. Phys. **5**, 451–452 (2009)
S. Ossicini, *L'universo è fatto di storie non solo di atomi - Breve storia delle truffe scientifiche* (Neri Pozza, Vicenza, Italia, 2012)
B. Owens, Long-term research: slow science. Nature **495**, 300–303 (2013)
J. Parravicini, Popper's nightmare. Org. J. Biol. Sci. **5**:21–25, 2022. https://doi.org/10.13133/2532-5876/17640
A. Pluchino, A.E. Biondo, A. Rapisarda, Talent versus luck: the role of randomness in success and failure. Adv. Complex Syst. **21**, 1850014 (2018)

K. Popper, *Conjectures and Refutations: The Growth of Scientific Knowledge* (Basic Books, London & New York, UK & USA, 1962)
E.S. Reich, *Plastic Fantastic - How the Biggest Fraud in Physics Shook the Scientific World* (Palgrave MacMillan, New York, USA, 2009)
R.J. Rosen, The slow-science manifesto: 'we don't twitter'. The Atlantic (2011). https://www.theatlantic.com/technology/archive/2011/07/the-slow-science-manifesto-we-dont-twitter/242770/
E. Rupp, Über die Interferenzeigenschaften des Kanalstrahllichtes. Sitzungsber. Preuss. Akad. Wiss. **25**, 341–351 (1926)
G. Salvemini, *Mussolini diplomatico*, 2a edn. (Laterza, Bari, Italia, 1952)
I. Sample, Nobel winner declares boycott of top science journals. The Guardian (2013)
R.H. Sanders, A historical perspective on modified Newtonian dynamics. Can. J. Phys. **93**(2), 126–138 (2015)
J.M. Santos, Quis judicabit ipsos judices? a case study on the dynamics of competitive funding panel evaluations. Res. Eval. **32**, 70–85 (2023)
R. Schekman, How journals like Nature, Cell and Science are damaging science. The Guardian (2013)
J.H. Schön, H. Meng, Z. Bao, Field-Effect Modulation of the Conductance of Single Molecules. Science **294**, 2138–2140 (2001a). Retracted on 1 Nov 2002
J.H. Schön, H. Meng, Z. Bao, Self-assembled monolayer organic field-effect transistors. Nature **413**, 713–716 (2001b). Retracted on 6 March 2003
E. Segrè, *From X-rays to Quarks: Modern Physicists and Their Discoveries* (W.H. Freeman & C., New York, USA, 1980)
A. Sokal, *Beyond the Hoax: Science, Philosophy, and Culture* (Oxford University Press, Oxford, UK, 2008)
J. van Dongen, Emil Rupp, Albert Einstein and the canal ray experiments on wave-particle duality: scientific fraud and theoretical bias. Hist. Stud. Phys. Biol. Sci. **37**(suppl.), 73–120 (2007)
R. Van Noorden, Hundreds of gibberish papers still lurk in the scientific literature. Nature **594**, 160 (2021)
S. Weinberg, *Dreams of a Final Theory* (Pantheon Books, New York, USA, 1992)
S. Zhang, K.H. Wapman, D.B. Larremore, A. Clauset, Labor advantages drive the greater productivity of faculty at elite universities. Sci. Adv. **8**, eabq7056 (2022)
J.M. Ziman, *Real Science: What it Is, and What it Means* (Cambridge University Press, Cambridge, UK, 2000)
T. Zoglauer, Truth relativism, science skepticism and the political consequences, in *Truth and Fake in the Post-Factual Digital Age: Distinctions in the Humanities and IT Sciences* (Springer Fachmedien Wiesbaden, Wiesbaden, Deutschland, 2023)

Chapter 11
Conclusion: The Scientific Sight as Multiple and Human

Science is not a monolithic body of doctrine. Science is a culture, constantly growing and changing.

Freeman Dyson

The foundations of physical science are complex and articulated just as the long and bumpy path determining its current connotations. Its starting point lies in the desire for knowledge. The practical answer to this desire involves all the human being: rationality, abstraction, practical sense, attention, imagination, passion, humility and pride, sociability, logic, the yearning for truth, the perception of beauty, the urge to understanding, the ability to plan and build, ambition, finitude, fallibility, contradiction, honesty and dishonesty, partisanship and probity. All these features can be found in the presuppositions and in the consequent construction of physics. However, its main strength probably lies in its wide adaptability: moved by the desire to increasingly deepen knowledge, physics has been constantly able to adjust, maintain, or change the methodological characteristics and schemes depending on the problem faced.

> (11.1) [K]nowledge needs a plurality of ideas, that well-established theories are never strong enough to terminate the existence of alternative approaches, and that a defence of such alternatives, being almost the only way of discovering the errors of highly respected and comprehensive points of view, is required.[1]

Nevertheless, this flexibility is not the result of arbitrariness and relativism; on the contrary, it is constantly guided by a *loyalty to reality* and by a humble listening to the data, thus to nature. It is the presence of reality data, i.e. the *experimental* nature of physics, that allows us to rediscover a coherence between all the elements that make up the discipline. The great variety of factors investigated in the previous

[1] Paul Feyerabend, Chap. 13 of Feyerabend (1993).

J. Parravicini, *The Foundations of Experimental Physics*,
https://doi.org/10.1007/978-3-031-55452-0_11

chapters can appear as a juxtaposed mass of dissonant objects without a comparison with natural phenomena: this comparison is like a thread running through initially unrelated elements and allowing the whole to compose in an astonishing unity.

The pre-eminence of objective and phenomenological data over ideal schemes is the foundation that allows modern physics to advance, always changing in its multiplicity and always remaining rigorous and faithful to its initial insights, i.e. the marriage between controlled and reasoned observation, known as *experiment*,[2] and the *mathematical language*.[3]

Given the powerful intuition of building this *scientia media*,[4] its form evolves through continuously responding to the demand for stable and objective knowledge, a desire always present in the human soul. Like a lake that overflows, the answer to this question touches every aspect of natural or social reality. Building knowledge is only a properly human activity. The power and limits of human knowledge is a topic that involves all men as it has got strong implications on profound and delicate human issues.

> (11.2) [Kant] thought that the problem "What can I know?" was one of the three most important questions a man could ask. Bertrand Russell [...] seems to side in this matter with Kant. And I believe that Russell is right when he attributes to epistemology practical consequences for science, for ethics, and even for politics. He points out, for example, that epistemological relativism, or the idea that there is no such thing as objective truth, and epistemological pragmatism, or the idea that truth is the same as usefulness, are closely linked with authoritarian and totalitarian ideas. [...] The situation is really very simple. The belief of a liberal -the belief in the possibility of a rule of law, of equal justice, of fundamental rights, and a free society- can easily survive the recognition that judges are not omniscient and may make mistakes about facts and that, in practice, absolute justice is hardly ever realized in any particular legal case. But this belief in the possibility of a rule of law, of justice, and of freedom, can hardly survive the acceptance of an epistemology which teaches that there are no objective facts; not merely in this particular case, but in any other case; and that the judge cannot have made a factual mistake because he can no more be wrong about the facts than he can be right.[5]

Our knowledge is never complete and thus it is imperfect; on the other hand, it is true and stable.

This is especially true in physics: its appeal and power lies in its generality, being able to investigate the largest class of natural phenomena, more than any other discipline, from the shape of the cosmos down to the most minute constituents of matter. If we deduced we know *everything* because of its breadth, we would have not understood the effective richness of the world in front of our eyes. Even if we limited this claim to the idea that physics (or the sciences more generally) theoretically grants us the possibility of knowing *perfectly and absolutely*, we would only get a closed and finite painting of the world and our finite mind would be able to encompass it all, at least asymptotically. Humanity, and scientists in particular, could fall into the

[2] Cf. Definition 4.2.

[3] Cf. Definition 2.1.

[4] See Sect. 4.1 and in particular the cit. 4.3 in Sect. 4.1.

[5] K. Popper, Introduction of Popper (1962).

presumption of being some divinity, opening up the possibility of prevarication and oppression by an authority that pretends to be the holder of perfect knowledge.

Our knowledge is true and stable, thus it is real; however, it is always incomplete, thus it is imperfect.

Nevertheless, if we deduced from this intrinsic incompleteness that our wisdom is illusory and real knowledge is not interesting, because it is ephemeral, or it is even impossible as there is no objectivity of reality, we would coherently lead to the death of knowledge. Popper warns that, as in the previous case, but more subtly, we would open the doors to the arbitrariness of an authority which, in the absence of criteria for obtaining objective knowledge, assumes the right to establish what to consider true.

The knowledge given by the natural sciences, and especially by physics as the most general and fundamental of them, is thus *true* and *reliable* even without being *perfect* and *absolute*. Consequently, there is no contradiction between *incompleteness* and *stability*, between the fact that the knowledge obtained through the scientific method is *incomplete* and always liable to be called into question and the fact that it is *stable*. We are able to grasp a large number of true features of reality without exhausting them all. It is a delicate position, which many can find difficult to accept, but it is the only one that takes into account the actual state of things. In this sense, Poincaré's[6] consideration is a profound truth. In spite of facing the incompleteness of scientific knowledge, its abuses and exploitation, it is not justified to refuge in systematic scepticism, or worse, indifference or superstition. In the same way, it is not justified to assume a scientistic position that passes off scientific knowledge as indisputable, because it is absolutely certain, definitive and all-encompassing, and so to judge as unreasonable and fanatical the *reasoned scepticism* of those who pose doubts and questions.

Given this delicate *status* of scientific knowledge, which is profound and solid even if imperfect and incomplete, it becomes clear why it is inappropriate to publicly speak about *science* and brandish like swords statements such as "I trust in *science*", "*science* says that", "we must believe *science*", often without further specifications. The word *science* may express completely different concepts. By *science* we may mean the scientific method, the phenomena object of its study, the data from experiments, scientific data in general, the scientific community, the *majority*, i.e. the *consensus*, of this community, the opinion of scientists, scientific *theories*, scientific *explanations*, scientific *knowledge*, scientific *research*, the *institutions* of scientific research, etc. Therefore, speaking about *science* without specifying its meaning is like providing experimental data without specifying the error: it is an undue oversimplification, a completely *non-scientific* statement.

It may perhaps be difficult for a scientist to always pay attention to all the assumptions here examined. Although it can sometimes be a heavy or cumbersome practice, this awareness is necessary and makes the practice of knowledge conscious as well as, at least in principle, profound and lasting for researchers, and good and fruitful for society.

[6] Quote 8.16 in Sect. 8.3.

On the other hand, the richness and the variety of the factors present in physics (and in all the mathematical and natural sciences) highlight how the strict separation between *scientific knowledge* and *humanistic knowledge*, between the mathematical, physical and natural sciences and the humanities is artificial. Certainly, there are differences in methods and techniques between the two categories, but analogous differences can also be found between the disciplines considered *scientific* in the strict sense: the experiments which an experimental physicist carry out in order to validate a theory are perhaps radically different from the demonstrations that a mathematician conducts in order to validate a theorem. Beyond the differences in techniques and methods, which are certainly profound, researchers of all disciplines are moved by that love for knowledge proper to all men. As matter of fact, all men have got a desire to know the truth of the surrounding things, whether or not they are researchers, scientists, or physicists; usually when men have got the chance to experience a rational and extensive knowledge, they fall in love with it. Then, when passion is found at the bottom of reason, true *wisdom* is experienced.

(11.3) *La vera comprensione nasce con l'amore sincero per la verità, il rispetto per la dignità dell'uomo, la convinzione che l'amore della sapienza può essere proposto ma non imposto. A questo amore ogni persona deve giungere liberamente attraverso la meditazione sulla propria esperienza e su tutta la propria vita che è unica e irripetibile; da questo punto di vista la comprensione non è relativismo, scetticismo, indifferenza rispetto alla verità, ma è piuttosto amore della sapienza profondo e libero, desiderio che tutti gli uomini ricerchino liberamente la sapienza per propria personale convinzione. [...]*

In fondo il compito a cui tutti noi, studiosi di scienze matematiche, fisiche e naturali e studiosi di scienze umane, siamo chiamati è la riscoperta del valore sapienziale delle nostre discipline, premessa necessaria per un armonico sviluppo di tutti i rami del sapere.[7]

True understanding comes with a sincere love of the truth, respect for the dignity of man, the conviction that love of wisdom can be proposed but not imposed. Everybody can freely reach this love by thinking about his own experience and about all his own life that it is unique and unrepeatable. This kind of comprehension is not relativism, scepticism or indifference to the truth but it is rather a deep free love of wisdom and the desire that all men freely search for wisdom because of their own personal conviction. [...]

The task that all we - scholars of mathematics, physics and natural sciences and scholars of human sciences - are called upon is simply the rediscovery of the value and wisdom of our disciplines, a necessary premise for a harmonious development of all the branches of knowledge.

References

E. De Giorgi, Lectio magistralis: "Il valore sapienziale della matematica", in *Ennio De Giorgi Selected Papers* (Springer, Berlin, Deutschland, 2006a)

E. De Giorgi, Lectio magistralis: "The sapiential value of mathematics", in *Ennio De Giorgi Selected Papers* (Springer, Berlin, Deutschland, 2006b)

[7] Ennio De Giorgi in De Giorgi (2006a), English translation from De Giorgi (2006b).

P. Feyerabend, *Against Method: Outline of an Anarchistic Theory of Knowledge*, 3rd edn. (Verso Books, London, UK, 1993)

K. Popper, *Conjectures and Refutations: The Growth of Scientific Knowledge* (Basic Books, London & New York, UK & USA, 1962)

Appendix A
Not Fundamental Units of Measure

See Table A.1.

Table A.1 The 22 SI not fundamental measurement units with their own names and symbols

Derived quantity	Name	Correspondence with basis SI units	Correspondence with other SI units
Plane angle	radian	rad = m/m	
Solid angle	steradian	$sr = m^2/m^2$	
Frequency	hertz	$Hz = s^{-1}$	
Force	newton	$N = kg\ m\ s^{-2}$	
Pressure, stress	pascal	$Pa = kg\ m^{-1}\ s^{-2}$	
Energy, work, heat quantity	joule	$J = kg\ m^2\ s^{-2}$	N m
Power, flux radian	watt	$W = kg\ m^2\ c^{-3}$	J/s
Electric charge	coulomb	C = A s	
Electric potential difference	volt	$V = kg\ m^2\ s^{-3}\ A^{-1}$	W/A
Capacity	farad	$F = kg^{-1}\ m^{-2}\ s^4\ A^2$	C/V
Electric resistance	ohm	$\Omega = kg\ m^2\ s^{-3}\ A^{-2}$	V/A
Electric conductance	siemens	$S = kg^{-1}\ m^{-2}\ s^3\ A^2$	A/V
Magnetic flux	weber	$Wb = kg\ m^2\ s^{-2}\ A^{-1}$	V s
Magnetic field*	tesla	$T = kg\ s^{-2}\ A^{-1}$	Wb/m^2
Inductance	henry	$H = kg\ m^2\ s^{-2}\ A^{-2}$	Wa/A
Celsius temperature	celsius degree	°C = K	
Light flux	lumen	lm = cd sr	cd sr
Luminance	lux	$lx = cd\ sr\ m^{-2}$	lm/m^2
Activity of a radionuclide	becquerel	$Bq = s^{-1}$	
Absorbed dose	gray	$Gy = m^2\ s^{-2}$	J/kg
Equivalent dose	sievert	$Sv = m^2\ s^{-2}$	J/kg
Catalytic activity	katal	$kat = mol\ s^{-1}$	

*Also called *magnetic induction field* or *magnetic flux density*

J. Parravicini, *The Foundations of Experimental Physics*,
https://doi.org/10.1007/978-3-031-55452-0

Appendix B
Some Logical Tools

B.1 Main Logical Symbols

In Table B.1 we show some of the main *logical symbols* usually employed in mathematical logic.[1]

Among the above logical operators, it is worth briefly illustrating the main features of the most used ones.[2] There are the *propositional connectors*:

- $\rightarrow$ means "implication", i.e. the proposition which precedes the symbol is a sufficient condition for the following proposition; it intuitively reproduces the expression "if…then…";
- $\leftrightarrow$ means "mutual implication", i.e. the proposition which precedes the symbol is a sufficient condition for the following proposition and vice versa; it intuitively reproduces the expression "…if and only if…";
- $\neg$ denies the validity of proposition which follows the symbol; it intuitively reproduces the "not" negation;
- $\wedge$ is the Boole *logical operator* "et" (or "and"); it intuitively reproduces the "and" conjunction;
- $\vee$ is the Boole *logical operator* "vel" (or "or"); it intuitively reproduces the "or" disjunction;
- $\veebar$ is the Boole *logical operator* "aut" (or "xor"); it intuitively reproduces the exclusive "or" disjunction.

Moreover, there is the symbol

- $\vdash$, meaning that the proposition following the symbol is the logical consequence of what it precedes it.

[1] Chapter 7 of Brewer (2022), Chap. 1 of Cole Kleene (1967), Copi et al. (2019).

[2] Cf. Chap. 1 of Enderton (2001).

J. Parravicini, *The Foundations of Experimental Physics*,
https://doi.org/10.1007/978-3-031-55452-0

Table B.1 List of the main logical symbols

Symbol	Name	How it is red	Explanation
$()$	Grouping	"brackets"	"the expression within ( ) should take precedence in calculations"
$\supset$ o $\rightarrow$ o $\Rightarrow$	Implication	"if P, then Q"	$P \supset Q \rightarrow$ "false if P is true and Q false; otherwise true"
$\equiv$ o $\leftrightarrow$ o $\Longleftrightarrow$ o $\sim$	Equivalence	"if and only if"	$P \equiv Q \rightarrow$ "P and Q are both true" "(...both false)"
$\neg$	Negation*	"not"	$\neg P \rightarrow$ "not P", "the opposite of P"
$\wedge$ o &	Conjunction	"both...and...", "and", "plus"	$x \wedge y \rightarrow$ "both x and y"; "$x + y$"
$\vee$	Inclusive disjunction	"one or both"	$x \vee y \rightarrow$ "either x or y or both"
$\veebar$	Exclusive disjunction	"either one or the other one (not both)"	$x \veebar y \rightarrow$ "x but not y"; "y but not x"
$\forall$	Universal quantifier	"for all"	$\forall x : P(x) \rightarrow$ "$P(x)$ is true for all x"
$\exists$	Existence quantifier	"exists"	$\exists x : P(x) \rightarrow$ "$P(x)$ is true for at least one x"
$\nexists$	Negative existence quantifier	"doesn't exist"	$\nexists x : P(x) \rightarrow$ "$P(x)$ isn't for any x"
$\in$	Belonging	"belongs to", "is an element of"	$x \in S \rightarrow$ "x is an element of the S set"
$\notin$	Not belonging	"doesn't belong to", "isn't an element of"	$x \notin S \rightarrow$ 'x isn't an element of the S set'
$:=$	Definition	"is defined as"	$P := Q \rightarrow$ "P and Q are logically equivalent"
$\therefore$	Logical consequence	"then"	$P \therefore Q \rightarrow$ "P, then Q"
$\vdash$	Turnstile	"demonstrates that", "whose logical consequence is that"	$P \vdash Q \rightarrow$ "P (syntactically) demonstrates Q"
$\nvdash$	Negative turnstile	"doesn't demonstrate that"	$P \nvdash Q \rightarrow$ "P doesn't prove Q"
$\models$	Double turnstile	"is a model of"	$P \models Q \rightarrow$ "P is (semantically) a model of Q"
$\top$	Tautology	Truth	$\top(P) \rightarrow$ "P is true under all conditions"
$\bot$	Contradiction	Falsity	$\bot(P) \rightarrow$ "P is false under all conditions"

*In some text the symbol $\sim$ (introduced by Peano and diffused by Russell) is used, which is employed, nevertheless, with several meanings in many contexts, so that its use can produce misunderstanding and inaccuracies

B.2 Outline of the Main Logical Tools

B.2.1 The Three "Laws of Thought"

Given an A proposition, we have the following *tree logical fundamental principles*; being their first statement due to Aristotle, they are also named *Aristotelian principles*.[3]

Identity Principle $A \rightarrow A$
Any statement implies itself, i.e. if a proposition is true, then it is true.
Non-contradiction Principle $\neg(A \wedge \neg A)$
No proposition can be both true and false at the same time.
Principle of *excluded middle* $A \vee \neg A$
Each proposition is either true or false, *tertium non datur* (there is no third possibility).

Although their status as effective *foundations* of the entire logic has been questioned in the modern age, their enormous importance is nevertheless recognized as fundamental elements for the construction of classical logic. In this context, the non-contradiction principle, the principle of the excluded middle, and the identity principle are of importance in descending order.[4]

B.2.2 Classical Syllogisms

Classical logic identifies 15 valid *formal categorical syllogisms*.[5] They are detected by Aristotle and systematized by medieval logicians.[6] Propositions are grouped in four typologies:

1. universal affirmative;
2. particular affirmative;
3. universal negative;
4. particular affirmative.

Each typology is identified by a letter, following a precise mnemonic scheme. Affirmative propositions are identified by the first vowels of the Latin word ADFIRMO ("I affirm"), while the negative ones are labelled by the vowels of the word NEGO ("I deny"). So we obtain

[3] Chapter 8 of Copi et al. (2019).

[4] ibid.

[5] Logic Overviews in Copi et al. (2019)

[6] Bernhard (2008), Copi et al. (2019), Parsons (2008).

AdfIrmo

$A : \forall x(P(x) \rightarrow Q(x))$. "For all x, if P holds, then Q will hold"; "All P is S".
$I : \exists x(P(x) \vee Q(x))$. "For (at least) one x, either P or Q holds"; "Some P is S".

nEgO

$E : \forall x(P(x) \rightarrow \neg Q(x))$. "For all x, P holds, then Q does not"; "All P is not S".
$O : \exists x(P(x) \vee \neg Q(x))$. "For (at least) one x either P or Q does not hold"; "Some P is not S".

A classical syllogism is made up of three declarative propositions, each of them belongs to one of the above kinds. The first two are the starting logical data, respectively, called *major premise* and *minor premise*; the third one expresses the logical conclusion of the premises. Usually, the propositions are ordered according to the following scheme, which is called *square of oppositions*.[7]

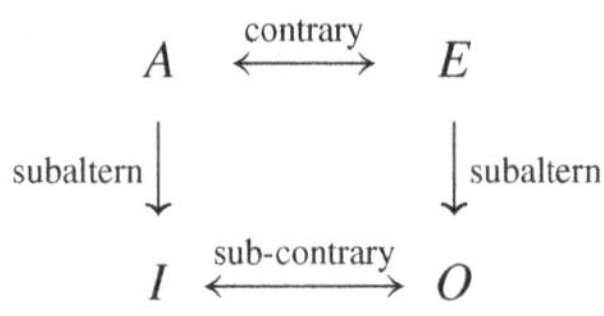

In the syllogism there are *three terms*: the predicate of the conclusion, the *major term*; the subject of the conclusion, the *minor term*; and the *middle term*, that is present in the premises but not in the conclusion. According to the mutual position of the terms, the valid syllogisms are grouped into four *figures*. In the medieval systematization there is a single word uniquely identifying the syllogism; this word is chosen so as to contain the three vowels identifying the kind of proposition to which each of the syllogism[8] statements belongs. The valid syllogisms, grouped by figure, in relation to the *square of oppositions*, are therefore the following ones:

1. BARBARA CELARENT DARII FERIO
2. CAMESTRES CESARE BAROCHO FESTINO
3. DATISI DISAMIS FERISON BOCARDO
4. CAMENES DIMARIS FREISON

The syllogisms of *first figure* are the simplest ones, and they are considered the fundamental ones that are assumed to be *true*. Those of the other figures can be traced back to previous syllogisms through syntactic manipulations called *conversiones*, in order to verify their effective validity.[9] This procedure, whose foundations are identified by Aristotle himself, constitutes the first example of *logical calculus*, which in modern times can be purely symbolically performed.[10]

---

[7] Bernhard (2008), Parsons (2008).

[8] Copenhaver (2014), Copi et al. (2019).

[9] In the choice of words identifying each kind of syllogism, the consonants provide information on the type of transformation to be performed.

[10] Church (1956), Copi et al. (2019).

B.2.3 Notes on Inference Rules

Pay attention to false inference rules.[11]

- While the *modus ponens* correctly applied is, following the (9.14),

  $$((A \to B) \land A) \to B,$$

  the incorrectly applied one gives

  $$((A \to B) \land B) \to A, \tag{B.1}$$

  which is called *fallacy of affirming consequence*. In other words, if A implies B and B holds (i.e. if it's true), we can't deduce that A is also true. Actually, this would be believing that a necessary condition is a sufficient condition. When we use the *modus ponens* to check a theory, we act according to (9.14):

  1. we assume the *hypothesis* that A (the theory) is true and we deduce its implications, among which the fact that A could be the cause of a phenomenon B that has not been observed yet;
  2. to verify the hypothesis, we investigate whether the phenomenon B happens (that is whether it's true);
  3. if B *doesn't* happen, that is $\neg B$, we can conclude that A is false, i.e. $\neg A$; vice versa, if B happens, we can *just say* that A is not false.

  Applying the *modus ponens* following the (B.1) would require the fallacious reasoning according to which

  1. we assume the *hypothesis* that A (the theory) is true and we deduce its implications, among which the fact that A could be the cause of a phenomenon B that has not been observed yet;
  2. to verify the hypothesis, we investigate whether the phenomenon B happens (that is whether it's true);
  3. if B does happen, we would *affirm* that A is true.

  This scheme is incorrect because, in order to "confirm" a theory, it is not sufficient to have *a single* experimental confirmation. It is instead necessary to get several ones, analysing different implications of it, and checking that no one of them be disregarded. Of course, this will never provide a *certitude* that A is true, instead, according to the *falsification* principle, it will ever provide more and more confirmations that it isn't false.
- While the *modus tollens* correctly applied is, according to (9.1),

  $$((A \to B) \land \neg B) \to \neg A,$$

[11] Chapter 2 of Copi (1973).

the incorrectly applied one is

$$((A \to B) \land \neg A) \to \neg B, \tag{B.2}$$

which is called *fallacy of denying premise*. In other words, if A implies B and A does not happen (i.e. it's false), we can't say that B is false. When we use the *modus tollens* in an experiment, we act according to (9.1):

1. we assume the *hypothesis* that A is, somehow, the cause of the phenomenon B;
2. to verify the hypothesis, we remove A *without acting on* B, that is we apply $\neg A$;
3. if the starting hypothesis is correct, is the experiment we must see that also B is lacking, i.e. $\neg B$.

Applying the *modus tollens* according to (B.2) would require the fallacious reasoning according to which

1. we assume the *hypothesis* that A is, somehow, the case of the phenomenon B;
2. to verify the hypothesis, we somehow remove B *without acting on* A, that is we apply $\neg B$;
3. if A does not happen, i.e. $\neg A$, then we deduce, incorrectly, that the starting hypothesis is correct.

The aforementioned reasoning is fallacious: if $A \to B$ is correct, removing B independently on A, even though A is lacking, there is no guarantee that $\neg A$ is due to $\neg B$.

Appendix C
Authors of Quotations

Anderson, Philip W. (1923–2020). USA theoretical physicist. Nobel Prize 1977 for Physics.

Bohr, Niels (1885–1962). Danish experimental physicist. Nobel Prize 1922 for Physics.

Cantor, Georg F.L.P. (1845–1918). German mathematician.

Carrel, Alexis (1973–1944). French physician, surgeon, and biologist. Nobel Prize 1912 for Medicine.

Courant, Richard (1888–1972). German mathematician.

De Broglie, Louis (1892–1987). French theoretical physicist. Nobel Prize 1929 for Physics.

De Giorgi, Ennio (1928–1996). Italian mathematician. Wolf Prize 1990 for Mathematics. Solver of *XIX Hilbert's problem.*

Dirac, Paul A.M. (1902–1984). English theoretical physicist. Nobel Prize 1933 for Physics.

Dyson, Freeman (1923–2020). English theoretical physicist and mathematician. Wolf Prize 1981 for Physics.

Einstein, Albert (1879–1955). German theoretical physicist. Nobel Prize 1921 for Physics.

Enriques, Federigo (1871–1946). Italian mathematicians, philosopher, and science historian.

Euclide (IV century B.C.–III century B.C.). Greek mathematician and philosopher.

Fermi, Enrico (1901–1954). Italian experimental physicist. Nobel Prize 1938 for Physics.

Feynman, Richard P. (1918–1988). USA theoretical physicist. Nobel Prize 1965 for Physics.

Feyerabend, Paul (1924–1994). Austrian science philosopher.

Frege, Gottlob (1848–1925). German mathematician and logician.

Galilei, Galileo (1564–1642). Italian physicist, mathematician, and philosopher.

Gell-Mann, Murray (1929–2019). USA theoretical physicist. Nobel Prize 1969 for Physics.

J. Parravicini, *The Foundations of Experimental Physics*,
https://doi.org/10.1007/978-3-031-55452-0

Gödel, Kurt (1906–1978). Austrian mathematician and logician. Author of *Incompleteness Theorems* named by him.

Heisenberg, Werner K. (1901–1976). German theoretical physicist. Nobel Prize 1932 for Physics.

Helmholtz, Herman von (1821–1894). German physicist, physician, physiologist.

Hilbert, David (1862–1943). German mathematician.

Kuhn, Thomas (1922–1996). USA physicist, philosopher, and science historian.

Lakatos, Imre (1922–1974). Hungarian science philosopher.

Landau, Lev D. (1908–1968). Soviet theoretical physicist. Nobel Prize 1962 for Physics.

Laplace, Pierre-Simon (1949–1827). French mathematician and physicist.

Laughlin, Robert (1950). USA theoretical physicist. Nobel Prize 1997 for Physics.

Lemaître, George E. (1894–1966). Belgian mathematical physicist, astronomer, and priest. First theorist of the *Big Bang* model.

Levi-Civita, Tullio (1873–1941). Italian theoretical physicist and mathematician. Inventor and developer, with G. Ricci-Curbastro, of *tensorial calculus*.

Lifšic, Evgenij M. (1915–1985). Soviet theoretical physicist.

Maxwell, James K. (1831–1879). Scot physicist and mathematician.

Neumann, Johan von (1903–1957). Hungarian theoretical physicist and mathematician.

Newton, Isaac (1642–1727). English physicist, mathematician, and philosopher.

Penrose, Roger (1931). English theoretical physicist and mathematician. Wolf Prize 1988 and Nobel Prize 2020 for Physics.

Podolsky, Boris (1896–1966). Russian mathematical physicist.

Poincaré, Henri J. (1854–1912). French mathematical physicist.

Polkinghorne, John (1930–2021). English theoretical physicist and philosopher.

Popper, Karl (1902–1994). Austrian science philosopher.

Prigogine, Ilya R. (1917–2003). Russian physical chemist. Pioneer of the complex system studies. Nobel Prize 1977 for Chemistry.

Ricci-Curbastro, Gregorio (1853–1925). Italian mathematician. Inventor and developer, with T. Levi-Citiva, of *tensorial calculus*.

Rosen, Nathan (1909–1995). Israeli-USA theoretical physicist.

Segrè, Emilio G. (1905–1989). Italian experimental physicist. Nobel Prize 1959 for Physics.

Tommaso D'Aquino (1225–1274). Italian philosopher and theologian.

Weinberg, Steven (1933–2021). USA theoretical physicist. Nobel Prize 1979 for Physics.

Whitehead, Alfred N. (1861–1947). English mathematician and logician.

Wiener, Norbert (1894–1964). USA mathematician. Inventor and developer, with C. Shannon, of *classic information theory*; father of *cybernetics*.

Wigner, Eugene P. (1902–1995). Hungarian theoretical physicist. Nobel Prize 1963 for Physics.

References

P. Bernhard, Visualizations of the square of opposition. Logica universalis **2**, 31–41 (2008)
W.D. Brewer, *Kurt Gödel. The Genius of Metamathematics* (Springer, Cham, Schweiz, 2022)
A. Church, *Introduction to Mathematical Logic* (Princeton University Press, Princeton, New Jersey, USA, 1956)
S. Cole Kleene, *Mathematical Logic* (Dover, Mineola, New York, USA, 1967)
B.P. Copenhaver, *Peter of Spain: Summaries of Logic - Text, Translation, Introduction, and Notes* (Oxford University Press, Oxford, UK, 2014)
I.M. Copi, *Symbolic Logic*, 4th edn. (MacMillian Publishing, New York, USA, 1973)
I.M. Copi, C. Cohen, V. Rodych, *Introduction to Logic*, 15th edn. (Taylor & Francis, New York, USA, 2019)
H.B. Enderton, *A Mathematical Introduction to Logic*, 2nd edn. (Harcourt Academic Press, San Diego, USA, 2001)
T. Parsons, Things that are right with the traditional square of opposition. Logica universalis **2**, 3–11 (2008)

J. Parravicini, *The Foundations of Experimental Physics*,
https://doi.org/10.1007/978-3-031-55452-0

Name Index

J. Parravicini, *The Foundations of Experimental Physics*,
https://doi.org/10.1007/978-3-031-55452-0

Analytical Index

J. Parravicini, *The Foundations of Experimental Physics*,
https://doi.org/10.1007/978-3-031-55452-0

GPSR Compliance

The European Union's (EU) General Product Safety Regulation (GPSR) is a set of rules that requires consumer products to be safe and our obligations to ensure this.

If you have any concerns about our products, you can contact us on ProductSafety@springernature.com

In case Publisher is established outside the EU, the EU authorized representative is:

Springer Nature Customer Service Center GmbH
Europaplatz 3
69115 Heidelberg, Germany

The manufacturer's authorised representative in the EU is Springer Nature Customer Service Centre GmbH, Europaplatz 3, 69115 Heidelberg, Germany. If you have any concerns regarding our products, please contact ProductSafety@springernature.com

Printed and bound by CPI Group (UK) Ltd, Croydon, CR0 4YY
01/10/2026
02225906-0006